“绿都北京”研究系列丛书

Green Beijing Research Series

北京三山五园地区绿道规划与设计研究

Planning and Design of Greenways in the Areas of the Three Hills and Five Gardens, Beijing

北京林业大学园林学院　主编

中国建筑工业出版社
CHINA ARCHITECTURE & BUILDING PRESS

图书在版编目(CIP)数据

北京三山五园地区绿道规划与设计研究/北京林业大学园林学院主编. --北京：中国建筑工业出版社，2018.4
(“绿都北京”研究系列丛书)
ISBN 978-7-112-22024-3

I.①北… II.①北… III.①道路绿化-绿化规划-研究-北京 IV.①TU985.18

中国版本图书馆CIP数据核字（2018）第060699号

责任编辑：杜　洁　李玲洁
责任校对：焦　乐

“绿都北京”研究系列丛书
北京三山五园地区绿道规划与设计研究
北京林业大学园林学院　主编
*
中国建筑工业出版社出版、发行(北京海淀三里河路9号)
各地新华书店、建筑书店经销
北京方嘉彩色印刷有限责任公司印刷
*
开本：787×1092毫米　1/16　印张：11¾　字数：387千字
2018年4月第一版　2018年4月第一次印刷
定价：99.00元
ISBN 978-7-112-22024-3
(31907)

编委会

本研究由城乡生态环境北京实验室支持

前 言

美国著名风景园林师西蒙兹（John Simonds 1913-2005）出版过一本文集Lessons(中文版书名为《启迪》)，书中有一篇文章记录了西蒙兹1939年到北平考察的经历。

在北京，西蒙兹拜访了一位祖上曾经参与规划了元大都的李姓建筑师，李先生非常赞赏他能到北京考察风景规划，并向西蒙兹简单地介绍了元大都的规划思想。“在这片有良好水源的平原上，将建设一个伟大的城市——人们在这里可以与上天、自然以及同伴们和谐共处”。“蓄水池以自然湖泊的面貌贯穿整个都城，挖出的土用来堆成湖边的小山，湖边和山上种植了从全国各地收集来的树木和花灌木”。“关于公园事宜和开放空间，可汗命令不能有孤立的公园。更准确地说，整个大都城将被规划成一个巨大而美丽的花园式公园，期间散布宫殿、庙宇、公共建筑、民居和市场，全都有机地结合在一起”。“从文献中我了解到北平被一些来此旅游的人称为世界上最美丽的城市，我不知道这是否正确。如果真是这样，那么这种美丽不是偶然形成的，而是从最大的布局构思到最小的细节——都是通过这样的方法规划而成的。”

北京的确如西蒙兹在文章中提到的李姓建筑师所说，是一座伟大的城市，也是一个巨大而美丽的花园式都城。

北京有着优越的地理条件，城市的西、北和东北被群山环绕，东南是平原。市域内有5条河流，其中的永定河在历史上不断改道，在这片土地上形成广阔的冲积扇平原，留下了几十条故道，这些故道随后演变为许多大大小小的湖泊，有些故道转为地下水流，在某些地方溢出地面，形成泉水。

北京又有着3000年的建城史。李先生提到的元大都已将人工的建造与自然环境完美地叠加融合在一起，到了清朝时期，北京城人工与自然的融合更加紧密完善。城市西北郊建造了三山五园园林群，西山和玉泉山的汇水和众多泉流汇纳在一起，形成这些园林中的湖泊，水又通过高梁河引入城市，串联起城中的一系列湖泊。许多宫苑、坛庙、王府临水而建，水岸也是城市重要的开放空间。城中水系再通过运河向东接通大运河。由此，北京城市内外的自然景观成为一个连贯完整的体系，这一自然系统承担着调节雨洪、城市供水、漕运、灌溉、提供公共空间、观光游览、塑造城市风貌等复合的功能。城市居住的基本单元——四合院平铺在棋盘格结构的城市中，但每一个四合院的院子里都有别样的风景，每个院子都种有大树，如果从空中鸟瞰，北京城完全掩映在绿色的海洋之中。

然而，随着人口的增加和城市建设的发展，北京的环境在迅速地变化着，古老的护城河早已消失，一起消失的还有城市中的不少湖泊和池塘。特别是快速城市化以来，北京的变化更为剧烈。老城中低矮的四合院被高楼大厦取代，步行尺度的胡同变成了宽阔的道路。老城之外，城市建设不断向周边蔓延，侵占着田野、树林和湿地，城市内外完整的自然系统被阻断，积极的公共空间不断消失，而交通设施的无限扩张，又使得城市被快速路不断地切割，城市渐渐失去了人性化的尺度、也渐渐失去了固有的个性与特色。

面对自然系统的断裂和公共空间的破碎与缺失等城市问题，作为风景园林、城市规划和建筑学的教育和研究者，我们看到了通过维护好北京现有的自然环境和公园绿地，利用北京的河道、废弃的铁路和城市中的开放土地，改造城市快速交通环路，建设一条条绿色的廊道，并形成城市中一个完整的绿色生态网络，从而再塑北京完整的自然系统和公共空间体系的巨大机会。

这条绿色的生态网络可以重新构筑贯穿城市内外的连续自然系统，使得城市的人工建造与自然环境有机地融合在一起；这个网络可以将由于建造各种基础设施而被隔离分割的城市重新连接并缝合起来，形成城市的公共空间体系；这个网络可以承载更加丰富多彩的都市生活，成为慢行系统、游览、休憩和运动的载体，也成为人们认知城市、体验城市的场所；这个网络还可以带来周边地区更多的商业机会，促进周围社区的活力；这个网络更是城市中重要的绿色基础设施，承担着雨洪管理、气候调节、生态廊道、生物栖息场构建、生物多样性保护的关键作用……

这套丛书收录的是我们对北京绿廊和生态网络构建的研究和设想。当然，畅想总是容易的，而实施却面临着巨大的困难和不确定性，但是我们看到，世界上任何伟大的城市之所以能够建成，就是从畅想开始的，如同元大都的建设一样。

在《启迪》中那篇谈到北京的文章最后，西蒙兹总结到：“要想规划一个伟大的城市，首先要学习规划园林，两者的原理是一样的”。

我们的研究实质上就是以规划园林的方式来改良城市，希望我们的这些研究成果也能对北京未来的建设和发展有所启迪。

2018 年 1 月

Forewords

The famous US landscape architect John Simonds once published a corpus named 'Lessons', and one of the articles in this book records the experience of Simonds's investigation to Beijing in 1939.

Simonds visited an architect surnamed Li whose ancestors once took part in planning of the Great Capital of Yuan in Beijing, and the architect admired that Simonds came to Beijing to study landscape planning. He also briefly introduced the planning thoughts of the Great Capital of Yuan to Simonds's group. According to architect Li, here on this well-watered plain, was to be built a great city in which man would find himself in harmony with God, with nature, and his fellow man. Throughout the capital were to be located reservoirs in the form of lakes and lagoons, the soil formed their excavations to be shaped into enfolding hills, planted with trees and flowering shrubs collected from the farthest reaches of his dominion. As for the matter of parks and open spaces, architect Li said the Khan decreed that no separate parks were to be set aside. Rather, the whole of Ta-Tu would be planned as one great inter-related garden-park, with palaces, temples, public buildings, homes and market places beautifully interspersed. He also added that he was led to believe that Peking (now present day Beijing) was regarded by some who have travelled here to be the most beautiful city of the world, which he could not know to be true. If so, it would be no happenstance, for from the broadest concept to the least detail – it was planned that way.

Just like what architect Li mentioned, Beijing is indeed a great city, also a grand gorgeous garden capital.

With superior geographical condition, Beijing city is surrounded by mountains in the west, north and northeast direction, and the southeast of the city is plain. There are 5 rivers in the city. Among them, the Yongding River has constant change of course in history, thus formed the vast alluvial fan plain here and has left dozens of old river courses. These old water courses then evolved into lakes with different scales, some even transformed into underground water streams and overflowed to the ground to form springs.

At the same time, Beijing has a history of 3000 years of city construction. As architect Li said, the Great Capital of Yuan has integrated the artificial construction and the natural environment perfectly. And when it came to Qing Dynasty, the integration of labor and nature is even more perfect in Beijing city. People built the 'Three Hills and Five Gardens' in the northwest of the city, so that the catchments of the West Mountain and Yuquan Mountain could join numerous springs together, and formed the lakes in these royal gardens. Then, water was introduced into the city through the Sorghum River, and thus a series of lakes inside the city are connected. Plenty of palatial gardens, temples and mansions of monarch were built in the waterfront, which makes the water bank an important open space for the whole city. The river system in the city heads for the east and connects to the Grand Canal, which makes the nature environment inside and outside the city into a coherent and complete system, which takes the charge of compound functions including the regulation of rain flood, city water supply, water transport, irrigation, providing public space, sightseeing function and shaping the cityscape. As the basic unit of urban living, Siheyuans are paved in the city with chessboard structure. Uniformed as they are in appearance, we can still see unique landscape and stories in each different courtyard. There are big trees thriving in each courtyard, as if they were telling the history of each family. If we have a bird's eye view from the air, Beijing will be completely covered in the green ocean.

However, with the population increase and the urban construction development, the environment of Beijing has been changing rapidly. The ancient moat has already disappeared, together with many lakes and ponds in the city. Beijing has changed even more fast and violent since the rapid urbanization. Low Siheyuans have been replaced by skyscrapers, and Hutongs of walking scale also became broad roads for vehicles. Apart from the Old City, the urban construction in Beijing has been spreading to the surrounding area, invading the fields, forests and wetlands. As a result, the holistic natural system both inside and outside the city is blocked, active public space is disappearing, and the unlimited expansion of transportation facilities make the city constantly cut by express ways. We cannot deny that the city has gradually lost its humanized scale, and it has also gradually lost its inherent personality and characteristics.

In face of the city fracture problems of natural systems and the broken public space, as landscape architects, urban planners, architecture educators and researchers, we see huge opportunities to maintain the existing natural environment and garden greenways, use the river courses, disused railways and open lands in Beijing to reform the city fast traffic roads and construct several green corridors in order to form a complete green ecological network in the city, and remold integrated natural system and public space system in Beijing.

This green ecological network can reconstruct the continuous natural system running throughout the city, so that the artificial construction of the city can be organically integrated with the natural environment. The network can connect and stitch the city divided by all kinds of infrastructure, and form a public space system in Beijing. What's more, the network can carry more colorful urban life styles and become the supporter of slow travel system, sightseeing, recreation and sports, and it will turn into a place for people to cognize and experience the city. It can also bring more business opportunities in the surrounding areas to promote the vitality of the communities in the neighborhood. Above all, the network is a significant ecological infrastructure in the city, which plays key roles in rain and flood management, climate regulation, ecological corridors, biological habitat construction and biodiversity conservation, etc.

This collection includes our researches and thinking of greenways and the construction of ecological corridor network in Beijing. It is without doubts that imagination is always easy, while implementation is always faced with great difficulties and uncertainties. But we can see that any great city in the world was finally built up based on the imaginations in the beginning, just like the construction of the Great Capital of Yuan.

In the article about Beijing from 'Lessons', Simonds concluded that: If you want to plan a great city, you need to learn to plan gardens first, for the principles of both are the same.

Essentially, our research is to explore a way to improve a city in the way of planning gardens, and we do hope that our research results may enlighten the future construction and development in Beijing.

Wang Xiangrong

January, 2018

目录 / contents

课题简介

Introduction of the Course

本课题是“绿都北京”系列研究之一，旨在把握“人居环境学科群”发展前沿，构建国内首创的以风景园林规划设计研究为核心，融合城市规划、建筑设计、园林历史、生态环境、植物景观、园林工程等多领域问题于一体的综合研究框架。

本课题聚焦于北京西北郊著名的三山五园地区。该地区地理环境条件得天独厚，自北京建城以来就成为维持城市生存最重要的水源地和自然生态屏障，清代盛期以来，随着皇家园林和行宫等的兴建而逐渐成为北京城最集中的自然、文化遗产聚集区，也是中国传统人居理念中城市与风景完美融合的代表，堪称世界级“自然和人类的共同杰作”。

新中国成立以来，在北京历次城市总体规划中，尽管对以颐和园、圆明园为代表的历史名园保护都提到了相当的高度，但“三山五园”却几乎从未被作为整体来考虑。近年来，随着北京城市建设的进一步扩张，该地区承担的城市职能愈发复杂，整体景观格局逐步解体，其遗憾不亚于对北京老城的破坏。如何充分发掘和积极保护该特殊地区的自然文化价值，以及在与城市发展协调的同时，如何逐步提升环境品质，已成为北京未来城市发展中极为复杂和重要的问题。

因此，本课题力图积极利用景观生态网络构建、城市历史景观认知、绿道规划建设等领域的前沿理论和科学分析方法，与现实情况紧密结合，从整体和局部两个层次，创造性地提出既有前瞻性、又具备可行性的系统性城市绿色空间提升、文化景观格局重塑方案。

Our subject is one of the 'Green Beijing Research Series' ongoing. We aim to grasp the frontiers of the human settlement environment subject, and then build a comprehensive research framework based on the research of landscape planning and design at first hand on China, integrating subjects of urban planning, architectural design, landscape history, ecological environment, vegetation landscape and garden engineering as a whole.

Our subject mainly focuses on the well-known 'Three Hills and Five Gardens' area in the northwestern suburb of Beijing. The region is richly endowed by nature with unique geographical environment, and it has become the most important water source and natural ecological barrier to maintain survival since the construction of Beijing began. With the constructions of royal gardens and palace museums since the prosperous period of Qing Dynasty, the area has gradually become the most concentrated area of natural and cultural heritage in Beijing. It is also the representative of the perfect fusion of urban and landscape in the traditional Chinese human habitat concept, which can be regarded as a world-class masterpiece of human and nature.

Since the establishment of People's Republic of China, although the protection of historic parks represented by Summer Palace and Yuan Ming Yuan has referred to a considerable height in the previous overall planning of Beijing, it cannot be ignored that the 'Three Hills and Five Gardens' area has almost never been considered as a whole. Recently, with the further expansion of urban construction in Beijing, the functions undertaken by this area are becoming more and more complex. We can see that the overall landscape pattern in this area is gradually disintegrated, which is regretful no less than the destruction of Beijing Old City. Problems about how to fully explore and actively protect the natural and cultural value of the area, and how to gradually improve the environmental quality have become extremely complex and significant in the future development in Beijing.

Therefore, we hope to actively use advanced theories and scientific analysis methods in the fields of landscape ecology network construction, city historical landscape cognition, greenway planning and construction, combining our research with actual situations, and then propose forward-looking and feasible plans of green space promotion and cultural landscape pattern reconstruction.

01 课题思考

STRATEGIC THOUGHT

北京是体现中国传统山水城市理念最重要的案例之一。三山五园的整体环境作为北京绝无仅有的城市文化景观，对其进行积极的保护迫在眉睫且意义重大。而面对这一极为特殊的研究对象，恰好是运用风景园林的研究视野、将相关多个专业的研究对象加以综合分析的极佳案例。然而由于该地段尺度巨大，且在功能类型、用地权属方面高度复杂，研究工作中的严密组织必不可少。

因此，本课题研究包含三个阶段的内容：一是从用地、交通、建筑、绿地与水系、历史遗产、非物质文化遗产及历史园林、历史水系七个角度入手，对该地区历史演变和现状条件进行专项深入调查，充分认知、科学评价其整体景观现状，提取其突出特色禀赋，充分考虑历史发展的脉络与上位规划的要求，从而完整、合理地制定“特色景观体系规划目标”；二是综合考虑现状问题与未来愿景，对该地区开放绿地格局进行整体提升规划，期望分别从城市、生态、文化等方面建立相互关联的要素体系；三是选择关键的节点地段，完成概念性设计，以实现这一特殊城市片区中绿道系统、生态网络和遗产廊道等的修复修补和完善提升。

Beijing is one of the most important cases reflecting Chinese traditional landscape city. As a unique urban cultural landscape in Beijing, it is imminent and significant to protect the overall environment of the 'Three Hills and Five Gardens' area. Because of the large scale and highly complexity in function type and land ownership, strict organization in the research is also essential.

Therefore, our research includes three stages. First, we should start with multiple viewpoints including land use, transportation, water system, architecture and garden history, so as to make in-depth investigation about region history and the present situation in this area. We should fully cognize and scientifically evaluate its overall landscape status, then extract its outstanding characteristic endowments, and fully consider the historical development as well as the requirements of upper plannings, so as to formulate characteristic landscape system plannings in a comprehensive and reasonable way. Second, we should consider the current situation and future vision comprehensively, so as to put forward the overall promoting plan for the open green space pattern in this area. We expect to establish an interrelated elements system from the aspects of city, ecology and culture. Third, we ought to select the key node locations and completing conceptual designs, so as to achieve the restoration and improvement of this area in aspects of greenway system, ecological network and heritage corridor.

基于实际地块的风景园林开放性研究
——三山五园地区绿道研究

An Open-ended Landscape Study on the Greenways in the Area of Three Hills and Five Gardens

林箐
Lin Qing

三山五园地区是北京市仅次于紫禁城和三海地区的文化遗产的集中地，是北京历史文化名城的重要组成部分。虽然近年的城市化发展已经将这个地区的大部分变为城市建成区，但它仍然是北京城区中建设强度较低、绿地比例较高的区域。三山五园地区不仅拥有重要的文化遗产，也是北京城市连接大西山的过渡区域，具有重要的生态意义。从风景园林的视角来看，该地区有许多可以改善的方面：历史文化遗产需要更好地保护，市民的生活需要更多的休闲场所，生态建设需要连通的绿色廊道，游客需要更好的游览服务和组织。我们选择这个地区进行绿道的研究，是希望通过这一过程探讨北京这座历史文化名城通过风景园林手段进行更新和环境改善的可能。这个课题中所面临的问题，也是新的社会背景下中国风景园林行业面临的诸多问题的一部分，非常具有代表性，例如：如何通过研究了解该地区的历史文化、自然特征和城市状况，从而确立恰当的景观发展思路？如何在纷繁复杂的城市网络中梳理出绿色的廊道，并建立人性化的慢行系统，保护该地区的文化特征，完善生态网络，促进绿色出行？如何在微观的层面通过设计解决具体的遗产保护、交通组织、景观提升和雨洪管理？

在历史和现状研究阶段，学生们以小组为单位，做了非常细致的调研和资料查阅。但也遇到了很多问题。因为规划面积比较大，现场调研不可能面面俱到，需要分出重点主次。由于很多学生在本科阶段只接触过中小型的场地设计，缺乏规划的概念，对于绿道的内涵也没有清晰的认识，对于课题要求的目标成果模糊不清，所以调查的时候针对性不强，需要重点调研的没有关注到，而不重要的罗列了许多。比如做现状绿地调研的同学，把重点放在现有公园绿地上，而对于最具有绿道潜力的各种附属绿地尤其是道路附属绿地匆匆带过，没有仔细调研；即便是现有的公园绿地，它的开放程度、出入口位置、周围的公交站点分布等非常重要的信息都遗漏了。其他各组的同学也都有类似的问题。经过课题组教师的启发、指导，经过多次补充调查，前期研究汇总出了非常扎实的成果。

在规划阶段，我们秉承开放的过程，希望每个小组能够提出自己的绿道选线方案，并阐述充分的理由。每个小组都试图寻找到合理的依据，并做了各种各样的分析。有些小组不仅逻辑严密，分析过程严谨，而且思考问题的方式非常独特，提出了令人信服的选线方案。也有些小组试图通过理性的分析来产生结果，但分析的对象和方法存在一定问题，后来在工作过程不断地矫正，最终调整为比较合理和可行的工作方法。具体方法的差异也导致了不同小组构建的绿道网络存在空间分布和尺度上的差异，带来了不同的结果，而这正是开放式研究课题想要达到的目标。

重点地段的设计是最接近学生们熟悉的风景园林设计的范畴。以往的设计课题，无论尺度大小、项目类型，总有不少学生只把视野局限在红线范围内，仅仅把它作为公园绿地，而不太考虑它与上位规划的衔接、与城市周边地区的融合以及应该承载的多种城市功能——如作为城市慢行系统的一部分、作为城市与自然联系的纽带、作为城市雨洪管理的绿色基础设施等等。经过上面两个阶段的铺垫，在这个阶段，对场地历史和现状的认识会比较自然地被结合在设计中，总体规划的成果也会贯彻在地块中。经过规划的环节之后再选择设计地块，更容易发现矛盾突出、存在问题而需要改进的地区。由于设计内容是城市绿道的节点，所以城市慢行体系的连通就成为一个重点，在这个过程中，必然要思考如何组织交通，如何调整城市用地，如何将分散的绿地联系起来，如何创造人性化的空间——如何用风景园林的手段来改善城市，解决城市问题，提升人们的生活品质，而不仅仅是在自我封闭的环境中凭空造景，孤芳自赏。

三山五园地区绿道这个研究课题，基于真实的地块，融合了多学科的方法和知识，采用了开放的规划设计过程，获得了一批既结合实际、又富有创新的成果。真实的地块能够锻炼发现问题和思考问题的能力；面临错综复杂的问题，必须运用多学科的知识进行分析思考，才能做出恰当的判断；而过程的开放能够提供很大的自由发挥空间，可以跳出实际项目的“窠巢”，找到独特的思路，提出创意性的解决方案（图 1）。

今天，各种与人类生存环境相关的问题日益严峻，每个人都能感受到身边环境的变化：城市离自然越来越远，而污染和热岛效应越来越严重；城市尺度越来越大，缺乏步行空间，城市的主角已经变成汽车而不是人；城市绿地从数量和内容上都难以满足需求，不成系统，可达性差；严重的城市内涝

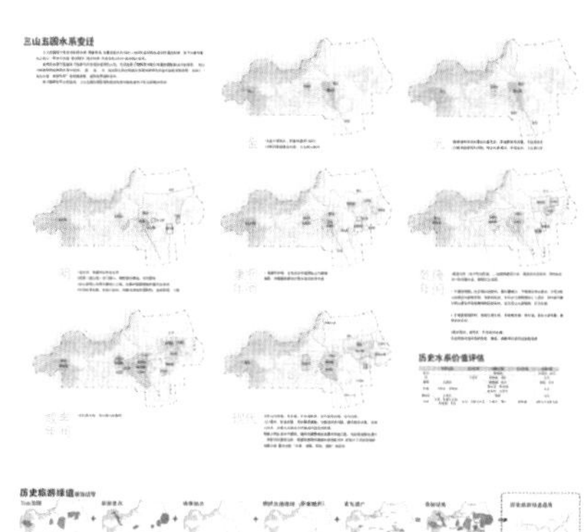

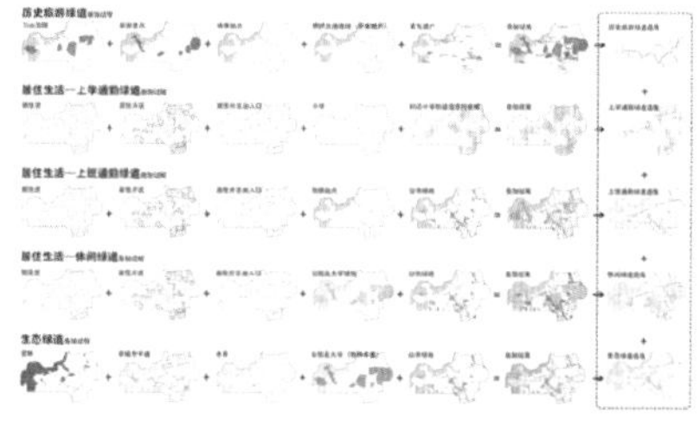

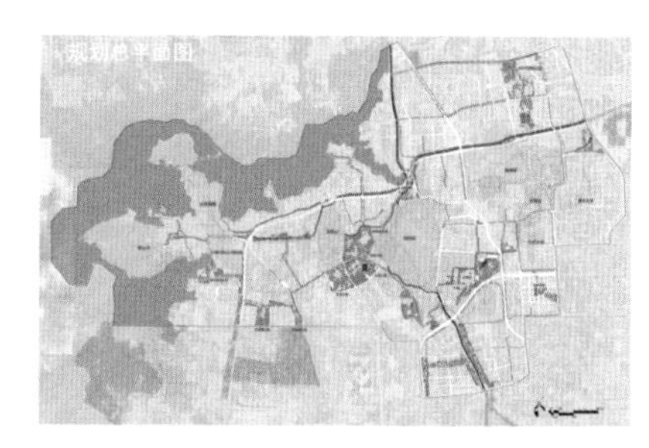

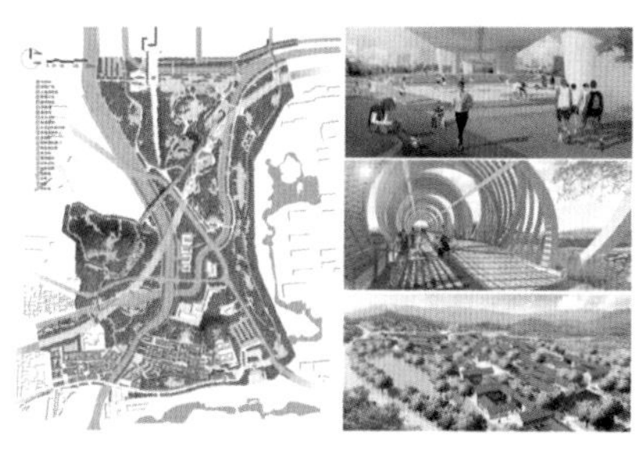

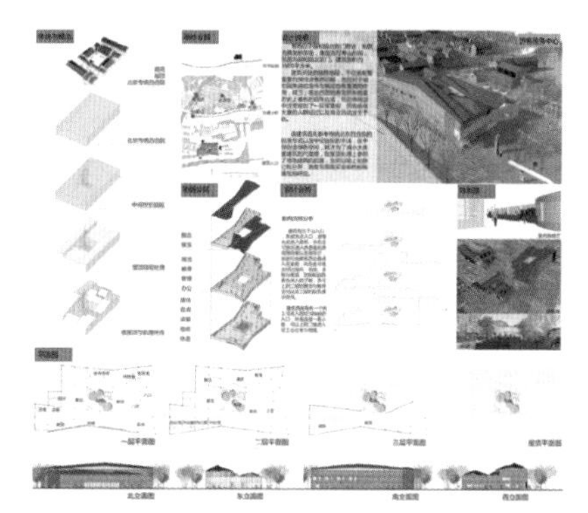

图 1　各阶段图纸体现了课题不同层次的要求——从研究、分析、规划到风景园林设计和建筑设计

与城市水系萎缩和地下水水位下降同时存在；城市无序蔓延，侵占乡村，蚕食自然，割裂水网绿脉……这些问题的解决都离不开风景园林师的参与。国内外的风景园林学者已经针对这些问题做了大量的研究和实践，提出了各种各样的观点和理论，所以才有这些年行业的热点话题——“绿色基础设施”、“景观都市主义”、“海绵城市”、“绿道”等等。

通过三山五园地区绿道这样一个综合性的课题研究，无论教师还是学生，都能够对这些理论有更深刻的理解，认识它们之间的内在联系。例如，本课题中，如选取香山地区作为重点设计地段，那么不得不面对的一个重要问题就是雨洪问题。由于这里是西山地区较大流域面积地表径流的一个汇流处，非常容易引起内涝，因此在连通绿道、梳理交通、提升景观、组织游览的同时，还要重点解决雨水的渗透、滞留和排放的问题。经过这个过程，“绿色基础设施”、“景观都市主义”、“海绵城市”、“绿道”就不仅仅是一个抽象的概念，而是和具体的措施、方法和方案联系了起来。这些理论实际上是对今天风景园林复合功能不同方面的强调。

今天，风景园林行业已经发展到了一个新的阶段，对风景园林人才提出了新的要求。风景园林师有越来越多的机会介入更广泛的人居环境建设中，并起到领导者和协调者的作用。新一代的风景园林师只有跳出公园绿地的藩篱，看到更广阔的城市、乡村和自然，才能把握时代赋予的机遇，将风景园林行业推向新的广度和深度。

备注：

图 1 作者：佟思明、王晞月、李璇、崔兹辰、李媛、徐慧、Mosita、莫日根吉。

（本文根据《风景园林》2016 年 01 期《务实、融合、开放——北林研究生风景园林设计课程改革尝试》删改而成）

The area of Three Hills and Five Gardens is one of the concentrations of cultural heritage, being next only to the Forbidden City and the Three Sea Area. It is an important part of Beijing as the famous historical and cultural city. Now this area remains low construction intensity and high proportion of green space. The Three Hills and Five Gardens not only has important cultural heritage but also is the transitional region connecting the city and the western hills in Beijing, which has great ecological significance.

Thus, we can find plenty of issues to improve from the view of landscape architecture. For example, historic and cultural heritage needs better protection, citizens need more leisure spaces, ecological construction needs connected green ways, and tourists need better tour service.

We pick greenway as our topic, so as to discuss the possibility of renewing this historical and cultural city and improving the environment at the same time. We think the problems we have to solve in this study represent plenty of challenges of landscape architecture development in China to a large extent.

In the stage of historic and present situation study, students join in groups to accomplish investigation and information assessment. We encourage students to advance their research from the view of urban planning and learn to distinguish primary and secondary issues at the same time, which enables them to sum up through prophase results.

In the stage of planning, each study group provides their greenway selection scheme and gives a full justification.

We can see obvious differences on greenway network and space distribution in different groups because of different research methods, and this is exactly the goal of our open research subject.

Then in the stage of key sector design, we combine the historic and present study with the design and especially attach great importance to the connection between the design and the master plan. We also put emphasis on the connection of urban slow-moving traffic system and think about how to improve our city through means of landscape architecture.

Nowadays, problems related to the living environment of human beings are becoming more serious than ever, and the solution of these problems is surely inseparable from the participation of landscape architects.

The study on Three Hills and Five Gardens enables us to deepen the understanding of theories in landscape architecture and urban planning, and turns out to be a comprehensive study with great reference significance.

跨专业协作　筑城绿一体

Multidisciplinary Cooperation Research for Building a Green City

刘志成
Liu Zhicheng

当前，风景园林设计行业的实践领域非常广泛，特别需要拓展研究领域与内容，培养具备多学科综合能力的复合型人才，探讨解决本学科所面对的综合性和复杂性问题的途径。为实现这一目标，课题选址在文化遗产高度聚集、城市用地条件比较复杂的北京三山五园地区，以“北京三山五园地区绿道概念性规划与重点地段设计”为题，展开研究，划定了81.33km^{2}的规划研究范围，用地规模较大，涉及内容综合性强，需要以风景园林学科的视角，综合运用城乡规划、建筑学、生态学等相关人居环境学科的研究成果，探讨场地规划与园林设计、生态修复与植被规划、工程技术与雨洪管理、城市设计与建筑设计等多方面内容，充分对接学科的前沿性话题与当下我国城市建设的重大课题，全方位拓展研究领域，多途径展开设计研究，广泛地探索城市问题的解决办法。整个课题研究包含两个阶段与层次：

1. 以规划为前导，以绿道构建研究为核心，形成城乡规划、建筑学、风景园林三位一体的研究机制：

本设计分为绿道概念性总体规划与重点地块设计两个阶段。鉴于研究的区域涉及内容复杂，现状调研成为规划与设计的重要前提。调研以分组、分专题工作的方式，从如下角度展开研究，包括：现状部分（用地类型、交通条件、绿地水系、文化遗产、建设强度与建筑风貌）、历史部分（水系演变、园林发展、非物质文化遗产）。重点工作是从历史演变和现状条件来看，综合分析这一地段的优势与问题。

绿道概念性总体规划阶段的核心任务是结合现状调研分析，探讨三山五园地区绿道的选线模式和路径规划。课题研究中允许对这一地区的绿地、水系要素的空间分布及城市建设安排进行必要的调整，以满足进一步保护历史环境、提升景观风貌、完善城市功能、合理促进社会经济发展及改善民生的目的。最终完成了区域绿道概念性规划，并进一步完成了水系统规划、慢行系统与服务设施规划、植被系统规划等专项规划。

综合性的研究内容需要多学科的教师团队与之相匹配，学院领导精心策划，抽调了园林设计教研组、园林历史与理论教研组、园林工程教研组、植物景观教研组、城市规划教研组和建筑设计教研组的八位教师，完成课题研究工作。在教学过程中，跨专业联合教学也为教师提供了很好的交流平台，增进了不同专业之间和不同教研组之间的相互了解，使教学过程不仅成为一个培养学生的平台，也成为一个教师探讨、研究规划与设计等相关问题，交流教学理念与教学方式的平台，形成了一个可以不断生长的平台，造就了进一步多途径开放教学的可能性。

教师跨专业协作，分别结合各自专业与研究领域，多层面、多视角地展开讨论，以不同的专业视角启发、引领学生们向更为广阔、更为深远的专业领域前行。在研究过程中，各学科之间相互支撑与合作，专业之间交流与碰撞，带来多元的思考、多样的选择与多维度的评价，使研究无论在广度还是在深度上，都能得到全面拓展，形成城乡规划、建筑学、风景园林三位一体的研究机制。

2. 多层次探讨，多维度创新，全面探究城市绿色空间构筑的内容与手段

风景园林规划与设计是一项综合性非常强的实践活动，不仅要求我们具备优良的专业素质、专业基础与专业技能，更需要具备创新精神与创新意识。研究内容已不仅限于城市绿地本身，需要多维度地探索，多层次地探讨最优的构建途径。

在规划前期，研究团队分别就城乡规划、城市设计、绿道规划三个层级展开主题探讨。在设计阶段，研究内容包括风景园林设计、风景园林工程与技术、生态修复、雨洪管理、植物景观、建筑设计六个专题，多层次、多维度地展开研究，强调专业基础研究的同时，注重研究深度的拓展，以国际化的视野、前沿性的思考探讨国内热点问题与社会需求，探索构建三山五园绿道的思路与途径。

研究团队以组为单位完成规划与设计任务，在查阅了大量的文献资料的基础上，展开系统研究。各位教师根据自己的专业特长，各抒己见，分别立足各自的专业领域，全面深化了对于场地的认知，理清了脉络，突出了重点内容，深入阐释绿道规划与城市开放空间的核心内容、相关规范与具体要求，畅谈绿道建设的意义与价值，多维度地展开讨论，全面拓展规划与设计思路，形成了良好的创新性研究氛围。各小组都将保护该区域的文化遗产作为规划与设计的重要内容，独立分析，分别从不同的视角切入，展开规划与设计，多维度创新，各具特色。

如有的小组将绿道的核心功能定位为完善三山五园地区皇家园林的文化遗产的保护体系，通过综合研究文化遗产、生态基底、基础设施、居民生活、旅游发展和空间特征六个因子，完成三山五园地区的绿道选线规划，并希望绿道能够进一步发挥传播历史文化、提升市民的文化自觉之目的。另一小组设计地块位于毗邻颐和园北宫门的青龙桥片区，以生态修复为主体，以“缝合”为概念，通过运用景观生态学理论，整合城市交通系统、构建生态廊桥与生态驳岸，对破碎的生境斑块进行“缝合”，实现青龙桥片区历史文化的繁荣、生态格局的健全和游憩功能的完善。团队以“水融园外园”为设计概念，运用 GIS 对场地进行了适宜性分析，研究确定了适合建设历史资源廊道、休闲游憩廊道、生态保护廊道的点和线，并结合潜力空间，形成绿道概念性规划。京密引水渠和清河夹合成的三角地段作为设计场地，采用 AHP 层次分析法对区域进行综合评价，确立生态型、都市型和文化型三类绿道，以综合评价结果确定需要连接的景观斑块，整合区域特色，形成绿道的规划，希望借助现有的生态资源，形成具有生态优异、文化特征突出、和谐、优美的都市空间。

敞开校门，积极邀请校外设计机构的设计师参

加阶段性研讨，是多层次、多维度探究的另一种体现。教学与研究过程中，教学团队的老师们先后邀请了北林地景、清华同衡规划院、EDSA、北京创新景观等设计机构的著名一线设计师和清华大学建筑学院教师参加阶段性评图。他们的到来，使学生们有机会与境内外的著名设计机构与优秀设计师面对面地沟通，讨论设计方案，直接了解成熟设计师对于具体项目的理解、认识、思维逻辑及设计机构的具体需要，对拓展设计思维的广度与深度都具有非常大的帮助。并且，这种交流也进一步加强了学院各个层面与各用人单位的相互了解，促成了专业教师与用人单位直接探讨、谋划专业人才培养方式。他们的建议与思考对进一步提升教学质量，更好地培养社会需要的优秀人才具有重要意义。

（本文节选自《风景园林》2016 年 01 期，原标题《交流启迪灵感 融合铸就创新——北林园林学院硕士研究生风景园林设计 STUDIO2 课程教学改革实践》）

Currently, landscape architecture has a wide range of practice field, it especially shows a great need for inter-disciplinary talents in order to extend its research field in face of the comprehensiveness and complexity of this subject. As a result, we locate our project in Three Hills and Five Gardens with highly aggregated cultural heritages and complex land-use conditions, and pursue our research under the topic of 'Conceptual Planning of Greenway and Main Section Design in Three Hills and Five Gardens'. With the research area of 81.33km^2, our project has comprehensive demanding. It requires researches from landscape architecture subject perspective and the application of research results of urban planning, architecture and ecology, etc. We also need to discuss issues such as site planning and landscape design, ecological restoration and vegetation planning, engineering technology and management of rain and flood, urban design and architecture design, etc , so as to call for cutting-edge topics and major topics of urban construction in China.

As a whole, our study contains two levels.

First, guided by planning, we take greenway construction as the core of our study and try to form a research mechanism including urban planning , architecture and landscape architecture. Our research process includes two steps: conceptual planning of greenway and main section design. During the teaching, we find interdisciplinary combined teaching a great communication platform for enhancing understandings between various subjects, which is very beneficial for our research.

Second, we try to carry forward our study from dimensional analysis and figure out the contents and means of urban green space construction.

Landscape architecture planning and design is a comprehensive and practical activity, so it not only demands excellent professional quality, but also a great sense of innovation. As a result, our research is not limited to the urban green space itself and we try to explore optimal approaches at a multidimensional degree.

In the earlier stage of the study, we considerate our site from three aspects: urban planning, urban design and greenway planning . In the stage of design, we unfold our research structured around six topics including landscape architectural design, landscape engineering and technology, ecological restoration, rain and flood management, vegetative landscape, and architectural design.

Key Researchers from different subjects provide a variety of ideas, and students are divided into different study groups to do researches from different perspectives with the background of cultural heritage in the district, such as research on cultural heritage protection system of royal gardens in Three Hills and Five Gardens, ecological restoration in Qing Long Qiao District, and environmental fitness analysis on Three Hills and Five Gardens, etc.

In the process of our study, designers and teachers from Beijing Beilin Landscape Architecture Institute, THUPDI, EDSA and Beijing Topsense Landscape Design took part in our periodicity review. Their suggests provided us a great help for our teaching and research.

“城市 - 风景 - 遗产”一体的研究框架及其在三山五园的应用构想

The "Urban-Landscape-Heritage" Integrated Research Framework and its Applications to the Landscape Regeneration of Three Hills and Five Gardens in Beijing

钱云
Qian Yun

1 研究视野的溯源及历史影响

对城市、风景、遗产一体进行研究的理念可追溯自中国上古时代聚落营造中“国野一体”的观念。在先秦时期礼文化的政治体制下，以分封为基础的国、野一体的城乡体系逐步形成[1-7]。秦汉以后两千多年的社会总体上继承了西周这一整套礼制制度的主要内容，使古代中国城市和周边郊野地区长期呈现高度一体的“统筹规划”[8]。在这一体系中，古人在聚落空间营造上高度强调对自然环境基底的尊重，“国”、“野”之间有机联络、功能互补，尤其是在聚落建设中对地形地貌勘察、水源保护、水灾防范等尤为关注。

这种“国野一体”的理念强调了人与天地之间、文化与自然之间的关系是连续的，这表现为中国传统城市与其周边自然环境高度的空间连续性，从而造就了诸多独特的、不可复制的城市形态和风貌。例如著名的历史文化名城南京素有“虎踞龙盘、负水带江”之城，其周边的三条山系与长江、秦淮河等水系格局直接影响了历代城市选址和布局，对古城的城墙轮廓、空间轴线、街道走向等都起到了重要的限定作用，造就了南京城依山就势、曲折蜿蜒的独特城市形态[9, 10]（图 1，图 2）。

同时，由于长期人类文化活动与自然环境的充分互动，“国野一体”中的“野”，即城市周边山水环境的内涵和范畴在历史发展中不断扩大，不仅限于各自然要素，也包括了大量的园林、寺庙、村庄、农田、墓葬等历史遗存或文化空间，这赋予城市周边山水环境以丰富的社会功能和文化内涵[11]。例如对泰山、黄河等名山大川的封禅、祭祀活动历代均为重要的国家大事之一，而各地府、县等周边的镇山、坐山等也都具有崇高的文化地位。这些历史遗产和文化空间要素往往沿驿道、运河、水系等线性空间分布，各节点间通过不同功能彼此串联形成

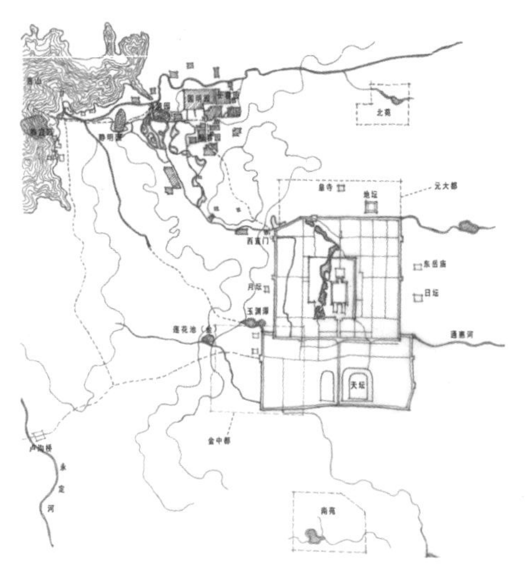

图 1　北京历史古城与周边环境要素相关的城市形态

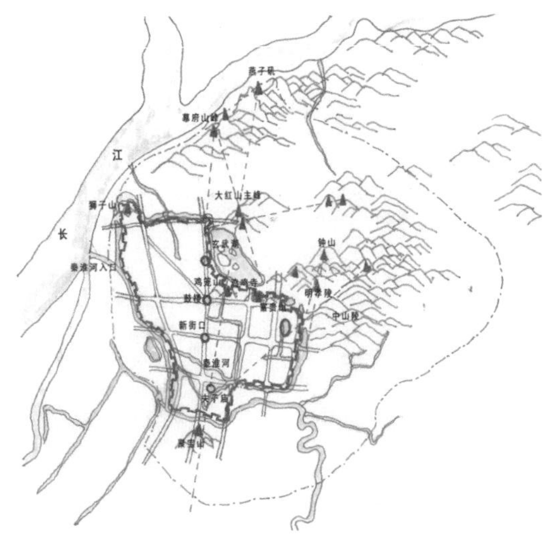

图 2　南京历史古城与周边环境要素相关的城市形态

网络，组成完整的文化景观体系，较全面地呈现出整个“国－野”体系在军事防御、社会组织、经济生产、祭祀礼仪、休闲游憩等诸方面、多层次的历史信息[11, 12]。

显而易见，这种基于“国野一体”空间范畴、饱含历史遗产的景观要素网络充分体现了东方文明在聚落营建上的理念，也为聚落整体形态的发展演变历程提供了较为可信的解释。因此，构建“城市－风景－遗产”一体的研究视野，致力于充分关注自然山水和历史文化遗产等景观要素的影响，有助于从“空间”和“时间”两个维度进一步拓展发掘“国－野”之间的互动影响，是人居环境学科领域充分继承和发扬中国传统文化思想的具体体现。

2 在城市特色塑造中的价值

在城市研究中，这一研究视野最重要的价值体现在城市特色风貌塑造研究中。工业时代以来，技术理性思想主导的大量“现代化”城市新区以“千城一面”的方式迅速蔓延，对传统城市周边的自然、遗产地段高强度的侵入，导致长久以来传统城市的景观格局体系日趋碎片化，不仅使景观视觉价值严重跌损，而且造成城市文化认同感的急剧下降，特色景观形态对人居聚落体系发展的解释性功能也大大衰退，这对于城市未来可持续发展探索造成了巨大的威胁。城市特色风貌的保护与重构也因此成为世界性的议题[13-15]。

在对“增长至上”、“人定胜天”思想滥觞深刻反思后，西方城市设计的发展走过了一条从否定传统到再次回归传统的轨迹。19 世纪末，卡米诺 • 西特指出欧洲传统城市中的人文主义精神正在失去，城市设计应努力塑造类似传统城镇的丰富、自然的空间形态[16]。埃德蒙 • 培根和简 • 雅各布斯则解释了传统城市中连续的步行体系和多样化社会生活对城市美学的影响[17, 18]。因此，向传统城市学习以把握人居环境发展中经久不衰的历史经验，已经得到了广泛共识[19, 20]。

众所周知，中国正在经历世界罕见的城市建设高速扩张期，也是城市特色风貌丧失尤为严重的时期。庆幸的是，当前诸多城市已经开始在城市特色景观格局营造方面展开研究，而事实证明，支撑城市发展的原生自然环境和真实留存的历史遗产，必将是城市形态中首要的、不可复制的特色所在。因此对中国城市中传统的“城市－风景－遗产”体系的再认识和重新构建，必将成为新时期城市特色风貌营造的核心任务。

3 在遗产保护实践中的价值

历史遗产保护与城市、景观领域的交叉融汇是随着其涵盖范畴、价值评估和保护策略等长期发展而不断明晰的。1964 年的《威尼斯宪章》首次提出了“历史地段”的概念，这一概念在 1976 年的《内罗毕建议》得以更为明确和广泛，即包括了“史前遗迹、历史城镇、老城区、老村庄、老村落以及古迹群”等[21]。2011 年，第 36 届联合国教科文组织大会提出了“城市历史景观”（Historical Urban Landscape, HUL）这一新概念，体现了诸多突破：首先是遗产的对象不仅限于人类建造的成果，而是强调要关注更为广阔的聚落外围环境；其次是不再强调将被保护的实体从城市整体环境中分离出来，而是致力于发掘其与外部要素的关联性；更重要的是，不再仅仅强调实物的保存，而是将历史信息保护与社会经济文化发展相结合，以现实的态度允许并引导城市历史景观在未来的“有机”生长[22]。具体而言，这一发展体现了遗产保护的区域化、网络化的趋势，用以应对城市快速扩张、人地关系危机和文化认同危机所带来的遗产孤立化、碎片化威胁[23, 24]。

随着诸多西方国家对遗产保护体系的不断完善，在遗产单体、历史保护区和城市周边整体景观体系的不同层次均提出了相应的保护措施。2000 年通过的《欧洲景观公约》，将景观作为文化、自然遗产多样性的体现和人们生活空间特性的基础，进行整体规划和管理[25-28]。日本也建立了以《历史风致法》为代表的较完善的景观保护法律体系[29, 30]。上述法规充分强调了城市及周边地区的自然、人文景观遗产体系的整体保护与当下建设发展的协调，使城市景观遗产成为城市战略资源的重要组成部分，对城市竞争力的贡献也日趋显著。

2005 年，国务院颁布《关于加强文化遗产保护的通知》，以“文化遗产”的提法逐步取代了过去“文物”的概念，并不断吸收国际遗产保护界的最新理念，越来越强调遗产保护的区域性、体系性，在具体实践上的举措也从无到有、不断探索[31, 32]。在 2002 年编制的《北京历史文化名城保护规划》中，针对旧城内的景观轴线、城郭轮廓、街道对景、建筑色彩、古树名木等保护措施细致、系统；但对外围环境只是散点式划出 2000 余处文物单位、西郊皇家园林等 10 片保护区以及莲花池历史水系等对象，并未对古城与其赖以生存的自然山水环境和历史聚落网络提出保护措施。而在 2010 年后的遗产保护规划实践中，杭州等城市依托其丰富的城市历史景观资源，探索性地提出了构建“遗产区域”、“遗产廊道”或“文化线路”的设想[33-37]，这体现了“城市－风景－遗产”一体的视角在遗产保护方面的有益尝试。

4 在风景园林营造中的价值

风景园林营造可上溯至古代的造园术。早期的园林因规模有限，与周边环境的相互影响和融合不甚显著，其营造逻辑主要满足审美和文化追求。西方工业革命以后，随着城市与周边自然环境的互动加剧，风景园林规划设计的范围迅速扩大，以奥姆斯特德为代表的现代风景园林师开始涉足较大尺度自然与城市景观的营造。自此，现代风景园林学科不再只关注被围墙所包围的、世外桃源式的风景，对待城市环境采用“佳者收之、俗者屏之”的思想也被改变，转而关注更宽泛的、包括各类城市公共空间等对象的景观系统，与生态科学与人文精神的互动越发深入[38-40]。

因此，当代风景园林规划设计的内容，已逐步被认为是对多层复杂系统及其叠加过程的研究、重塑。这一复杂系统包括反映地貌塑造过程的自然系统、源于自然基底并依赖生物自然生长的农业系统以及人工显著改变原始地貌并具备系统功能的聚落系统。各个自然层和文化层长时间形成了由下而上堆积形成的过程，每一层都为后来一层提供了空间上的环境累积[41]。而风景园林规划设计除了注重空间，还必须考虑时间维度，从景观系统的生成发育和动态演变中把握其研究重点，聚焦三个子系统叠加时的关联。具体而言，风景园林规划设计就是在选定地域背景下，剖析景观系统的生产、生活、生态属性和人文艺术内涵，注重自然、农业和聚落系统各自形态的生成机制和互动影响，进而探讨其更新途径及在面对工业化强力冲击下的自我修复和可持续生长策略[42-49]。

因此，“城市－风景－遗产”一体的研究视野的引入，反映了风景园林学科从主要关注“建造活动”转向更加关注“层级体系的演变过程”的进展。而在强调“可持续发展”的时代背景下，对城市、历史和生态的解读，共同为城市景观格局的重塑、再生提供了依据，也强化了风景园林的学科生命力。

5 对北京三山五园地区的思考

北京是体现中国古代山水城市理念最重要的案例之一。其中，西北郊的“三山五园”体现了中国传统人居理念中与山水的交融，也是历史遗存高度聚集的地区[11,50-52]。燕山余脉、永定河冲积扇和大片稻田，造就了这一地段蜿蜒丰富的视觉景观、丰沛的水源及丰茂的植被。依托这些自然条件，多个皇家园林在自然景观上互借互成，西山作为视觉大背景，香炉峰、玉泉山、万寿山等自然山体分别成为不同园林的主要借景点或控制点，共同形成一个“移天缩地在君怀”的眺望体系[53, 54]。在功能上，五个核心园林各执其责：静宜园、静明园主要承担了涵养水源和山居游赏的功能，畅春园为皇太后住所，圆明园集园居、理政、游憩于一身，颐和园初为城市的调节水库，后代替圆明园承担园居、理政功能；此外还有大量配套服务设施分布于各个园林之间及周边，如京西稻田为皇室提供稻米，八旗营房是皇家安全防卫体系，周边的私家园林群供与园林中的皇帝日常议政的官员居住[55]。在如此集中的山水地域内，历史遗产保存之完整、信息之丰富、影响之深远，在世界范围内绝无仅有（图3）。

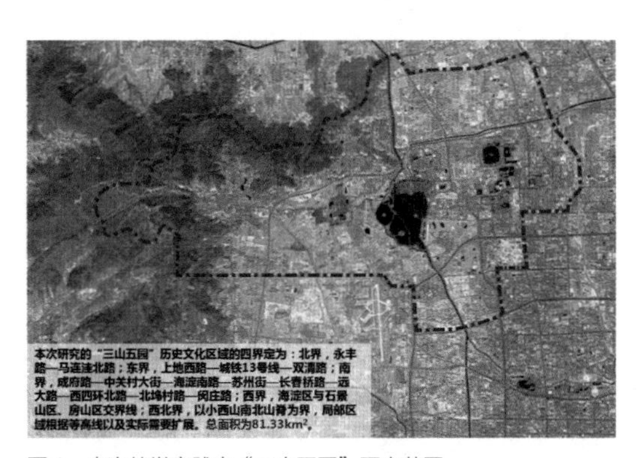

图3 本次教学实践中“三山五园”研究范围

时至今日，三山五园已完全融入北京城市建成区中。大量城市建设使其整体景观格局逐步陷入“孤岛化、割裂化”。而各园林之间的田野、兵营、水系、历史村镇等则更加受到忽视，逐渐湮没在城市的居住、文教和商业区中，完全失去了景观和功能上的“有机连接”作用。

在北京历次城市规划中，“三山五园”几乎从未被作为整体来考虑[52,56]。当前，宏观景观格局受到的威胁主要表现在两个维度上：一是五环路等大型基础设施的建设，加剧了自然和历史要素之间的割裂局面；二是从三维的角度，以中关村西区为代表的大量高层建筑，直接破坏了多个历史场景的天际线和自然历史要素之间丰富的视觉联系网络。有学者指出，“三山五园”整体景观格局的逐步解体，其遗憾甚至不亚于当年对北京古城的摧毁[52]。而长期以来由于“城市－风景－遗产”一体视野的缺乏，绝大多数的民众甚至相当数量的专业人士，却对此熟视无睹，或正毫无知觉地参与到这一场新的浩劫中来。

毫无疑问，“三山五园”的整体环境作为北京独一无二的城市文化景观，对其进行积极的保护迫在眉睫且意义重大。而面对这一极为特殊的研究对象，恰好是运用“城市－风景－遗产”一体的研究视野、将相关多个专业的研究对象加以综合分析的极佳案例。然而由于该地段尺度巨大，且在功能类型、用地权属方面高度复杂，展开研究的基础调查工作量相对庞大。因此笔者期望将相关研究工作分为三个阶段：第一阶段是充分认知、科学评价其整体景观风貌现状，提取其突出特色禀赋，充分考虑历史发展的趋势与未来（上位）规划的要求，从而完整、合理地量身订制出“特色景观体系规划目标”。第二阶段则期望分别从城市（生产与生活）、风景（生态与环境）、遗产（历史与文化）三个方面对这一目标进行具体解读，每个方面分别建立三个要素体系。其中重点强调的是，基于城市与风景要素的关联互动，建立“轮廓眺望体系”；基于风景与遗产要素的关联互动，建立“遗产廊道体系”；二者共同作为构建“城市特色景观格局”的核心内容。第三阶段则是选择这一体系中重要的节点地段，完成局部设计探讨（图 4）。

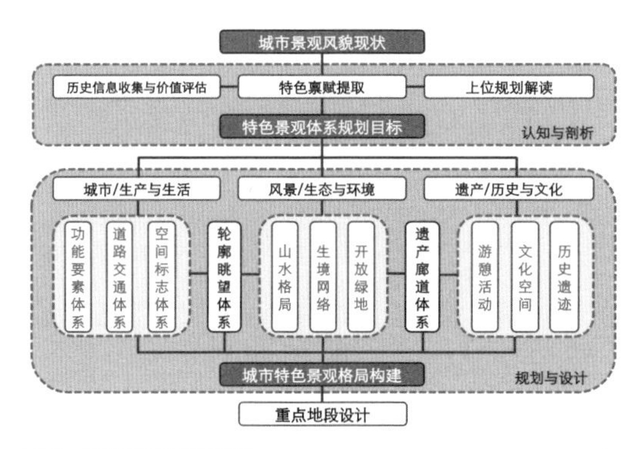

图 4　研究工作框架构想

总之，整个研究过程中充分强调多要素整体结构的“完整性”和“连续性”，在此基础上对空间的多样性、景观尺度的控制、遗产环境的塑造等，尽量提出创造性的解决方案。

6 结论与展望

自古以来，中国传统城市营造中，以其独特的方式表达了对自然山水、历史文化的高度尊重，从宏观尺度形成了诸多如“三山五园”等堪称“自然和人类的共同杰作”的文化景观遗产。因此对于当代的规划设计师，无论从塑造城市特色风貌、促进遗产保护系统网络化、还是探索特色景观体系的可持续发展路径方面，运用“城市－风景－遗产”一体的宏观视角都已成为研究分析中克服以偏概全、“盲人摸象式”错误倾向的必然选择。北京林业大学园林学院开展的以“三山五园特色景观体系重塑”为主题的探讨，为这一研究视野在实际问题中的运用进行了初步尝试，相关构想为今后展开深入、系统的研究奠定了有益的基础。

参考文献：

[1] 张杰，邓翔宇．论聚落遗产与文化景观的系统保护 [J]. 城市与区域规划研究，2008(3):7-23.
[2] 卜工．文化起源的中国模式 [M]. 科学出版社，2007.
[3] 贺业钜．考工记营国制度研究 [M]. 中国建筑工业出版社，1985.
[4] 董鉴泓．中国城市建设史 [M]. 北京：中国建筑工业出版社，1989.
[5] 贺业钜．中国古代城市规划史 [M]. 北京：中国建工出版社，2004.
[6] 汪德华．中国城市规划史纲 [M]. 南京：东南大学出版社，2005.
[7] 李学勤．周礼注疏 [M]. 北京：北京大学出版社，1999.
[8] 张自慧．礼文化中的人与自然之和谐观 [J]. 贵州社会科学，2005(5):69-70.
[9] 姚亦峰．基于自然地理格局的南京古都景观研究 [J]. 建筑学报，2007(2).
[10] 金汤．南京城市山水景观状况与保护 [J]. 规划师，2003(4):88-90.
[11] 邓翔宇．中国传统城市周边地区系统性保护研究 [D]. 北京：清华大学，2010. 规划研究，2008(3):7-23.
[12] 张弓．中国古代城市设计山水限定因素考量——以承德为例 [D]. 北京：清华大学，2006.
[13] 汪德华，王景慧．对城市特色问题的认识 [J]. 城市规划，1989(2):17-19.
[14] 张继刚．城市景观风貌的研究对象、体系结构与方法浅谈——兼谈城市风貌特色 [J]. 规划师，2007(8):14-18.
[15][美] 刘易斯•芒福德．城市发展史：起源、演变和前景 [M]. 宋峻岭，倪文彦，译．北京：中国建筑工业出版社，2005:21.
[16][奥] 卡米诺•西特．城市建设艺术 [M]. 仲德崑，译．南京：东南大学出版社，1990.
[17][美] 埃德蒙•培根，黄富厢．城市设计 [M]. 朱琪，译．北京：中国建筑工业出版社，2003.
[18][加] 简•雅各布斯．美国大城市的死与生 [M]. 金衡山，译．江苏：译林出版社，2012.
[19] 张杰，张弓，等．向传统城市学习——以创造城市生活为主旨的城市设计方法研究 [J]. 城市规划，2013(3):26-30.
[20] 张杰，吕杰．从大尺度城市设计到“日常生活空间”[J]. 城市规划，2003(9):40-44.
[21] 联合国教科文组织世界遗产中心，国际古迹遗址理事会，国家文物保护与修复研究中心，中国国家文物局．国际文化遗产保护文件选编 [M]. 北京：文物出版社，2009.
[22][荷] 罗•范•奥尔斯．城市历史景观的概念及其与文化景观的联系 [J]. 韩锋，王溪，译．中国园林，2012(5):16-18.

[23] 史晨暄 . 世界遗产保护新趋势 [J]. 世界建筑 ,2004(6):80-82.
[24] 郑颖，杨昌鸣 . 城市历史景观的启示——从“历史城区保护”到“城市发展框架下的城市遗产保护”[J]. 城市建筑 ,2012(8):41-44.
[25] 张松，蔡敦达 . 欧美城市的风景保护与风景规划 [J]. 城市规划 ,2003(9):63-66,70.
[26] 肖笃宁，曹宇 . 欧洲景观条约与景观生态学研究 [J]. 生态学杂志 ,2000(6):75-77.
[27] 赵明，张松 . 城市景观的保护与塑造——以法国里昂的规划实践为例 [C]. 规划 50 年——2006 中国城市规划年会论文集：风景与园林绿化 ,2006.
[28][日] 西村幸夫，等 . 城市风景规划——欧美景观控制方法与实务 [M]. 张松，蔡敦达，译 . 上海科学技术出版社 ,2005.
[29] 张松 . 日本历史景观保护相关法规制度的特征及其启示 [J]. 同济大学学报 ,2015(3):49-58.
[30] 相秉军，杨自安，顾卫东 . 中日传统城市景观的保护、再生与创造——以京都和苏州为例 [J]. 现代城市研究 , 2000(5):39-44.
[31] 刘祎绯 . 我国文化遗产认知的空间扩展历程 [J]. 建筑与文化 ,2015(2):128-129.
[32] 刘祎绯 . 我国文化遗产认知的时间扩展历程 [J]. 建筑与文化 ,2015(6):132-133.
[33] 刘祎绯 . 文化景观启发的三种价值维度：以世界遗产文化景观为例 [J]. 风景园林 ,2015(8):50-55.
[34] 王景慧 . 文化线路的保护规划方法 [J]. 中国名城 ,2009(7):10-13.
[35] 吕舟 . 文化线路构建文化遗产保护网络 [J]. 中国文物科学研究 ,2006(1):59-63.
[36] 俞孔坚，李伟，李迪华 . 遗产廊道与大运河整体保护的理论框架 [J]. 城市问题 ,2004(1):28-31,51.
[37] 朱强，李伟 . 遗产区域：一种大尺度文化景观保护的新方法 [J]. 中国人口 • 资源与环境 ,2007(1):50-55.
[38] 孙筱祥 . 风景园林 (Landscape Architecture) 从造园术、造园艺术、风景造园——到风景园林、地球表层规划 [J]. 中国园林 ,2002(4):7-12.
[39] 王向荣，林箐 . 现代景观的价值取向 [J]. 南京林业大学学报 ,2002(4):43-49.
[40][英] 伊恩 .D. 怀特 .16 世纪以来的景观与历史 [M]. 王思思，译 . 北京：中国建筑工业出版社 ,2011.
[41] 侯晓蕾，郭巍 . 场所与乡愁——风景园林视野中的乡土景观研究方法探析 [J]. 城市发展研究 ,2015(4):80-85.
[42] 俞孔坚 . 景观：文化、生态与感知 [M]. 北京：科学出版社 ,1998.
[43] 钱云，庄子莹 . 乡土景观研究视野与方法及风景园林学实践 [J]. 中国园林 ,2014(12):31-35.
[44] 郑曦 . 鉴湖、西湖、湘湖——钱塘江下游地区三大著名湖泊的景观演变与城市化发展启示 [J]. 中国园林 ,2014(11):69-73.
[45] 韩炳越，沈实现 . 基于地域特征的风景园林设计 [J]. 中国园林 ,2005(7):61-67.
[46] 林箐，王向荣 . 风景园林与文化 [J]. 中国园林 ,2009(9):19-23.
[47] 王向荣，韩炳越 . 杭州“西湖西进”可行性研究 [J]. 中国园林 ,2001(6):11-14.
[48] 王向荣，韩炳越 . 社会 . 生态与艺术的融合——绍兴市镜湖景区概念性规划 [J]. 中国园林 ,2004(11):20-24.
[49] 王向荣，韩炳越 . 资源保护、历史延续与景观再生——杭州湘湖保护与开发启动区块规划 [J]. 中国园林 ,2005(1):13-19.
[50] 郭黛姮 . 三山五园：北京历史文化最辉煌的乐章 [J]. 北京联合大学学报 ,2014(1):58-60.
[51] 刘剑，胡立辉，李树华 . 北京三山五园地区景观历史性变迁分析 [J]. 中国园林 ,2011(2):54-58.
[52] 阙镇清 . 再失一城——北京西北郊皇家园林集群：三山五园在城市化过程中的没落 [J]. 装饰 ,2007(11):16-20.
[53] 祝丹 . 北京颐和园景观与“三山五园”的构成关系 [J]. 大连民族学院学报 ,2013(3):291-295.
[54] 张杰，熊玮 . 清代皇家园林规划设计控制的量化研究——以圆明三园、清漪园为例 [J]. 世界建筑 ,2004(11):90-95.
[55] 王其亨，张龙，张凤梧 . 从颐和园大他坦说起——浅论圆明园和颐和园历史功能的转换 [C].《圆明园》学刊 ,2008(8):150-154.
[56] 周景峰，汤羽扬 . 京郊历史文化保护区保护问题的思考 [J]. 工程建设与设计 ,2005(4):40-43.

（本文节选自《风景园林》2016 年 01 期《“城市 - 风景 - 遗产”一体研究视野的价值与应用—以“北京三山五园景观格局重塑”教学探索为例》）

1 The Origin of the Research and its Historical Influence

Our research on urban landscape and heritage can be traced back to the concept of 'urban-rural integration' since the ancient times in China, which formed unified planning of urban and rural areas in our country.

The concept emphasized the connection between human and the universe, as well as culture and nature. And it is especially expressed as the spatial continuity of Chinese traditional urban area and the natural environment surroundings, which helps to create unique urban morphology and unrepeatable urban features, such as the historical and cultural city of Nanjing.

Thanks to the long-term interaction of human cultural activities and natural environment, landscape environment around cities possesses more connotations and larger categories, including a great amount of historical remains and cultural space such as gardens, temples, villages, tombs, etc.

This endows natural environment with rich social functions and cultural connotation, just like what lofty cultural status Mount Tai and the Yellow River have.

Apparently, the construction of historical heritage and landscape elements which is based on the urban-rural integration concept fully embodies the settlement construction idea of the oriental civilization, it also provides us with reasonable explanations on the evolution progress of settlement morphology. So the 'urban-lanscape-heritage' integrated research framework helps us to study the interaction of urban and rural area from time and spatial dimensions.

2 The Value in the Molding of Urban Characteristics

The most important value in our study embodies in the research on modeling urban features. A large number of same-patterned modern new districts has led to the fragmentation of traditional urban landscape pattern, so urban features protection and reconstitution has become one of world issues.

After profound reform and introspection, western countries return to tradition again in their urban design process. They learn from traditional cities to get historical experience about the development of human settlements environment. Cities in China also begin to carry out researches on the construction of urban landscape pattern, natural environment and historical heritage protection. These make our study a core task of urban features construction nowadays.

3 The Value in the Practice of Heritage Protection

Along with the evolution of category determination, value assessment and protection strategy, historical heritage protection are becoming more integrated with urban planning and landscape architecture. Concepts have been brought up such as 'Historical Area' and 'Historical Urban Landscape'. Firstly, we realize that heritage protection should pay more attention to the environment around urban settlements. Secondly, we should try to explore the relation of the protected entity and the overall urban environment. What's more, we should focus on combining historical information protection with economic and cultural development.

Plenty of countries have already put forward corresponding protection measures such as 'European Landscape Convention' and 'Japanese Features Law'. In 2005, the State Council of China promulgated 'Notice on Strengthening the Protection of Cultural Heritage', which emphasized the regional and the systematicity of heritage protection.

4 The Value in Landscape Architecture

Landscape design in early times mainly satisfy the aesthetic and cultural pursuit. Thanks to industrial revolution, modern landscape architecture concerns landscape system of a broader range as well as the interaction of ecological science and humanism.

Now, landscape architecture is considered as research on multi-level complex system and its additive process, including natural system, agricultural system, settlement system, etc. Except for concerns about space, the subject encourages exploring dynamic evolution of the landscape system from time dimension.

As a result, the 'Urban-Landscape-Heritage' integrated research framework represents the transform from 'construction activities' to 'the evolution process of hierarchy system' of the subject.

5 Thoughts on the 'Three Hills and Five Gardens' Area in Beijing

Beijing is one of the most important case of ancient Chinese landscape city ideas. The 'Three Hills and Five Gardens' area embodies the natural landscape in Chinese traditional concept of human habitat, it is also an area of highly aggregated historical remains. Five core gardens in this area have different functions, and a large number of supporting service facilities are located in the surroundings.

Till now, a great quantity of urban construction has made the overall landscape pattern split. Elements in landscape are ignored an drowned in huge residential and business quarters.

All previous urban planning in Beijing failed to take the 'Three Hills and Five Gardens' area as a whole in consideration, which makes this area faced with threats including the separation of natural and historical elements, and the destruction of the skyline of historical blocks.

So it is out of question that the positive protection of the 'Three Hills and Five Gardens' is imminent. We suggest dividing the research work into three stages. First, we should accomplish scientific evaluation of the overall features in this area and setting goals for characteristic landscape system planning. Second, establish different factor systems from the view of urban, landscape and heritage. Third, we can pick main nodes in the systems to explore design of joints.

In a word, we should put great emphasis on the integrity and continuity of the overall structure with multiple elements.

6 Conclusion and Expectation

For modern planners and designers, it has become an inevitable choice to improve our life and environment from a macro-perspective, combining urban, landscape and heritage. So our research on remodeling the landscape system of the 'Three Hills and Five Gardens' will surely lay a useful foundation for the follow-up study.

城市视野与研究精神

Urban Perspective and Research Spirit

李倞
Li Liang

1 前言

面对快速城市化发展，以及由此产生的巨大环境压力，风景园林师正在依托自身专业特色，不断地拓展学科边界，实践领域也在随之发生巨大变化。正如当今在风景园林领域的许多热点理论——“景观都市主义”、“生态城市主义”等一样，“城市主义（Urbanism）”都是一个重要的落脚点。尤其在中国，城市化发展和相关问题的缓解将是未来风景园林最重要的实践方向之一。为了应对这种需求，风景园林师需要建立一种“城市视野”，能够从风景园林的视角来看待当今势不可挡的城市化趋势，培养一种对未来城市化问题的敏锐嗅觉，探索风景园林在城市化过程中可能扮演的角色和发挥的潜力，并能够采取一些具有理想化和创造性但又具有未来合理性的设计解决方案，作为一种实验性的探讨。

在当今学科领域拓展合作不断增多、知识呈现爆炸性增长的趋势下，风景园林师在未来设计中可能遇到的问题将会日益复杂，很可能会超越自身现有的知识体系，需要具有一种基于设计的研究能力，并使这种以问题为导向的设计研究成为一种设计习惯。风景园林师可以在设计中体现一种“研究精神”，能够依托设计实践，针对遇到的新问题迅速获取知识，不断地整合和更新现有的知识体系，并将其运用到设计之中，从而为提出创造性的设计解决方案提供可能。

此次课题研究所选择的北京三山五园地区，位于城乡边缘地带，正在面临包括生态、历史、社会、经济等一系列问题，可以作为北京复杂城市化过程的一个缩影进行研究。该区域一直饱受快速城市化发展的困扰[1-2]，在未来面临巨大的发展压力和潜力，是我们拓展城市研究视野、培养设计研究能力的一个非常合适的选择。

2 设计视野拓展

在课题研究一开始，我们就发现许多学生的设计视野比较狭窄，往往只关注红线内的区域，对风景园林相关研究领域的关注也比较少，这造成了大家在解决“三山五园”的复杂区域问题时产生了不少困扰。因此，我们开始反复强调设计的视野问题，引导大家不要把场地当作是一个有边界的公园，而是将三山五园地区的规划设计放到一个城市层面，不单纯是关注一条街道、一个建筑或某一个领域，而是要依托一个更加宏观的视野去梳理场地中的多种复杂联系，发掘设计的潜力。

为了拓展学生的设计视野，我们在研究题目的设计上也包含了两个不同的尺度层次——约 $81hm^2$ 的规划区域和内部一块约 $40hm^2$ 的设计范围（$40hm^2$ 的设计范围需要学生根据规划研究自己划定）（图 1）。在区域规划阶段，学生需要开展更

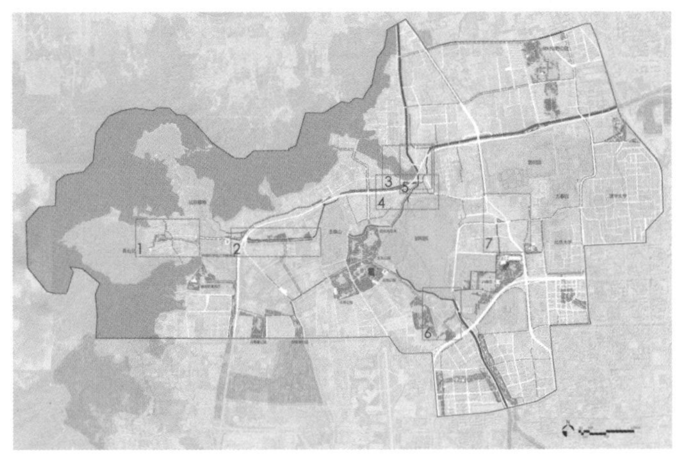

图 1 三山五园规划范围及详细设计片区

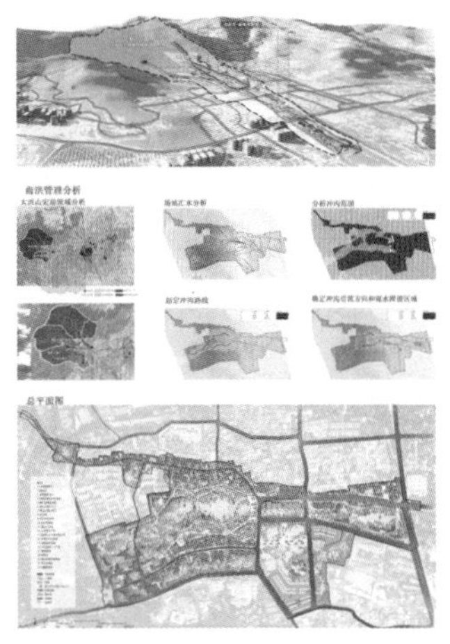
图 2 香山买卖街片区设计

大范围的规划研究，有的甚至必须扩展到整个城市；而在设计阶段，学生则要求与自己之前完成的规划成果进行衔接，自觉地将区域视野纳入设计研究的范畴之中。

视野拓展的另一个方向就是超越自身的学科领域，培养一种多学科综合的设计思路。课题的指导教师团队包含了城市规划、建筑、园林规划、设计、施工、植物景观等多个领域，学生团队也包含了风景园林、城市规划和建筑学方向。除了风景园林相关领域的合作，学生在设计中也往往需要根据场地需求，运用历史文化遗产、水生态系统、生物多样性、农业生产等多个领域的相关理论和技术成果开展规划设计。

3 设计与研究结合

设计是一个以问题为导向的研究过程。我们发现许多学生往往更加注重对设计形式的推敲，而忽视对设计过程逻辑的研究。因此，在此次课题研究中，除了注重设计的形式美感和空间体验，我们更加强调展现一种清晰而富有逻辑的设计过程，学生需要在设计过程中不断明确所要解决的核心问题，逐步推进相关的设计研究。

研究的第一部分内容是完成文献和现场研究，对三山五园地区的历史发展过程和现状场地条件进行全面整理。学生被分为 7 个专题研究小组，分别从现状用地条件、交通系统、绿地系统、水系统、文化遗产、建设强度与建筑风貌、历史水系演变、历史园林发展演变、非物质文化遗产等 9 个层面开展现状基础资料的整理和研究工作。这些专项基础资料的研究整理本身也成为整个课题的重要成果之一。

在引导学生对相关理论的研究和运用过程中，我们反复强调研究的严谨性，注重从理论向设计方法的转化，进而提出兼顾技术创新和实际操作能力的设计解决方案。在 81km^2 的区域绿道系统规划研究中，学生从最初的只是完成一条连贯的自行车道（目前国内对“绿道”概念理解的普遍误区），逐渐转化为构建一条能够发挥多种复合功能的区域绿色网络，并通过总结相关研究成果，逐步探索出符合“三山五园”区域特色的城市绿道构建方法。在场地设计阶段，学生也会针对场地问题，结合设计目标，开展一系列的专项设计研究，包括以“海绵城市”为目标的 LID 低影响开发、雨水生态管控、湿地水净化技术；以历史园林保护和再生为目标的文化遗产保护、历史区域更新、历史遗产廊道；以动植物保护为目标的景观生态学、城市生物多样性、生物迁徙廊道设计等，使最终的设计成果成为能够凝聚设计研究的载体。

4 多元化设计方向

风景园林是一个具有创新能力的媒介，也是一个容许差异的“多元化”与“多样性”的代名词。在课题研究中，我们不是希望带领学生去完成一个熟练的成熟设计方案，而是希望能够激发学生的热情，引导学生自己去发现问题，并通过研究，提出具有创造力和实验性的解决方案。教师在其中发挥的作用主要是发现学生方案的特色，通过引导和建议帮助学生推进研究，并最终提出一个独特的、具有合理性的设计成果。

在场地设计阶段，课题并没有给大家限定相同的设计场地，而只要求了研究面积，把场地选择也作为一项研究内容，希望大家能够结合规划寻找最能激发设计灵感、实现规划理想的关键区域开展设计。最终，7 个研究小组分别选择包括香山买卖街片区、青龙桥片区、颐和园北宫门片区、香泉环岛——颐和园西门带、颐和园西门片区等场地。由于每一个设计小组所选择的场地不同，设计思路也会存在差异。例如，在香山买卖街片区，设计方案重点解决了交通拥堵和社区发展混乱的问题，创造了更加舒适的香山入口环境和富有活力的商业街区，并因为其作为香山最重要的自然排水廊道面临极大的洪泛隐患，通过数据分析大胆地对绿地和建设用地进行了调整，使场地能够满足雨洪管理的生态功能（图 2）；青龙桥片区由于被五环路等 6 条城市道路、京密引水渠和清河所切割，设计方案提出了“BIG‘O’”绿环的概念，将周边支离破碎的土地、滨河空间、慢行交通和生态廊道重新串联起来，并重点沿“绿环”设计了一系列公共空间节点，使整个“绿环”成为能够凝聚整个区域的公共活力中心（图 3）。学生所选择的这些设计区域基本覆盖了整个三山五园地区最有意思和最重要的节点，并因为场地的差异性呈

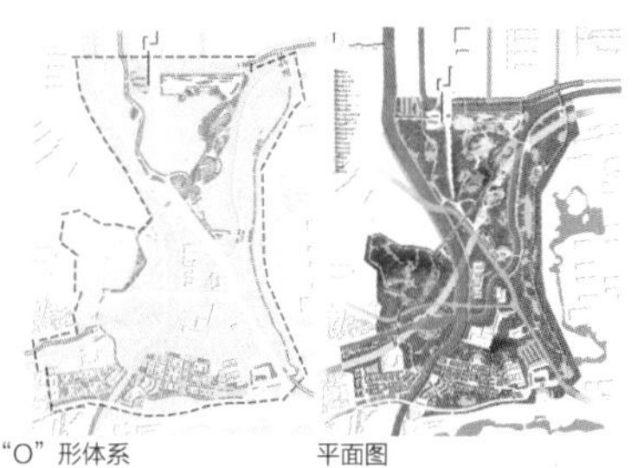

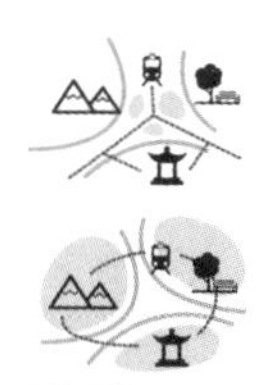

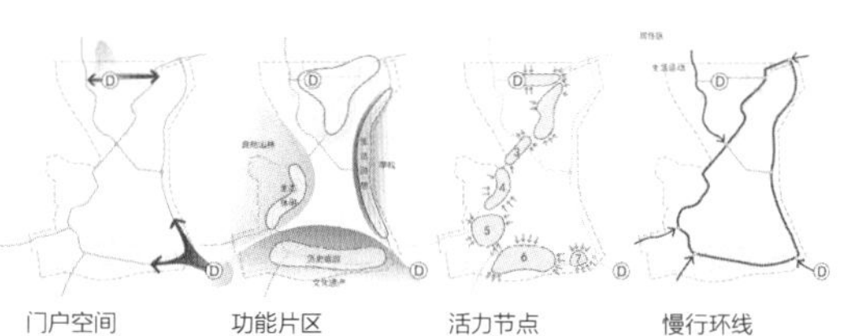

图 3　青龙桥片区设计

现出不同的发展方向，为区域的多元化发展创造了多重可能性。

5 结语

三山五园地区的研究需要建立一种城市视野来聚焦这片处于城市化风暴中的城乡渗透区域，由于异常复杂而需要进行设计研究探索创新解决方案，并最终为充满活力和机遇的区域发展探讨多元化的方向。

在课题研究过程中，我们也发现了几个非常关键的环节：首先是将视野放大，开始关注城市，寻找问题的环节。研究不仅需要看任务书的要求，更需要结合调研，自己去发现，并通过思考准确判断需要解决的关键问题；其次是研究思维的培养以及与设计结合的环节。大家需要根据问题逐渐理出一条清晰的研究线路，推进研究的不断深入，并能够真正地将研究转化为设计解决方案；第三是方案的深化和优选环节。研究必须在众多的解决途径中，不断通过质疑、取舍、整合、强化，逐渐形成一个最适合场地、最合理有效和具有创造力的解决方案。

备注：

图 1、3 由崔滋辰、王晞月、佟思明、李璇、李媛、莫日根吉、徐慧绘制；图 2 由尚尔基、李娜婷、周珏琳、王训迪、孙津、高琪、吴晓彤绘制。

参考文献：

[1] 阙镇清．再失一城——北京西北郊皇家园林集群：三山五园在城市化过程中的没落 [J]. 装饰 ,2007,(11)．

[2] 詹姆士科纳．绪论——复兴景观是一场重要的文化运动 [C]. 詹姆士科纳．论当代景观建筑学的复兴．北京：中国建筑工业出版社 ,2008:2.

（本文节选自《风景园林》2016 年 01 期《城市视野与研究精神》）

1 Foreword

Under the background of rapid urbanization and great environmental pressure, landscape architects are trying to broaden the field of the subject. Just like the hot spot theories in landscape area such as 'Landscape Urbanism' and 'Ecological Urbanism'. 'Urbanism' is also an important foothold. Especially in China, problems about urbanization development are going to be one of the most important practical directions in landscape architecture. In order to respond to such demands, landscape architects need to set up an 'urban vision' and take some solutions with idealization and creativity as well as rationality. They have to show a kind of 'research spirit' in their design, and keep on integrating and updating their knowledge system, then apply them into designs.

In our research subject, we choose the 'Three Hills and Five Gardens' area as our site, which is located in the urban-rural fringe area. Facing with a series of problems including ecology, history, society and economy, the area can be considered as an epitome of the complex urbanization process in Beijing.

2 Expansion of the Visual Field in Design

In order to avoid students from designing with a narrow field of vision, we put repeated emphasis on broadening the view of design and making our design an urban-level issue. Therefore, we are able to card a variety of complex relationship in our site.

Two layers of different scales are considered on our research subject, which include an 81 km^2 planning district and a 40 hm^2 design site. In the stage of regional planning, students have to carry out their researches in a larger range, some even extend to the whole city. While in the stage of design, students are required to integrate the regional view into their researches and designs.

We hope to develop a design idea of multidisciplinary synthesis, therefore the faculty advisors in our subject come from multiple fields including urban planning, architecture, garden planning, design, engineer, vegetative landscape, etc, so do our students.

3 Combination of Design and Research

In our research subject, we not only put an emphasis on the beauty in form and special experience of our design, but also lay stress on a clear and logical design process.

When helping students to apply relevant theories into their designs, we drum in the rigor of research and aspire for designs giving consideration to both technical innovation and practical operation ability.

4 Diversified Design Directions

Through the research, we hope to bring up solutions with creativity and practicality. So teachers aim to find different characteristics in students' design and help to carry forward their researches, so as to put forward unique and reasonable design results.

Design sites are not limited, thus each group's idea is different. The sites students choose have already covered the most interesting and important nodes in the 'Three Hills and Five Gardens' area, and the difference of these sites provide possibilities for the pluralistic development of the region.

5 Epilogue

As we can see, the 'Three Hills and Five Gardens' area is located right in the urban and rural region, so we have to set up a view of urban development to explore innovative solutions. We find several key links in our research process. First, we successfully broaden our horizon to pay attention to the urban level. Second, we pay more attention to the combination of the research thought cultivation and design capability. Third, we also focus on deepening students' ideas and optimizing the schemes.

北京“三山五园”整体性研究思考

The Research Cognition of “Three Hills and Five Gardens” Region in Beijing

朱强　张云路　李雄
Zhu Qiang　Zhang Yunlu　Li Xiong

1 引言

“三山五园”在历史上是指包括畅春园、圆明园、香山静宜园、玉泉山静明园、万寿山清漪园（颐和园）这五座皇家园林组成的核心区、核心区内部穿插的过渡区（村庄、私家园林、农田、水利设施等）和周边的军事防御区组成的一个皇家园林综合功能区[1]，它由一个庞大而完整的生态、人文和社会系统交织而成，有着非同寻常的历史价值。

“三山五园”在第二次鸦片战争后遭到了战争的致命打击，其庞大的聚落网络也随着清王朝的覆灭而解体，随后便在长年中遭受了较为严重的破坏。新中国成立之后，党和政府加大了对于该地区的保护，2012 年的北京市党代会首次将“三山五园”历史文化片区建设列入首都历史文化名城的保护项目，《北京城市总体规划（2016-2035）》首次提出三山五园地区整体保护。

作为北京近郊一个重要的集自然与人文于一体的历史文化片区，无论是在自然科学还是人文科学界，“三山五园”都已经成为研究的热点，其现实意义在于：①文化遗产的数量多、密度大、质量高，保护意义重大；②作为北京西北郊历史文化公园的一部分和市区西北部重要的生态基础设施，既是休闲游憩的场所，也是抑制城市无序扩张的绿色隔离空间；③由于地处城市与农村的过渡地带，该区域内开放空间的开发以及文教科研区的合理规划建设可促进社会、经济等方面的健康发展。

目前，由于三山五园地区的历史格局受到了近年来快速城市建设的冲击，已经融入城市之中，因此其主要面临的威胁与挑战主要来自内部，如何科学地处理历史园林及周边环境的保护与城市建设的关系、人居环境改善与生态文明建设的关系，成为当今迫切需要研究的课题，本文针对区域新的问题与发展需求，对“三山五园”整体性发展提出新的研究思考。

2 研究目标

结合风景园林学科的特点，本文尝试提出应对“三山五园”问题和发展需求的风景园林学科三大研究目标：

第一，针对“三山五园”片区整体发展中的实际问题，风景园林学科尝试建立一种以人与环境的协调为中心、以自然与人文为研究对象的新的研究框架，强调三山五园地区整体性发展对于区域的社会、文化、经济的重要意义。

第二，风景园林学科更加强调学科之间的综合，成为在“三山五园”研究领域中整合不同专业与研究领域的平台，最终以一个整合不同专业的模式来探索其中的问题，从而提供更加科学全面的解决思路。

第三，将“三山五园”作为一个生态、历史、文化功能综合区域进行研究，更深入地辨识、描述和分析“三山五园”在历史发展中的空间变化规律和特征，系统地分析城市化过程与园林的相互关系与互动机制，为未来该区域保护与建设提供依据和参考。

3 研究内容

3.1 生态基础设施研究

由于“三山五园”地处北京中心城区西北部的城市建成区与西山之间，风景园林学科应紧密结合历史地理与规划建筑等学科，在宏观层面，提出行之有效的措施和办法，避免出现“三山五园”中历史名园成为生态孤岛的情况持续恶化；应结合城市的发展，为区域内绿地系统的完善与更新提供宏观层面的指导，同时应在园外物质空间与视觉空间的开发建设中提供很好的指导和管控，避免出现违背生态发展的大小规模的建设，尤其注重天际线的保护。在微观层面，风景园林学科可能会着重于区域内诸多单体园林的外环境自然风貌的恢复（如万寿山、玉泉山之间的两山区域）以及新建公园绿地的建设（如北坞公园）。

3.2 区域整体发展演变特征研究

风景园林学科应突破静态的“三山五园”的研究思维，加强该地区尤其是清代之前的历史研究，从而填补历史空白。当前城市的发展变化迅速，以动态的视角研究“三山五园”，可以突破原本以清代为主要研究的时间段，从而构建场地历史、现在与未来的全面发展关系，寻找区域发展演变的规律和特征，可以对该区域的未来发展规划提供科学的指引。

3.3 评价体系研究

风景园林学科应遵循定性与定量相结合的原则，在继续坚持原有定性研究的基础之上，还应该加强对“三山五园”进行定量分析，从而能够更加客观地研究该区域的历史发展轨迹并且预测未来的发展趋势，也能为影响该区域发展的不同要素建立科学关联性，为更加深入地研究其作用关系提供依据。

3.4 绿色资源保护和合理利用研究

首都城镇化建设和气候变化尽管在一定程度上改变了三山五园地区的历史格局，但同时也为它带来了发展机遇：风景园林学科应充分发挥绿地资源的生态效益，在现有条件的基础上对该地区的生态环境进行一定程度上的重塑，包括将原有稻田改造为具有历史风貌的现代郊野公园，如北坞公园中还保留了京西稻的历史记忆。近年来西山城市森林生态系统修复、大西山景观规划、“三山五园”绿道建设和“三山五园”历史文化景区规划等研究实践已经开展，尤其是作为三山五园地区重要的生态屏障的西山森林，因受到技术和社会等因素的影响任重而道远，风景园林可为这些实践中绿色资源的保护和利用提供理论指导。

4 结语

“三山五园”不但属于历史与现在，同样属于未来。对于科学界而言，如何综合多个学科的智慧，在有效保护现有历史文化资源的基础上，正确处理历史园林及周边环境与城市建设的关系、人居环境改善与生态文明建设的关系，已经成为迫在眉睫的问题。

当前，风景园林学科、规划建筑学科、地理历史学科，已经对“三山五园”进行了大量的研究、也取得了十分丰硕的成果（图 1）。但站在历史的新起点上会发现：一方面，“三山五园”作为一个庞大的历史文化片区体系，存在着整体研究上的一些空

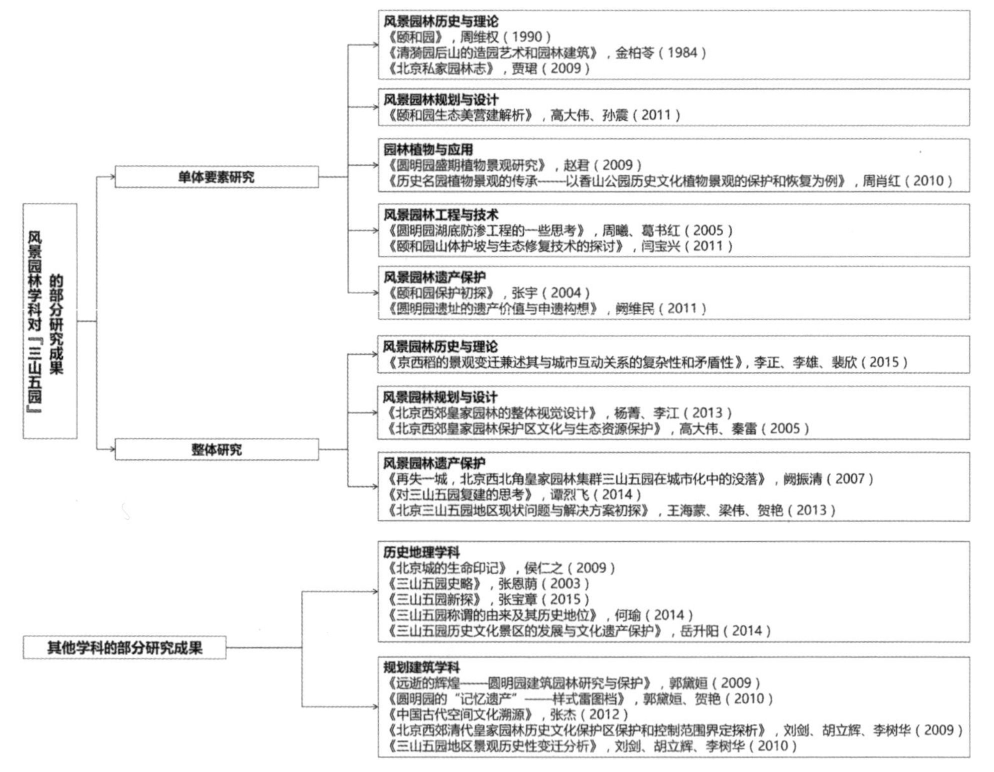

图 1　基于风景园林学科的三山五园片区现有研究体系

白，这些空白或许对于整个“三山五园”来说具有至关重要的意义，都需要学者们持续地关注和挖掘；另一方面，风景园林学科应该多尝试突破传统的研究方法，借鉴其他学科的研究思路和方法，特别应综合吸收森林生态学的理论，开拓新的思路，将过去、现在和未来进行恰当的对接，如思考古典园林中的山水关系、空间理法、建筑营造、文化意象表达等等对于当代风景园林规划与设计的启示，也可以思考文化遗产与人居环境的关系等问题，对于未来该片区的整体发展提供不同的思路。

1 Introduction

The 'Three Hills and Five Gardens' area is one of the comprehensive functional areas for royal gardens in Beijing. It is made up of large and complete systems including ecology, humanism and society, so it is of extraordinary historical value.

The area has suffered fatal blows since the Second Opium War. However, after the founding of People's Republic of China, the Communist Party and the government have increased the protection of the area.

Now, the area has become a hot spot for researching, which practical significance lies in three facts. First, there are cultural heritages with large quantity, high intensity and high quality. Second, it is not only a space for recreation, but also a green space for restraining the disorderly expansion of the city. Third, the development of the open space in the region and the rational planning and construction of the cultural and educational research areas can help promote the sound development of the society and the economy.

2 Research Goals

We are trying to put forward three major research goals for the landscape architecture subject towards the development needs of the 'Three Hills and Five Gardens' area.

Firstly, we should establish a new research framework which is centered on the coordination of human and environment, taking nature and humanity as the objects of the research, and emphasize the overall development of the area.

Secondly, we should integrate platforms of different specialties and research fields, so as to provide more scientific and comprehensive solutions.

Thirdly, it is necessary to carry out researches taking the 'Three Hills and Five Gardens' as an integrated area of ecological, historical and cultural functions, and then systematicly analyze the relationship and interaction mechanism between urbanization process and the gardens.

3 Research Contents

3.1 Study on Ecological Infrastructure

We should take effective measures to avoid the historic parks from becoming ecological islands any more, and perfect the green system in the region, with particular emphasis in the protection of the skyline. At the micro level, landscape architects should focus on the recovery of the natural features of the external environment of many single gardens in the region.

3.2 Study on the Evolution Characteristics of the Overall Regional Development

Landscape architects should strengthen the historical study of this area, especially those before Qing Dynasty. By studying the 'Three Hills and Five Gardens' from a dynamic perspective, we can find laws and characteristics of regional development.

3.3 Study on Evaluation System

We should strengthen the quantitative analysis of the area, study the historical development and predict the future development trend more objectively, so as to provide reliable basis for further researches on its relationship.

3.4 Study on the Protection and Rational Utilization of Green Resources

Landscape architects should give full play to the ecological benefits of green resources, and reconstruct the ecological environment of the region on the basis of existing conditions, including modern country parks, forest parks and greenway constructions.

4 Epilogue

For us, it has become an imminent problem to figure out issues like how to integrate multiple disciplines of wisdom and correctly handle the relationship between historical landscapes and urban constructions on the basis of effective protection of historical and cultural resources. The balance of the improvement of human settlement environment and the construction of ecological civilization should also be paid attention to.

On the one hand, as a huge historical and cultural system, there are plenty of blanks in the research on the 'Three Hills and Five Gardens' area, which may have great significance for the whole area, so it is necessary for scholars to continuously pay attention to the issue and excavate beneficial solutions.

On the other hand, we landscape architects should try to break through the traditional research methods and especially absorb the theory of forest ecology comprehensively, so as to provide different ideas for the future development of this area.

古代北京城市水系规划对现代海绵城市建设的借鉴意义

The Experience of River System Planning in Ancient Beijing for The Modern Sponge City Construction

王沛永
Wang Peiyong

在北京西山的三山五园地区这个特殊的地带，历史上分布着大量涌泉、湖淀、河道，成为皇家贵族的主要园林及风景游览地。同时也保留着劳动人民充分利用这样的山水环境进行劳作、并与洪涝灾害作斗争的痕迹。它更是京城人民赖以生存的水利、漕运、饮用的水源地。现在面对这个多样复杂的历史、人文和风景遗产地，人们又开始以生态环境、休闲游憩的目光重新审视此地，赋予它更多的内涵。

如果从雨洪管理的视角来观察三山五园地区，我们可以发现当前城市建设对三山五园地区是忽视的，大量的城市道路、工厂、居民区的建设破坏了本地雨水的自然循环与流动，自然的生态功能在降低。研究古代北京城的城市规划与水体关系，可以发现古人非常重视城市与水系的密切关系，构建了北京城优美的人居环境，对现代海绵城市建设也有借鉴意义。

1 北京水系变迁史

1.1 水系变迁概况

北京的历史是一部人与水互动的历史。在蓟城时期就开凿了车箱渠引水。金代开凿了护城河、金口河、长河，使高粱河与闸河（今通惠河）相连。元代建元大都时开金水河，引白浮泉水，接通惠河。明代在元大都基础上改建都城，外护城河与内护城河相连，挖内外金水河，绕紫禁城凿护城河。明都城内建设了大量河网水系，排水明沟，在防洪防涝方面发挥了巨大作用。明代定都北京前，河道堵塞水系退化，水患次数增多，达到平均 6.5 年一次，建城后经水系治理，降为 12.5 年一次，效果显著。清代北京西郊的一系列园林建设，维持了大量河湖湿地，对北京抵抗西山洪水起到重要作用。与明代相比，清代建都北京的 267 年间，共有 5 次较大的水患，平均 53 年一次（图 1）。

1.2 城市与水系变迁的关系

明北京城在元大都城址的基础上南移以避险，清朝沿袭明代都城和皇城。北城墙南移有利于抵御西山洪涝带来的淹城风险。后又开拓都城，最终形成“凸”字形布局。从金中都、元大都到明北京，城墙屡次变化，原因之一是避免雨洪威胁，进而产生水系的变迁，导致城市形态的变化。

2 明清北京城市水系规划经验

2.1 完整的水网建构

从金到清所有的水利建设都在致力于构建完整的水网体系，包括低洼地，池塘，湖泊，河道水渠，护城河，城外的自然河流系统。完整水系网络的构

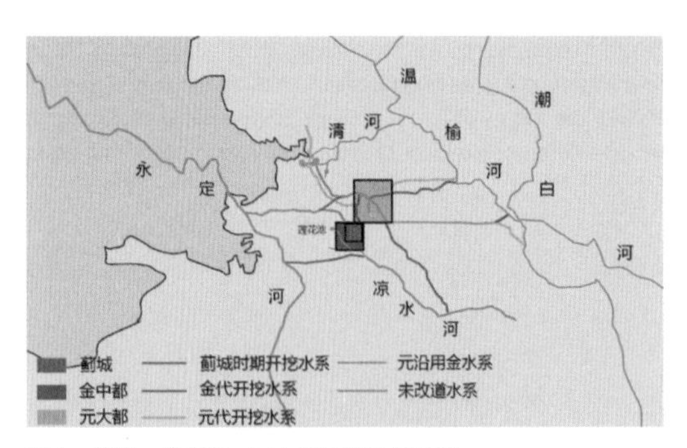

图 1　蓟城、金中都、元大都与附近水系图

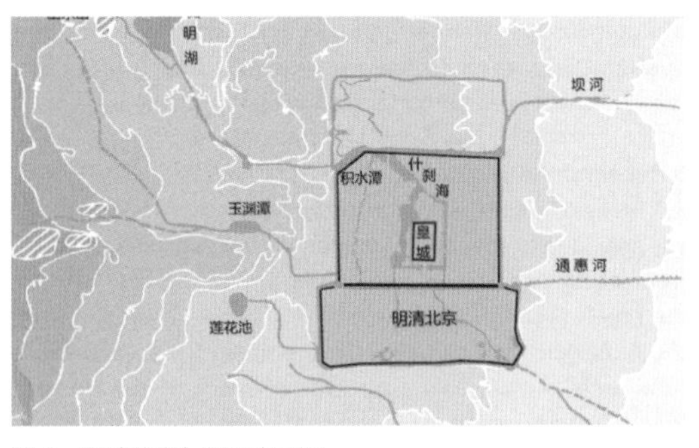

图 2　明清北京与附近水系图

建不仅满足了漕运，为生活和娱乐提供了充足的水源，而且也在城市防洪排涝方面起到了巨大的作用。暴雨时雨水顺排水沟渠进入城市内部的湖泊、池塘，再由河道引水出城到城外护城壕池，再排到城外的自然河流中。雨水在完整的水网中缓缓流过，增加雨水下渗的机会，雨水进入地下参与自然循环。

从元大都起北京城就开始修建城内排水系统，主要包括城内明渠暗沟、城壕、城内河道。据史料记载，明朝启年间，内外城壕约 40.47km，加上紫禁城的城壕共长 44.27km。城内排水干渠河道总长约 64.27km，城内河道密度为 1.07km/km^2。到了清乾隆年间，内城大沟约 97.70km，小沟约 313.92km，内城面积约为 35.5km^2，所以内城大沟密度为 2.75 km/km^2，小沟密度为 8.84 km/km^2。城内排水系统完备，城外有大量的河网水系可以进行雨洪调蓄，共同组成一个保证雨洪安全的系统（图 2）。

2.2 功能完善的水系统

完整的水网结构建立之后，还要求水体有完善的功能。城市低洼地，池塘汇聚雨水，就近收集滞留雨水。人工修建的排水沟渠是地面上的排水系统，保证雨水就近排放。各种引水河道水渠，湖泊为满足城市生活、宫苑给水、漕运用水，沟通串联起整个完整水系。暗沟主要用于人口密集区域，多在雨洪排放的起始段设置，路边的排水沟渠相勾连，引导雨水进入城内排水河道。这些河道具有很强的蓄水作用，足以抵御暴雨径流带来的威胁。

3 明清西郊水系治理与利用

清代时期北京城内的内涝次数减少，一方面永定河治理比较完善，河道决堤的次数减少，另一方面西山雨洪也得以治理，外水围城的机率降低了。北京西郊的湿地水体一直是主要的水源地，也是洪泛威胁的源头之一。清代皇家园林群“三山五园”的建设是水系治理与人居环境建设的典范，其经验与做法值得我们今天学习。

北京西郊水草丰美、风景秀丽，从元代开始就有皇家贵族的别墅园林，明清时期更是皇家园林的集中地。这里除了园林水体、河网，还有大量的沼泽地、水稻田，在北京城市水系构成中占据重要的地位，是湿地集中的地带。清代皇家贵族利用这里的水系形势，建造了大量“水景”园林，以河湖吐纳雨洪。西郊园林群的建设遵循了自然河湖相连的原则，利用水系低洼或宽阔处，扩建水面造园，是对河湖水系的完善和发展。尤其是昆明湖由瓮山泊扩大而成，为西北郊园林群造景水源，也为京城用

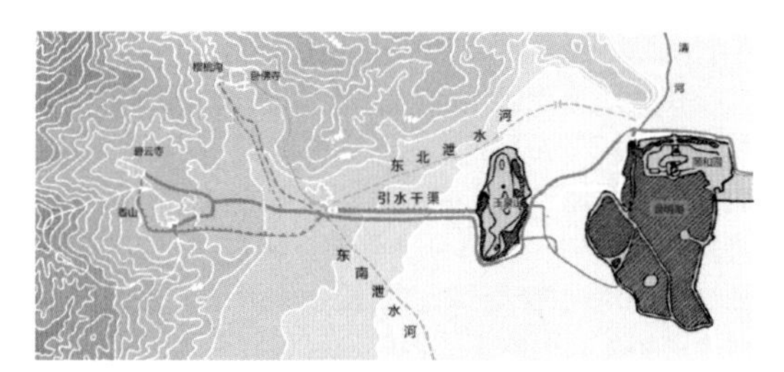

图 3　西郊引水渠和排洪渠位置图

水周边田地灌溉提供了充足且可调蓄的水源保障。由于对皇家园林持续的维护管理，使得这些河湖水面不间断地发挥防止雨洪灾害的作用。

园林内湖体吸纳了西山汇流的大量洪水，也承载了水系周围的地表排水。由于水体具有足够的蓄水容积，所以对洪涝灾害具有巨大的缓冲作用，在一定程度上保证了北京城的雨洪安全。雨洪过程结束之后，湖体吸纳的雨水可以缓慢释放，补充下游园林系统和河道系统的水系水量，发挥了巨大的海绵作用。

为了保证昆明湖在干旱时期的蓄水量，从香山碧云寺、樱桃沟及玉泉山修建了石质的引水干渠直达昆明湖。同时为了抵御山洪，减少过量的雨洪对昆明湖的冲击，修建了东北和东南两条泄水河，将超量雨洪引入清河和高粱河，绕开了昆明湖及玉泉山皇家园林群，保证皇家园林的安全，也使得这一地带水系之间的功能更合理（图 3）。

4 对现代海绵城市建设的借鉴意义

现代城市扩张导致城市内及城市近郊的河湖水系发生了巨大的变化，一些河道和池塘被填埋，从而产生了排水不畅、内涝现象。人们逐渐认识到要解决城市内涝问题，不能仅靠容积和排水速度有限的管道进行排水，应综合运用管网、河道、低洼地等综合设施，建设海绵城市。应从城市水系的重新整治入手，统筹城市水体循环，恢复一些河网低洼地，梳理城市河道。应综合利用蓄、分、导、渗等措施对城市水系进行整治。

4.1 增加城市内外的蓄水空间

在降雨过程中通过蓄积一定量的雨水，使得雨水不至于快速下泄，可以保证降雨与排水错时进行，有利于削减洪峰。清代北京城外西郊的皇家园林群建设，客观上保留了大量的河湖湿地，在源头蓄积雨水，不致使雨洪迅速接近北京城，即为源头控制。城市水系应具备与其城市规模相适应的蓄水容积，且应互联互通，不然即使有完整的水网结构，也会因排水不畅而产生内涝。

4.2 分洪是保证雨洪安全的重要措施

在雨洪产生的源头处将过量的雨水分流出去，引入远离城市的河流，可以避免损失。北京西山雨洪出山处修建了东北泄水河和东南泄水河，日常小降雨时河道将雨水引至昆明湖，而强降雨发生时，通过北泄水河将部分雨水直接引到清河，通过东南泄水河将部分雨水引到西郊，绕过北京城后排入城南河道。这是兼顾了排水安全与日常使用的最佳方案，对现代城市的河网水系治理提供了范例。

4.3 增加城市内的排水沟渠密度

现代城市主要依靠地下管网将雨水排入河道，城市内部排水仅有主干河渠，支渠减少，微型河道消失，河网密度大大下降。然而管道对雨水的排放能力是有限的，无法及时将雨水导出，暴雨来临时存在很大的内涝风险。

城市建设应有意识地增加一些自然的排水沟渠，尽可能多地保留城市原有自然排水汇流的渠道，并在其两侧控制一定宽度的绿地，形成排水河道的同时也增加了城市绿地。这些排水沟渠应连成网络，并连通城市较大的河道，成为城市自然水系的组成部分。

4.4 各种措施促进雨水下渗

促使更多的雨水下渗进入土壤，可以形成地表水与地下水之间的良性自然循环，对城市水生态环境起到良好的促进作用。古代北京通过西郊的皇家园林中大量的水系河道，城内大量的排水沟渠与湖面，共同管理雨水，调节雨洪水量，维持了良好的自然景观与水生态系统。

5 结语

古代北京的水系目前仍然在发挥着排洪、景观作用，我们同时也看到当前城市发展偏离了“天人合一”的传统，过分依赖管网排水，河湖的防洪功能被削弱了很多。现代城市规划应先分析自然水文过程，通过对水系的规划整理，保证水系网络完整性，水体功能的完善。在满足水域面积、河网密度的情况下营造兼顾游赏和调蓄雨洪功能的城市空间。城市园林绿地的布局在满足居民游憩休闲的基本功能前提下，可在排水河网两侧进行绿化建设，在改善城市环境条件的同时形成优美的绿地景观，构建城市特色人居环境，对改善城市环境，调节小气候，尤其是对城市可持续发展有重要影响。

在海绵城市建设的背景下，我们应继承和发扬古代北京城市水系规划与建设的经验，变管网排水为综合治理，增加水道、湿地的占地面积，恢复河网生态功能。以水定城，景观优先，水绿结合的城市规划方法和建设模式应该重新被定为城市建设的优先模式，并发挥它巨大的价值。

参考文献：

[1] 侯仁之，北京城的生命印记 [M]，北京：生活 · 读书 · 新知三联书店，2009.3

[2] 吴庆州，中国古代城市防洪研究 [M]，北京：中国建筑工业出版社，2008.12

（本文删改自《园林》2015 年 07 期《古代北京城市水系规划对现代海绵城市建设的借鉴意义》）

A great amount of springs, rivers and lakes are located in the 'Three Hills and Five Gardens' area, therefore the area has become a great scenic spot for royal gardens and natural sceneries. It is also the water source for Beijing residents to live on. Facing such a complex area with historical heritage, cultural heritage and natural heritage, today people begin to reassess this area in the view of ecology and recreation.

Currently, massive construction of urban road system, factories and residential have destroyed the recycling system of local rainwater, and have reduced the ecological function of our environment. What we are trying to do is explore the relationship between the river system and Beijing city, further more to provide reference for the construction of sponge city in the future.

1 The Transition of Water System in Beijing

1.1 General Situation

The history of Beijing revolves around the interaction of man and water. In Ming Dynasty, people constructed plenty of drainage ditches and completed the river network, which played a huge role in the aspects of flood control and water logging. In Qing Dynasty, a series of gardens were constructed in the western suburb of Beijing, they also helped maintaining a large number of river and lake wetlands and played an important role in Beijing's resistance to the flood in the Xi Mountain.

1.2 The Relationship Between City and the Transition of Water System in Beijing

The rulers of Qing Dynasty followed the capital of Ming

Dynasty and built the Forbidden City. Then they expanded the capital area to form a convex shape.

2 Planning Experience of Urban Water System in Beijing in Ming and Qing Dynasty

2.1 Completed Water Network Construction

From Jin Dynasty to Qing Dynasty , all the water conservancy constructions are devoted to building up a completed water network, including low-lying lands, ponds, lakes, rivers, moats and natural river systems outside the city. Completed water network not only meets the need of water transport, but also provides sufficient water source for livelihood and entertainment, it also plays an important part in flood control and water logging.

People began to build the drainage system within the city since Yuan Dynasty. Completed urban drainage system and water system make up a mature system to ensure the safety of rain and flood.

2.2 A Well-function Water System

Perfect function is needed in water system. For example, low lands and ponds are able to gather rainwater, built drainage ditches are well-function drainage system, and various kinds of water diversion channels connect the whole water system in the city.

3 Water System Management and Utilization in the Western Suburb of Ming and Qing Dynasty

Thanks to the governance of Yongding River and the control of the rain and flood in Xi Mountain, the number of water logging in Beijing obviously decreased in Qing Dynasty.

We can also say the construction of the 'Three Hills and Five Gardens' royal garden group is the model of water system management and human living environment construction, which experience is worth learning today.

The western suburb of Beijing occupies an important position in the water system of Beijing. In is also the concentrated area of wetland. By using the water system, the royal aristocracy of Qing Dynasty built a large number of waterscape gardens in order to resist rain and flood.

The rivers in these gardens absorbed a large amount of floods in Xi Mountain, and it carried the surface drainage around the whole water system as well, which exerted a huge sponge effect.

At the same time, in order to resist torrential flood and reduce the impact of excessive rain on Kunming Lake, people built two rivers in the northeast and southeast, so as to make the function of the northeast and southeast, so as to make the function of the water system in this region more reasonable.

4 The Reference Significance to the Construction of Modern Sponge City

The expansion of modern cities has led to great changes in the river and lake water system, no matter in the cities or in the suburbs. As a result, people gradually realize that to solve the problem of urban water logging, we should not only drain by drainage pipes with limited volume and draining speed, but also use comprehensive facilities such as pipe network, river course and low lands to build sponge city.

4.1 we should increase the storage space inside and outside the city. The accumulation of certain amount of rain can keep rainfall from fast discharge and guarantee that rainfall and drainage are carried out in different periods of time.

4.2 flood diversion is an important measure to ensure rain and flood safety. It is possible to avoid loss by diverting the excess rain and introducing it into the rivers far away from the city.

4.3 it is beneficial to increase the density of drainage ditches in cities. Urban constructions should consciously increase natural drainage ditches and maintain the original channels in the city as many as possible, it is also necessary to control the greenbelts with a certain width on both sides of the rivers.

4.4 we can learn from the royal gardens in the ancient western suburbs of Beijing that infiltrating more rainwater into the soil can help from a benign natural circulation between surface water and groundwater, which can play a good part in promoting urban water environment.

5 Epilogue

Today, the water system of ancient Beijing still plays an important role of water drainage and landscape effect. We should also see that the flood control function of rivers and lakes today have been weakened a lot by relying too much on the drainage of pipe network. The natural hydrological process should be analyzed first in modern urban planning. We should also guarantee the integrity of water system network and the improvement of water function through water system planning and management, and then carry out greening constructions on both sides of the drainage river network.

In a word, we should inherit and carry forward the experience of the planning and construction of the urban water system in ancient Beijing, and beautify our environment under the construction modes of 'Landscape Priority' and 'Water-Greening Combination'.

02 专题研究

SPECIALIZED STUDIES

对三山五园地区开展全面深入的调查研究是进行规划设计研讨工作的基础。根据课题研究目的和场地实际情况，拟定的专题研究分为如下七个方面，包括五个现状专题研究，两个历史专题研究。

用地现状专题研究主要对地段内各类土地使用进行普查统计，分析各类土地的数量、分布特征，探讨现状与生态、历史保护之间的矛盾以及各类用地之间的矛盾等。

交通现状专题研究主要对地段内各类公共交通方式和线路进行调查，分析各类公共交通与用地、历史遗产和绿地水系之间的关系，提出改善交通尤其是慢行交通的可行性。

建筑现状专题研究主要对地段内各类建筑的功能、建设强度、高度、风格、开放性等进行调查，分析其与历史遗产、绿地水系之间的关系与矛盾，提出未来可更新改造的建议等。

绿地与水系现状专题研究主要对地段内各类绿地的数量、分布、植被状况、开放度等进行调查，分析其与历史遗产和其他功能性用地的关系与矛盾，提出未来可恢复、拓展及更新的潜力区段。对地段内各类水系的线路、流向、水质、岸线形式及开放度等进行调查，分析其体系完整性、雨洪容纳能力和景观特征，提出未来可恢复、拓展及更新的潜力区段。

历史遗产现状专题研究主要对地段内各级、各类文化遗产的数量、分布、使用和保护现状等进行调查，分析其分布特征和保护措施缺陷，提出未来进一步加强保护及合理利用的措施。

非物质文化遗产及历史园林专题研究除对地段内为数众多的各类皇家、私家园林的发展脉络进行梳理外，还对相应的传统文学、美术、表演艺术等多方面的非物质文化遗产发展及其载体分布进行调查，提出现状发展的缺陷及未来发展的方向。

历史水系专题研究主要对地段内水系的数量、类别、分布特征等从时间上进行梳理，总结其发展演变的规律及主要的影响要素，提出现状发展的缺陷及未来发展的可行方向。

Carrying out comprehensive and deep-going investigation on the area of 'Three Hills and Five Gardens' is the basis for the discussion of planning and designing. According to the research purpose and the actual situation of the area, the thematic study is divided into seven aspects, including five current status and two historical studies.

Specialized Studies on Existing Land Use
We conduct census and statistics of various land use in this area, analyze the quantity and distribution characteristics of all kinds of land, and discuss the contradiction of the current situation, ecological and historical protection, as well as the contradiction between various land use.

Specialized Studies on Existing Traffic
We mainly investigate various public transportation modes and routes in this area, analyze the relationships including that between all kinds of public transportation and land use, as well as that between historical heritages and water systems, in order to put forward the feasibility of improving traffic, especially slow traffic.

Specialized Studies on Existing Buildings
We mainly investigate the functions, intensity, height, style and openness of all kinds of buildings in this area, analyze their relationship and contradiction with historical heritages and the water system, then put forward suggestions for the renewal and transportation.

Specialized Studies on Existing Green-Water Systems
We mainly investigate the quantity, distribution, vegetation status and openness of various kinds of green space in this area, analyze its relationship and contradiction with historical heritage and other functional lands. We also investigate the routes, flow direction, water quality, coastline form and opening degree of various river systems, analyze the integrity of the system, the capacity of rainfall and flood and the landscape characteristics, and finally put forward potential recovery areas that can be restored, expanded and renewed in the future.

Specialized Studies on Existing Heritage
We mainly investigate the quantity, distribution, usage and protection status of cultural heritage at all kinds and all levels, analyze its distribution characteristics and the defects of protection measures, and then put forward measures for further strengthening protection and reasonable utilization.

Specialized Studies on History of Garden and Intangible Heritages
In addition to the development of various types of royal and private gardens in this area, we also have to investigate the development and carrier distribution of intangible cultural heritage such as traditional literature, fine arts and performing arts, and then put forward the shortcomings of the current development and point out the future development direction.

Specialized Studies on History of Water Systems
We mainly sort out the quantity, category and distribution characteristics of the water system from time to time, summarize the laws and main influencing factors of its development, and then put forward the shortcomings of the current situation and feasible directions for the future.

用地现状专题研究

Specialized Studies on Existing Land Use

尚尔基、李娜亭、周珏琳、王训迪、孙津、高琪、吴晓彤

Shang Erji/Li Nating/Zhou Juelin/Wang Xundi/Sun Jin/Gao Qi/Wu Xiaotong

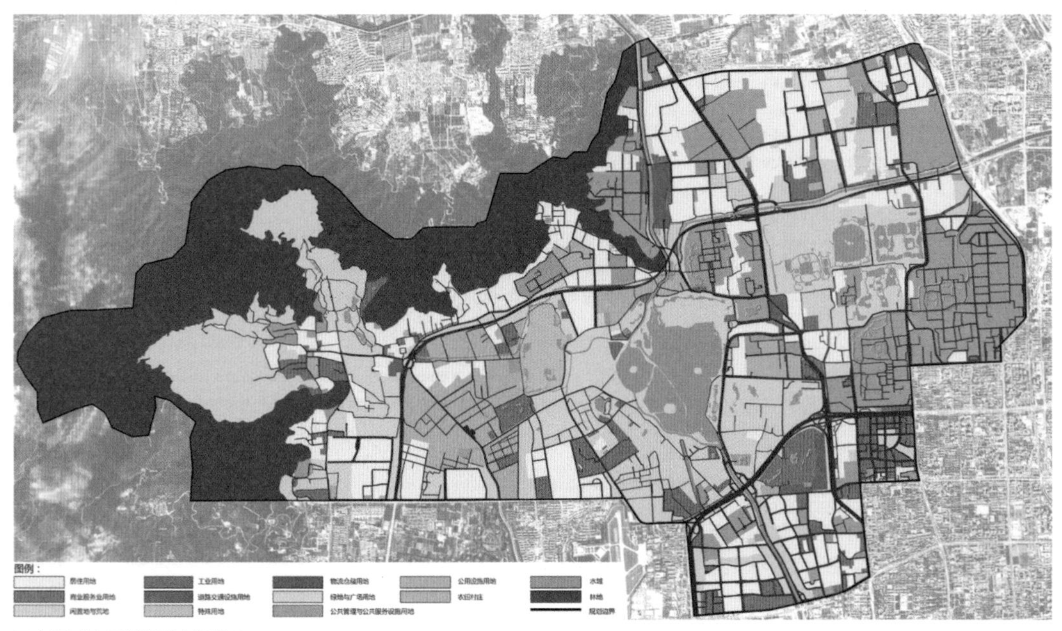

三山五园地区现状用地分析图

Land Use Categories

三山五园地区总面积为 81.33km²，建设用地面积 63.32km²，非建设用地面积为 18.01km²。其中绿地与广场用地总面积为 2163.4hm²，居住用地总面积为 1175.2hm²，公共管理与公共服务用地总面积为 994.7hm²，商业服务业设施用地总面积为 428.5hm²，工业用地总面积为 9.1hm²，物流仓储用地总面积为 5.6hm²，交通设施用地总面积为 1531.7hm²，市政公用设施用地总面积为 24.4hm²。农林用地总面积为 1434.6hm²，特殊用地等其他非建设用地总面积为 365.8hm²。

具体各类用地分析如下：

用地类型面积统计与国家标准对比分析表

Land Use Proportions Compared with National Standard

大类	类别名称	总面积（hm²）	比例	国家标准	对比	备注
G	绿地	2163.4	34.2%	10% ~ 15%	高	历史名园集中
R	居住用地	1175.2	18.6%	25% ~ 40%	低	城中村分布广
A	公共管理与公共服务用地	994.7	15.7%	5% ~ 8%	高	高等院校集中
B	商业服务业设施用地	428.5	6.8%	—	—	中关村地带为商业中心，其余散布
M	工业用地	9.1	0.1%	15% ~ 30%	低	—
W	物流仓储用地	5.6	0.1%	—	—	—
S	交通设施用地	1531.7	24.2%	10% ~ 30%	—	越靠近城中心路网结构越清晰
U	市政公用设施用地	24.4	0.4%	—	—	北京市供水水源、南水北调工程保护地
合计		6332.6				

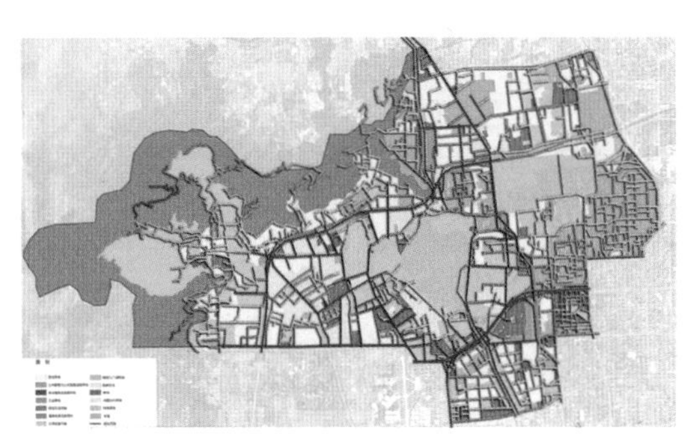

全局整合度与用地类型
Spatial Integration by Space Syntax and Land Use

全局整合度和建筑密度
Spatial Integration by Space Syntax and Building Density

建筑容积率对比分析
Spatial Integration by Space Syntax and Floor Area Ratio

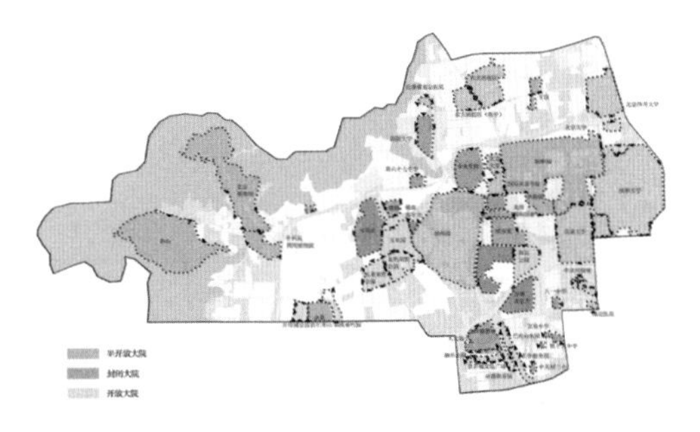

重点单位分布
Distribution of Key Work Units

（1）绿地与广场用地：占用地面积 34.2%，其分布以历史园林为核心，周边散布点状和带状绿地，绿地总量大，环境质量高。大院式绿地较多，但开放性不高，而缺乏连接性。

（2）居住用地：占用地面积 18.6%，东南部集中分布，西北部零星分布，一、二、三、四类居住用地均有覆盖。三、四类居住用地（主要为城中村）规模虽不大，布局分散，穿插在各类用地之间，阻碍景区之间的联系，严重影响城市风貌。

（3）商业用地：占用地面积 6.85%。高科技产业园区用地集中在地段东南侧，其余低端商业用地分布散乱，且服务质量较差；部分商业用地与公共交通关联度不高，与其他类型用地的联系弱；或与城中村用地混杂、环境质量低下、配套设施不完善有关。

（4）道路交通设施用地：占用地面积 24.2%。构成网状格局，东南部路网密度大且规整，西北部路网密度小且凌乱；四环、五环快速路相对封闭，道路两侧交流与联系较弱，尤其制约五环路北侧片区的发展。

（5）公共管理与公共服务用地：占用地面积 15.7%。该类用地总量相对较大，高校等文化设施数量多，公用服务设施较全，但各类服务设施分布不平衡；科研教育用地内环境质量好，建设强度不大，校园开放程度有限。除科研教育用地外，西部公共管理与公共服务用地偏少，设施类别不够丰富。

（6）农林用地：园地数量充足，大多集中分布；基本处于封闭管理，与其他用地衔接较差。

地段内有大量“封闭型”单位大院，主要包括驻军、高级居住区、高校、公园等。根据其对公众开放程度，又分为封闭型、半开放型及开放型三类。

城中村内部现状问题
Urban Villages with Poor Living Conditions

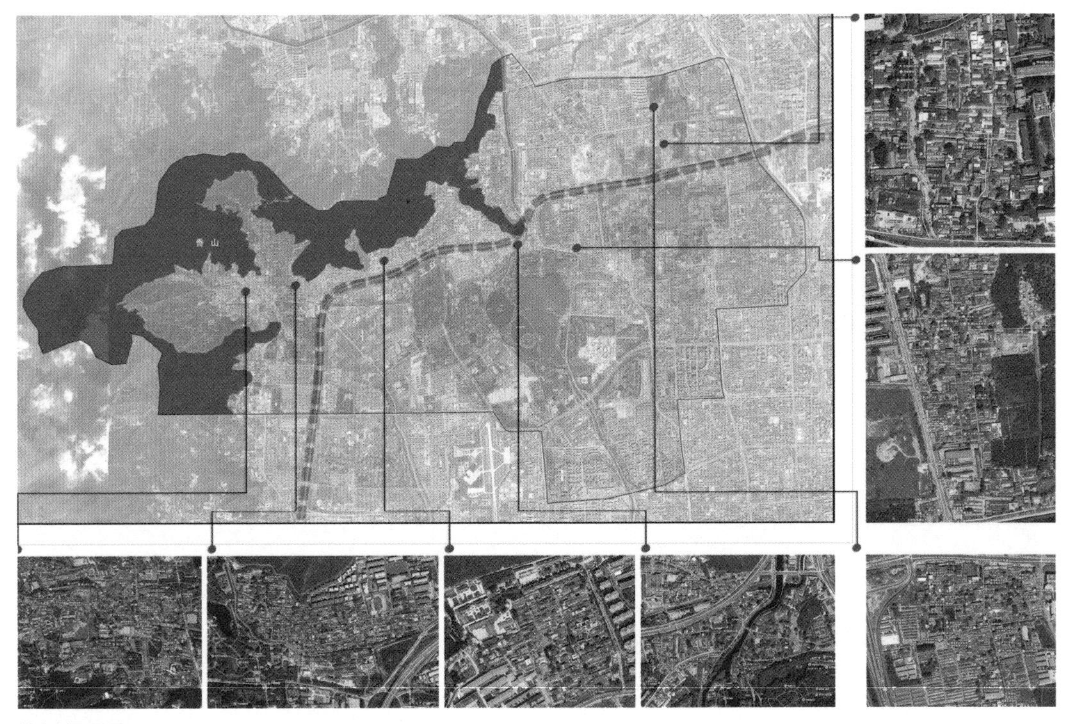

城中村卫星图
Locations of Urban Villages by Satellite Map

针对用地现状及存在问题，提出了 8 个方面的规划建议与相应措施：

（1）加强各分散绿地的关联，完善绿网结构，建立完整的绿地系统（尤其加强历史名园与周边绿地的联系），为绿道提供更舒适的环境。

（2）完善绿地内的基础设施，为绿道规划提供基础。

（3）改造城中村，充分利用存量建设用地，通过原址改造和搬迁安置，引导居住用地向集约利用方向发展，为绿道选线提供空间。

（4）在西部集中布置商业用地，增强集聚效应，完善路网结构，加强商业用地与公共交通的关联性，带动绿道发展。

（5）加大对公共管理与公共服务用地建设的投入力度，努力营造良好的社会氛围，带动绿道发展。

（6）带动田园风光展示带的开发利用，引入绿道，促进其发展。

（7）调整交通用地，在保持快速路独立性的同

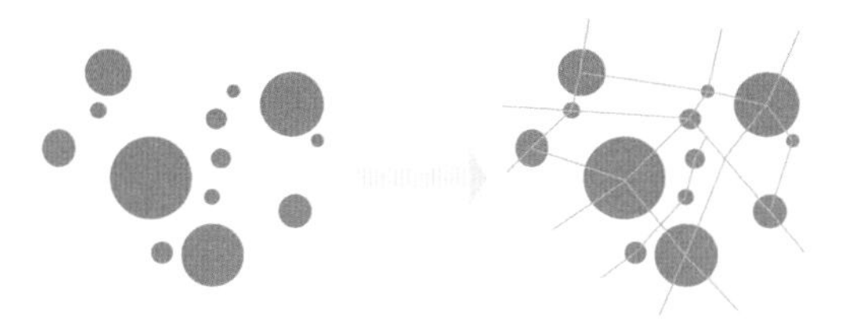

加强各分散绿地间的关联
Strengthen Links between Green Spaces

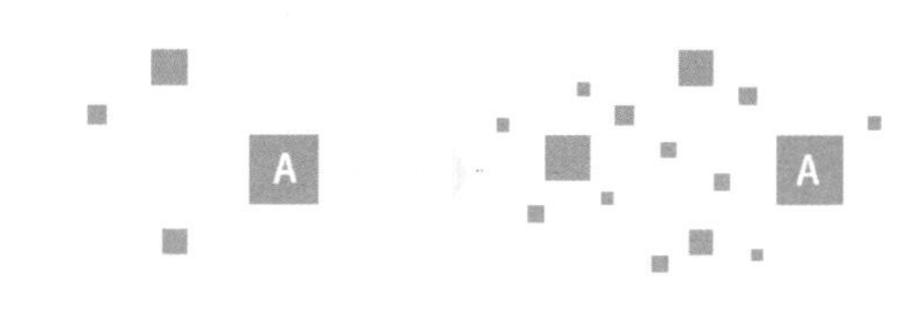

加大公共管理与公共服务用地的投入力度
Increase Investment to Public Sectors

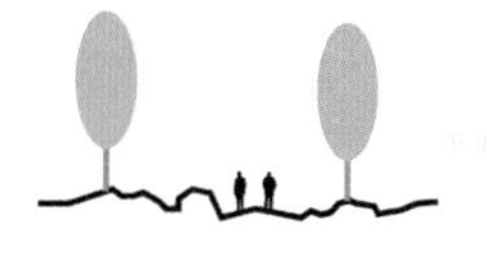

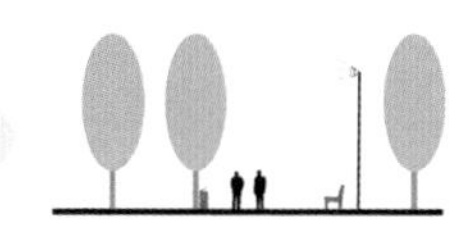

完善绿地内的基础设施
Improve Infrastructure within Green Spaces

加强快速路两侧用地之间的联系
Enhance Connections across Urban Expressways

改造城中村
Renew Urban Villages

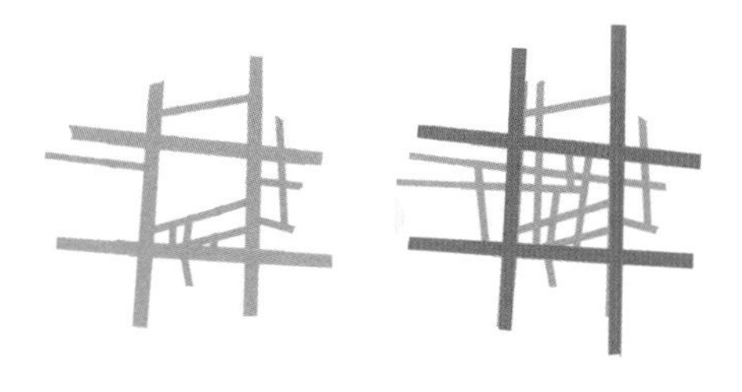

完善路网结构
Improve the Road System

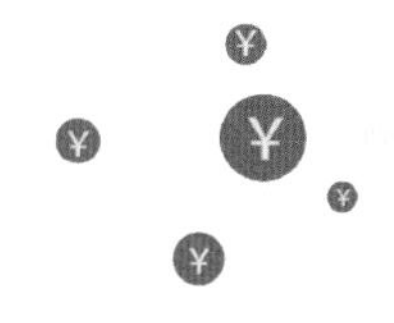

西部集中布置商业用地
Increase Commercial Land in the West Part

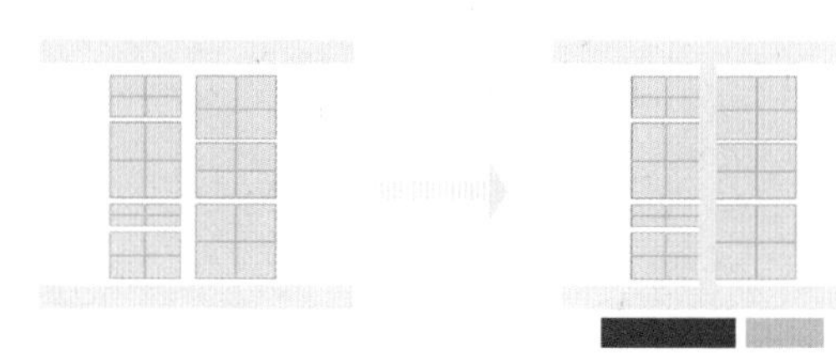

开发利用田园风光带
Develop Rural Landscape Belt

时，加大快速路两侧用地之间的联系，通过改造与快速路平行的路段（包括香山路、北四环西路辅路等），增加地下穿越等方式，可通过绿道连接快速路两侧，提高绿道的连通性。

（8）提高路网密度，完善路网结构，发挥交通区位的优势，发挥最大经济效益，为绿道选线发展提供更多可能性。

交通现状专题研究

Specialized Studies on Existing Traffic

刘加维、吴明豪、王茜、刘童、张琦雅、夏甜、韩冰
Liu Jiawei/Wu Minghao/Wang Xi/Liu Tong/Zhang Qiya/Xia Tian/Han Bing

本专题对于交通现状的调研分别以现状路网、交通服务体系、实时路况、立交现状、水陆交通及避灾通道六项进行展开，确立游憩、生态、文化等多元目标为导向，探讨其交通格局和特征，得出结论并提出建议，为三山五园地区的未来绿道选线及整体开发保护提供参考。

该地区内，五环快速路割裂了南北片区。在高架桥的位置则提供了绿道选线经过的可能。万泉河快速路割裂了其东西两侧地块，在西苑桥和西苑北桥能够提供直接的穿行机会。

在综合各种因素的基础上，对于慢行空间体系建立了评价系统，并以此为基础评估了区域内现有的慢行体系并进行分级。评价指标分为道路自身因素、设施因素以及绿化因素。由此得到的评分结果划分慢性空间的等级分别为：9~10 分，A 级慢行空间；7~8 分，B 级慢行空间；5~6 分，C 级慢行空间；3~4 分，D 级慢行空间；1~2 分，E 级慢行空间。

通过评价分析可以得知，该地区慢行空间总体建设情况较好，但仍存在着管理不佳、占道情况严重、绿化有待提升等问题。最不理想的慢行空间一般靠近城乡结合部，机动车和非机动车混行是最主要的问题。然而上述地段往往周围绿地充足，有较大发展和改造为优质慢行空间的潜力。主要包括树村路、旱河路、门头村路等。

该地区公共交通便利性一般：包括四条地铁线路（两条建成，两条在建），没有 BRT 公交线路通过；普通公交线路较为发达，基本覆盖了从西边的香山

现状路网分析图
Existing Road System

现状快速路分析
Existing High Speed Roads

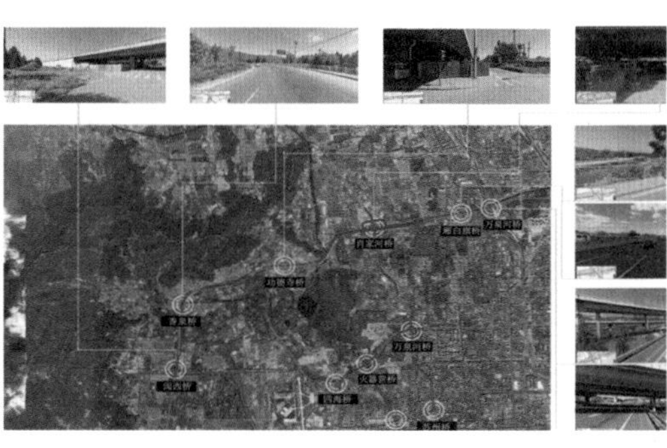

现状立交桥分析
Existing Overpasses

到东侧的上地、蓝旗营、海淀黄庄以及从北侧的百望山森林公园到南侧的远大路，其中西苑作为疏通东西、南北交通的重要枢纽。

以 TOD 公交导向开发以及社区型公交社区模式为评价基础，对地区内的公交地铁接驳现状进行评价。公交地铁的接驳覆盖范围：四环内能够满足，五环西侧玉泉山以西覆盖较为稀疏，北侧个别居住区和商业区处覆盖度稍显欠缺。

根据实时监控分析，在工作日拥堵情况最多发生在环路立交桥位置以及部分商务写字楼集中的区域（例如中关村等）；同时也出现在高校和景区汇集的区域（清华大学、北京大学及圆明园地段）。拥堵发生的时候，拥堵路段较长、影响的范围较大。拥堵情况多发生在上下班时段，主要拥堵的交通类型为通勤交通 。

根据实时监测的结果发现，在周末，除中关村等拥堵情况依旧十分严重的地段，总体拥堵路段相对较短，影响的范围比较小。其余立交桥等交通节点位置的拥堵情况比工作日相对减轻了许多。形成鲜明对比的是，通往香山、植物园等郊区重要景点的香山路在丰户营处成为周末最为拥挤的路段之一。此路段的拥堵持续时间长，并且情况也比较严重。

按照道路的等级及在防灾救灾中道路的救援等级分为：①城市级防灾救灾通道，用来联系灾区与非灾区、城市局部（受灾）地区与其他城市，以及各防灾分区。主要为城市快速路和主干路，如五环、四环等。②区级防灾救灾通道，灾害发生及灾害稳定后内部运送救灾物品、人员等。有效宽度不小于 15m。疏散道路避开危险源。确保消防车、大型救援车辆的通行。③社区级防灾救灾通道，功能以人员避难疏散为主，辅助连接防灾通道。有效宽度在 8m 以上。每一街区至少应具有 2 条以上的避难道路。

在不同等级的防灾避灾通道上，串联起主要的医院作为重要的医疗救护点，并结合现有的作为应急避难场所的公园绿地、街旁绿地、小区游园等，形成完整的防灾避灾体系。

通过以上分析，可以得出本专题的研究结论：

（1）现状路网外部绿地分布横向以四环、五环两侧各 100m 的绿化带为基础；纵向以京密引水渠和万泉河为基础。内部则由于现状快速路、主干路对绿道建设的影响主要体现在道路对原有绿地体系的切割，切断了原有绿地之间的联系，不利于生物之间的交流和生态多样性的形成，所以整体性较差，需要加强颐和园、圆明园、香山以及西山森林公园之间的联系。

（2）交通服务体系：以 TOD 理论为评价标准，对现状公交、地铁布局进行分析评价，发现四环内站点服务半径能够覆盖全部区域，五环西侧玉泉山以西覆盖较为稀疏，北部地区个别居住区和商业区处覆盖度稍显欠缺。大部分地铁站点接驳较好，但万安公墓、颐和园南门和万泉桥河三站接驳较为不便；最不理想的慢行空间大多靠近城乡结合部，机动车和非机动车混行是最主要的问题，但上述地段周围绿地充足，有较大发展和改造为优质慢行空间的潜力。城区慢行空间总体建设情况较好，但考虑到该区土地性质不易更改，提升改造潜力不大，建议绿道选线应绕行。停车场建设应综合考虑周边用地情况、路网承载能力，分等级规模建制，力图基本满足停车需求。

（3）三山五园地区交通拥堵现象多呈现季节性潮汐状态，在周末节假日期间，拥堵状况较为严重。日常拥堵情况多发生在环路立交处，还有部分写字楼集中的区域。

（4）该地区立交桥数量较多，主要分布在五环、四环和万泉河快速路沿线；场地内还有许多座供行

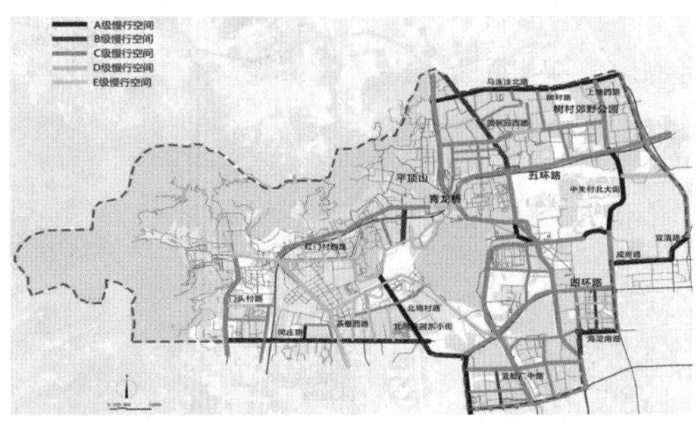

慢行交通空间分析
Existing Slow Traffic System

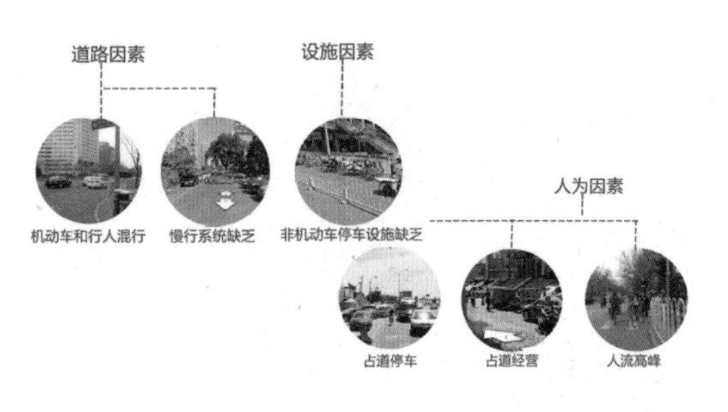

拥堵路段分析示意图
Frequently Congested Roads

防灾避灾通道及救援点示意图
Existing Fire and Disaster Evacuation Paths and Rescue Spots

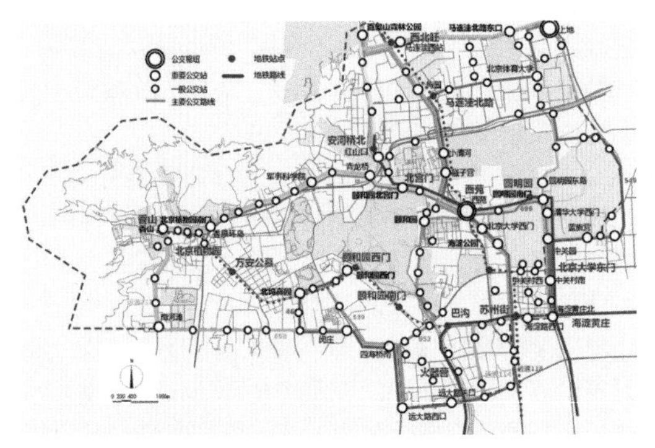

公交地铁现状分析图
Existing Bus and Rail Stops

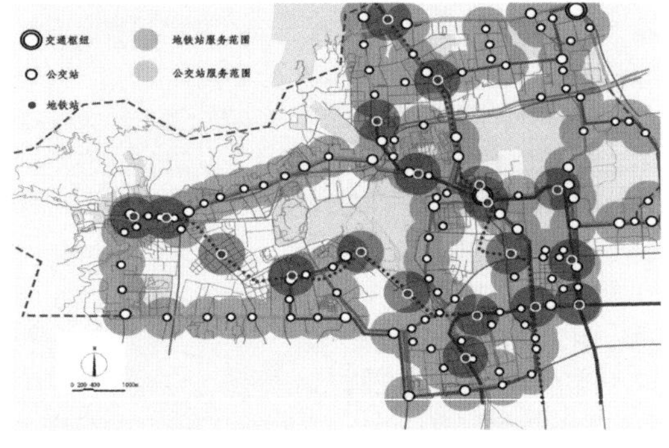

公交地铁接驳分析图
Existing Public Transit Hubs

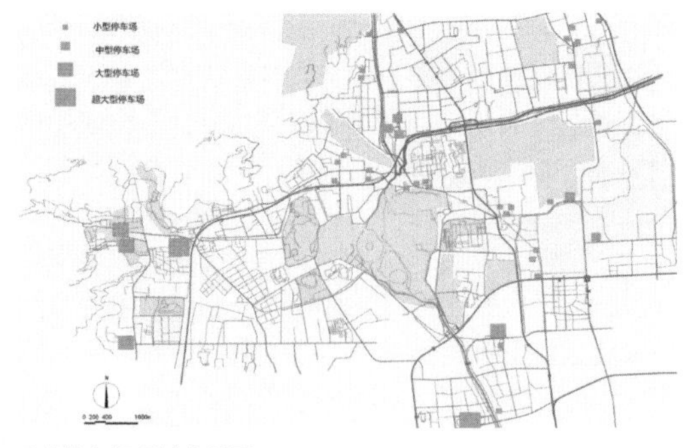

公共停车场现状分析图
Current Situation of Public Parking-lots

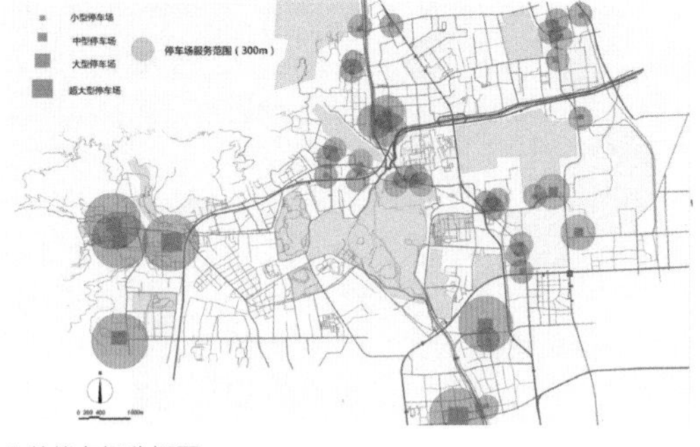

公共停车场分析图
Current Analysis of Existing Parking-lots

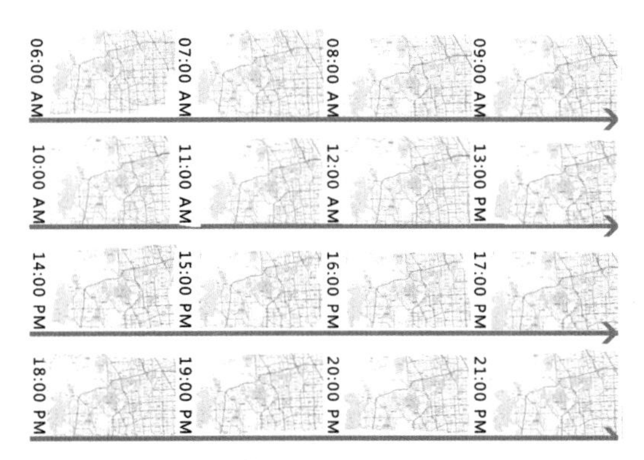

工作日实时监控交通现状
Traffic Conditions on Working Days

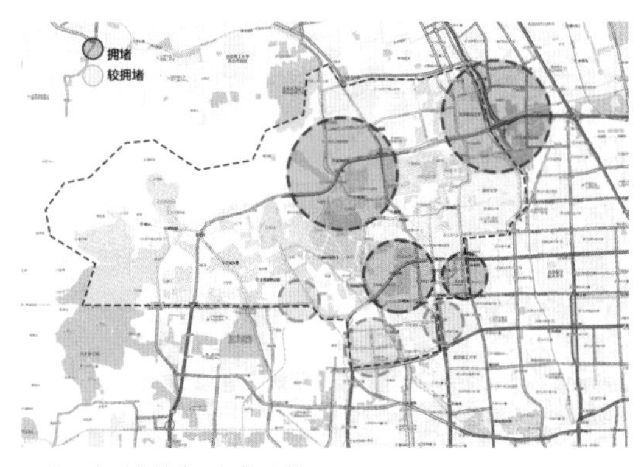

工作日实时监控交通拥堵现状
Traffic Jams on Working Days

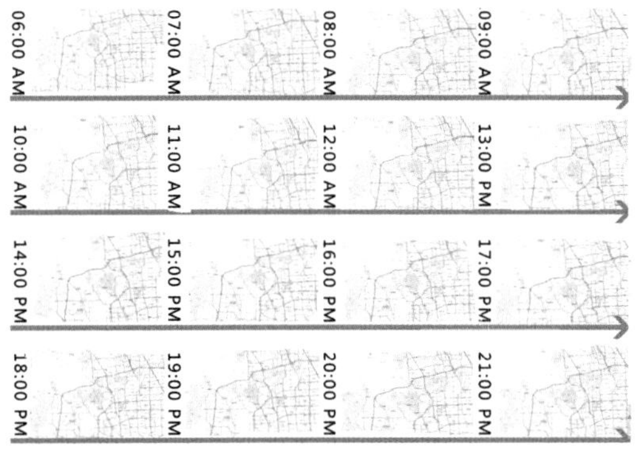

周末实时监控交通现状
Traffic Conditions on Weekends

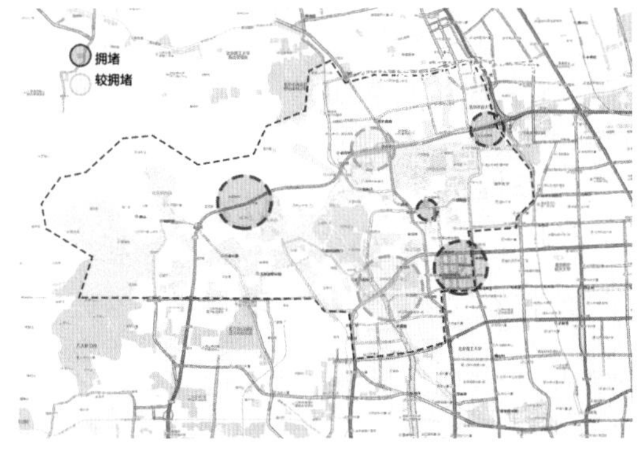

周末实时监控交通拥堵现状
Traffic Jams on Weekends

人和慢行交通使用的过河桥梁和过街天桥，使用状况良好。

（5）依据北京城市绿地系统的结构特点，结合地质条件及现状因素分析评价，发现地区防灾避灾通道设置、应急避难场所布局能有效覆盖全区，满足地区防灾避灾需求，道路两侧绿化和照明效果较好。

三山五园地区文化底蕴丰厚、景观特征鲜明、城市功能复杂多样，它包括大量人文与自然结合的历史文化遗产和景观遗产，因而规划中首先应该以遗产保护、生态保护和景观风貌保护为前提，在未来绿道规划选线中从实际出发，在保护好历史文化资源的基础上，考虑环境整治诉求，改善地区交通环境条件，积极引导产业发展，合理规划和利用现有水线、路线、轨道线，串联起大片的郊野生态绿化和点缀其中的园林遗产，尽可能地使城市建设与古典园林与其外围历史景观形成良好互动的可持续发展。

建筑现状专题研究

Specialized Studies on Existing Building

王博娅、丁小夏、郑慧、张司晗、郭沁、刘健鹏
Wang Boya/Ding Xiaoxia/Zheng Hui/Zhang Sihan/Guo Qin/Liu Jianpeng

本专题研究中，考虑到研究的需要和调查的可操作性，把居住区、单位大院、街区等作为单位空间，一共划分了 285 个单位空间，并对地块进行编号。本次调查不对单个建筑进行评价；绿地不在调查范围内；以卫星图和现状作为基础资料，绘制成 CAD 图纸，方便进一步研究。

针对上述 285 个调查对象，选定建筑性质、建筑风貌、容积率、建筑密度、建筑高度、建筑色彩等 6 项作为调查因子，通过实地踏勘、现场采访、遥感技术进行调查，并通过数据统计分析总结单位空间的现状情况。

建筑性质按照其使用性质把该场地的建筑物分为居住建筑、工业建筑、园林建筑和其他。建筑风貌指建筑设计中在内容和外貌方面所反映的特征。容积率是总建筑面积和建筑用地面积的比值。建筑密度指项目用地范围内所有建筑的基底总面积与规划建设用地面积之比。建筑高度指建筑物室外平面至外墙顶部的总高度。建筑色彩是城市景观中的主体部分，直接影响了城市色彩的美。

单位大院性质与其用地性质相一致，相应的建筑性质有居住建筑、行政办公建筑、文教建筑、科研建筑、医疗建筑、商业建筑、体育建筑、旅馆建筑、交通建筑、工业建筑和其他类别。

建筑风貌受不同时代的政治、社会经济以及材料、技术的制约，呈现出多种多样的风貌。居住建筑中，郊区或者城中村依然是以普通居住建筑为主。20 世纪 70 年代之后，受西方现代主义建筑思想的影响，出现了一系列现代风格的居住建筑。20 世纪 90 年代，部分开发商崇尚“欧陆风格”和市场运作的不规范，“欧陆风格”卷土重来。而今，越来越多的人提倡把传统的文化内涵和建筑精髓配以现代的生活方式，从而引领新的中式生活，新中式风格的别墅应运而生。公共建筑中，大部分是具有现代风格的建筑，由于建筑出现的时代背景和设计师的经历，出现部分折中主义的公共建筑，还有少数传统建筑是由古典园林直接改建的。工业建筑全部为现代风格的建筑。园林建筑主要以传统建筑为主。

根据建筑容积率分类标准和现场实际调查验证，把容积率分为 < 0.8，0.8 ~ 1.2，1.2 ~ 1.5，1.5 ~ 2.2，3.3 ~ 4.0，> 4 的标准分类整合。根

“三山五园”历史风貌图
Historic Picture of "The Three Hills and Five Gardens"

单位空间划分编号图
Index of Plots

京密引水渠
Jingmi Diversion Canal

城市建设现状
Existing Urban Buildings

西山鸟瞰图
Bird's Eye View

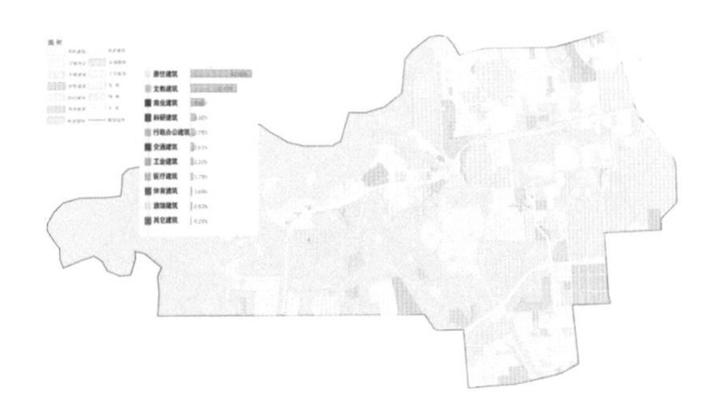

居住建筑
Residental architecture

商业建筑
Commercial architecture

文教建筑
Cultural architecture

建筑性质现状调查图
Categories of Existing Buildings

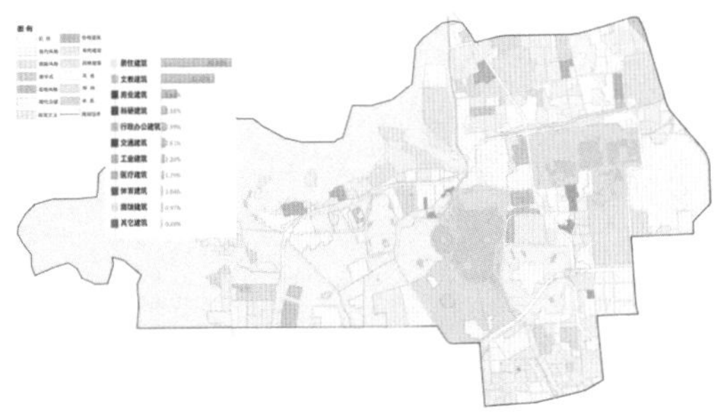

新中式建筑
Neo-Chinese architecture

民居
Folk house

现代风格建筑
Modern architecture

建筑风貌现状调查图
Architectural Styles of Existing Buildings

容积率小于 1
Plot ratio less than 1

容积率大于 4
Plot ratio more than 4

城中村
Village in a city

容积率现状调查图
Floor Area Ratio by Plots Buildings

密度 15% ~ 30%
Density 15% ~ 30%

密度 30% ~ 50%
Density 30% ~ 50%

密度 50% ~ 70%
Density 50% ~ 70%

建筑密度现状调查图
Building Density by Plots

据单位空间性质的不同，建筑密度分为 < 15%，15% ~ 30%，31% ~ 50%，51% ~ 70%，> 70% 等几个等级。根据建筑高度相关规定，把建筑分为低层建筑（1 ~ 3 层）、多层建筑（4 ~ 6 层）、中高层建筑（7 ~ 10 层）、高层建筑（10 层~100m）和超高层建筑（100m 以上）等几类。

宏观布局上，居住建筑分布较广，作为整个场地的"面"出现。文教建筑主要分布在东部地区，四环和五环之间，作为场地的特色出现。商业建筑、行政办公建筑、医疗建筑、体育建筑、旅馆建筑、交通建筑等满足人们基本生活需求的建筑，以东西方向的四环、五环，南北方向的京密引水渠为轴，串联分布。整体形成"两横、一纵、五片区"的布局模式，方便人们生活。

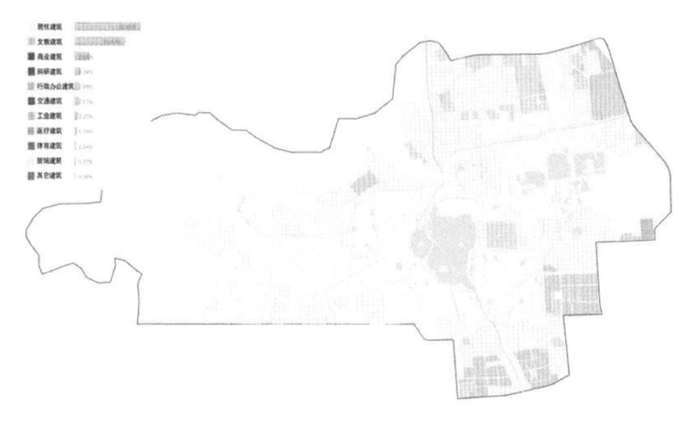

低层建筑
Low-rise building

多层建筑
Multi-story building

高层建筑
High-rise building

建筑高度现状调查图
Height of Existing Buildings

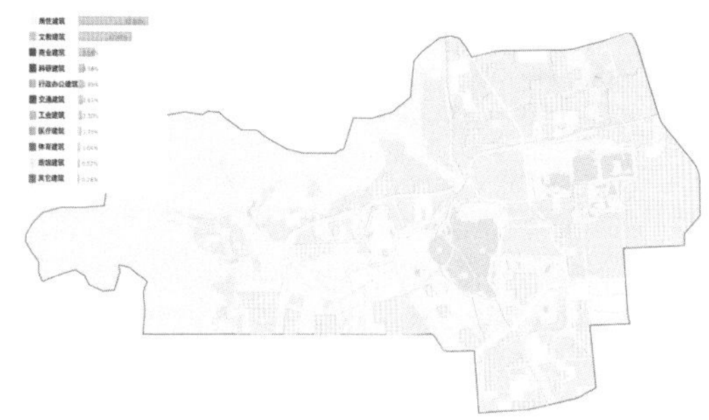

黄色系
Yellow color system

红色系
Red color system

白灰色系
White and grey color system

建筑色彩现状调查图
Colors of Existing Buildings

中观布局中，居住建筑多集中在东部，西部开发程度低。商业建筑中中关村西区作为北京市大型的公共建筑组团，规模较大。而服务于场地内的商业建筑分布较均匀，说明该地区基本生活较方便。同时，商业建筑多临近主路，交通便利，分布合理。

微观设计层面，快速路四环、五环周边的红门村、丰户营小区等居住建筑，影响较大。

各建筑风貌情况，普通居住建筑最多，约为36%，现代风格的公共建筑约占22%，现代风格的居住建筑约占15%，其他风格建筑比例较小。居住建筑中，郊区或者城中村依然是普通居住建筑，例如肖家河、六郎庄等地区。

容积率在0.8以下的地区所占比例最大，约为36%，分布在中部和西部地区，多为待开发的城中村或已建成的别墅区。容积率较高地区集中分布在东北、东南地区，例如万柳地区和中关村西区。部分地区容积率大于4.0，多为位于高密度住宅区的商业建筑。建筑密度在50%～70%的单位空间所占比重最大，为41%，主要分布在五环以外。建筑密度在15%～30%的次之，30%～50%之间比重也较大。而建筑密度小于15%和大于70%的大院都仅占2%。西北部山脚下的村中地区建筑密度大，甚至超过70%。肖家河地区的城中村建筑密度也很大。部分别墅地区建筑密度非常小。大部分区域的建筑密度在30%左右，比较正常。

该地区建筑以低层和多层为主，多层建筑比重最大为48%，低层建筑占29%，超高层建筑仅占1%。从整体分布来看，低层建筑分布在西部地区，高层和超高层多分布在东部地区。东南部和东北部的高层建筑对场地形成了围合之势。中高层多分布在东北部，多层建筑在场地中分布较均匀。肖家河、西苑地区有较多的低层建筑，主要是城中村。东西方向主要的视廊上大部分为较低建筑，四环以内的中关村西区和万柳地区的建筑高度较高。在东西方向的视觉走廊上，颐东苑小区的建筑高度突出，阻挡了视线、破坏了视廊。清华科技园为超高层建筑，在城市空间上比较突出，山脚下的中国林业科学研究院亦是如此。而后营村、树村等城中村由于偏低的高度，在城市空间中显得较为局促。

整体来看，场地中建筑以红色系和白灰色系为主。各色系分布不成体系，色彩多样，布局较乱。中观角度上，历史风貌保护片区内，大部分控制较好，同时出现部分不和谐因素。在颐和园周围的色彩重点控制区内以白灰色系为主，少量红色系出现。圆明园周边的重点控制区颜色较为混杂。中关村西区的色彩统一但和周边不协调。微观角度上，香山公园附近的黄色系和红色系建筑突出，与整体建筑的白灰色系形成较强烈的对比。

绿地与水系现状专题研究

Specialized Studies on Existing Green-Water Systems

何二洁、罗亚文、朱友强、甘其芳、孟宇飞、施晨、施莹、张俞

He Erjie/Luo Yawen/Zhu Youqiang/Gan Qifang/Meng Yufei/Shi Chen/Shi Ying/Zhang Yu

三山五园地区位于北京西北郊生态楔形廊道，是北京西郊皇家园林区和北京“宜居城市和历史名城”的重要标志。生态保护是绿道规划最为重要的功能，绿地水系是体现生态保护最为核心的要素，因此基于三山五园地区绿地水系要素的专项研究，对于北京三山五园地区绿道概念性规划起到了重要的指导意义。

1 现状绿地

三山五园地区位于北京西北郊连接中心城区的楔形绿地，该楔形绿地北接西山余脉，南插北京老城，对于缓解北京城的生态环境恶化起到了重要的作用。该地区现状绿地总量较大，除大量城市绿地外，还含有大量山体、耕地、农业采摘园等类型的生态用地。在城市绿地中，公园绿地用地总量居首。城市绿地类型同比北京其他区域类型最为丰富，除综合性公园、社区公园外，还含有植物园、历史名园、郊野公园等类型的公园。三山五园地区也是高校和机关大院的聚集区，相配备的高质量附属绿地对于该地区生态保护起到了一定的作用。三山五园地区总体的绿地格局可以划定为香山、颐和园、圆明园三大斑块，由西山嵌入居住用地的楔形绿地是其一大特点。四环、五环城市快速路的切割使得该地区绿地格局呈现断裂化的空间结构，用地类型呈现出环状梯度模式，降低了绿地之间的连通度。

绿地开放性分析图中分析了该片区内全部绿地的开放性与封闭性，封闭性绿地主要为专类公园，且主要分布于三山五园东西轴线上，隔断了南北向绿地的连通，是影响绿地连接性的重要因素。区域内具有较好的景观视点，万寿山、玉泉山、平顶山、西山及香山是主要的视点，绿道的选线将考虑这些重要的视线范围。

2 现状水系

三山五园地区含有多条水系连通城郊和城区，现状有水的河道主要为京密引水渠、清河、万泉河、

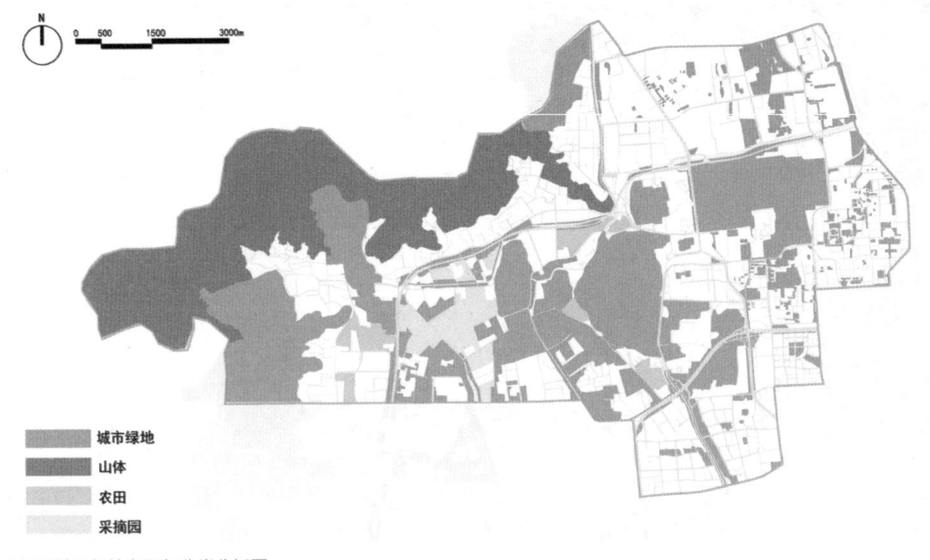

三山五园地区绿地布局与分类分析图
Existing Green Land by Categories

城市绿地布局及分类
Existing Urban Green Land

绿地开放性分析
Openness of Green Land

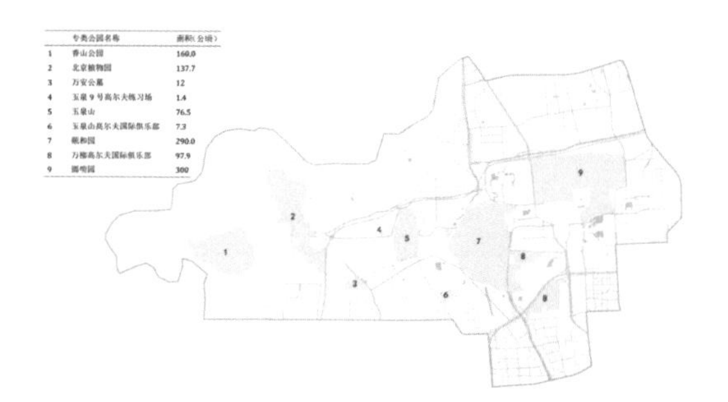

	专类公园名称	面积(公顷)
1	香山公园	160.0
2	北京植物园	137.7
3	万安公墓	12
4	玉泉9号高尔夫练习场	1.4
5	玉泉山	76.5
6	玉泉山高尔夫国际俱乐部	7.3
7	颐和园	290.0
8	万柳高尔夫国际俱乐部	97.9
9	圆明园	300

专类公园布局及分类
Existing Specialized Parks

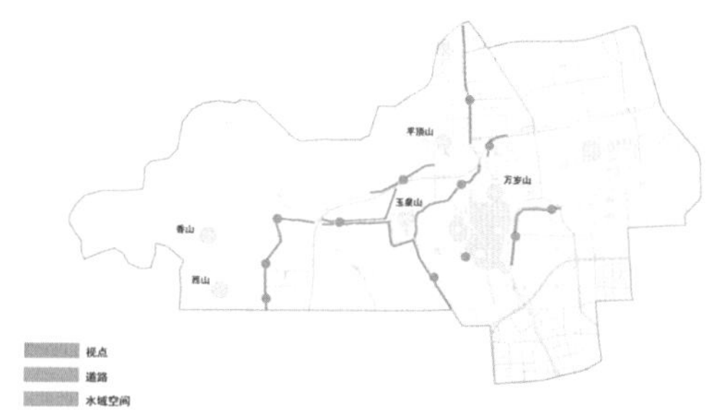

视线分析
Analysis of Sight Lines

绿地与水系总布局
Existing Green and Water Area

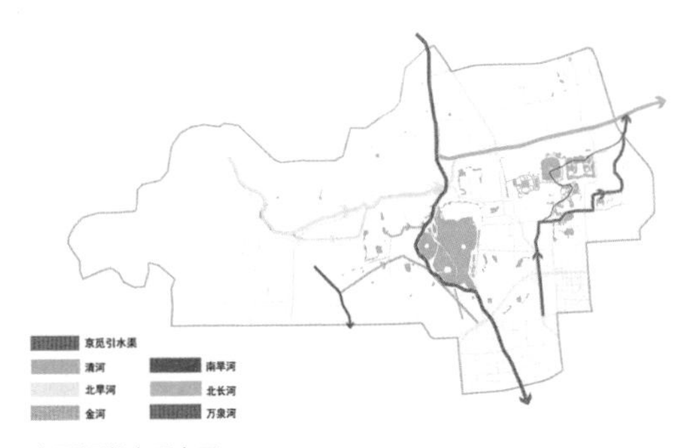

主要河湖水系布局
Existing Water System

水系与城市交通网络分析
Analysis of Water-Urban Networks

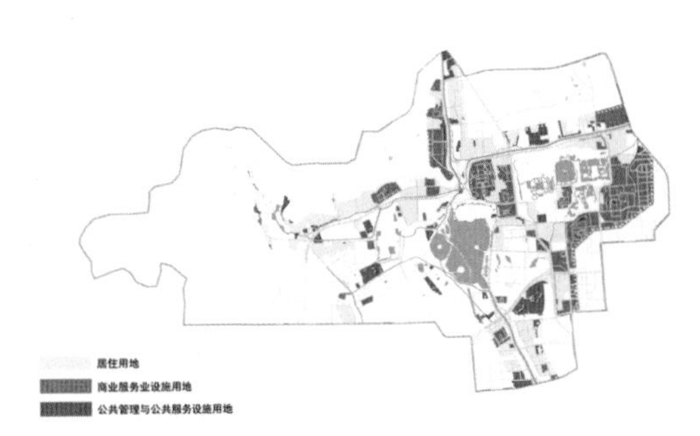

水系周边用地类型分析
Land Use of Waterfront Areas

河道功能表
The List of Rivers

河道名称	起止点	河道功能	河道长度(km)	水质现状
京密引水渠	百望山方向-团城湖 团城湖-玉渊潭方向	水源河道 风景观赏	9.3	II类、昆玉III类
清河	安河闸-清河闸	防洪排污 风景观赏	5.2	IV类
万泉河	巴沟村北路-汇入清河	防洪排水 风景观赏	7.2	V类
北旱河	樱桃沟-安河闸	防洪排水 风景观赏	4.8	无水
北长河	玉泉山-安河闸	防洪排水 风景观赏	3.2	IV类
南旱河	万安公墓-闵庄路	防洪排水 风景观赏	1.4	无水
金河	南旱河与万安东路相交处-金河退水闸	防洪排水	3.6	无水

南旱河、北长河，现状无水的河道有北旱河、南旱河及金河三条河道。该地区的河道的主要功能多为防洪排污与风景观赏，其中京密引水渠是北京城市供水的主要河道，也是北京最长的人工渠道。清河是北京北部的主要排污河，水质为Ⅳ类水。

水系与城市网络分析图中通过对水系周边的城市交通网络分析，河道岸线与城市交通网络整体连接性较好，但部分水系因为封闭绿地或居住区的阻隔与城市的连接性较差。其中，北旱河、万泉河及京密引水渠南部与城市网络有较好的连接，北长河可达性较差，需要增加与城市道路的连通。水系周边主要用地性质经分析得知，河道周边的主要用地为居住用地、商业服务设施用地、公共管理与公共设施用地。同时河道周边主要的服务人群为居民、上班族、游客等。周边服务人群的分析是绿道选线的重要影响因子。

三山五园地区现状水系主要包含以下问题。在水系与绿化方面，水系的防洪功能占主导地位，牺牲了水系的其他功能；河道河床硬化严重，生态功能退化；河流景观过于人工化，简单乏味；除京密引水渠外，水质普遍较差。在水系与城市关系方面，河道沿线缺乏开放空间，道路噪声及扬尘影响滨水景观游憩体验。在水系与文化方面，河流沿线极其缺乏体现文化特色的景观和标志物，散落在河流沿线的文保单位没有得到良好保护。

3 问题与策略

结合绿地水系的各部分研究，总结出三山五园地区绿地水系现状问题并提出相应策略：

问题一：西山山体与城市相割裂。西山片区，是三山五园片区独特的自然优势，楔形绿地插向城市内，但是城市道路阻隔了楔形绿地与城市的联系，没有形成很好的绿地网络。

策略：恢复西山山体植被，将自然渗透入城市，建立道路生态廊道，打通道路间的生态廊道。

问题二：公园绿地分布不均衡。整个三山五园研究范围内，休闲游憩功能的公园绿地主要集中分布于中西部，南北部分布极少。

策略：通过土地置换等方式增加公园绿地量，特别是综合性公园的数量。

问题三：历史文化景区之间功能缺乏连通性。集中分布在香山、玉泉山及颐和园之间，虽然自然资源优越，景观具有连续性，但游憩功能的连通性较弱。

山体与城市关系
Mountain-Urban Relations

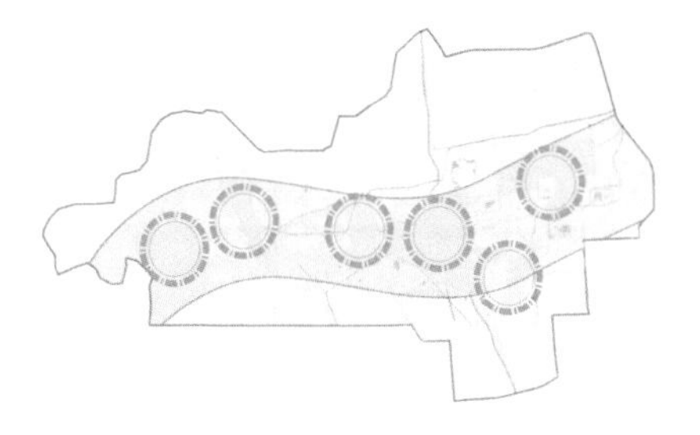

城市绿地布局
Existing Green Structure

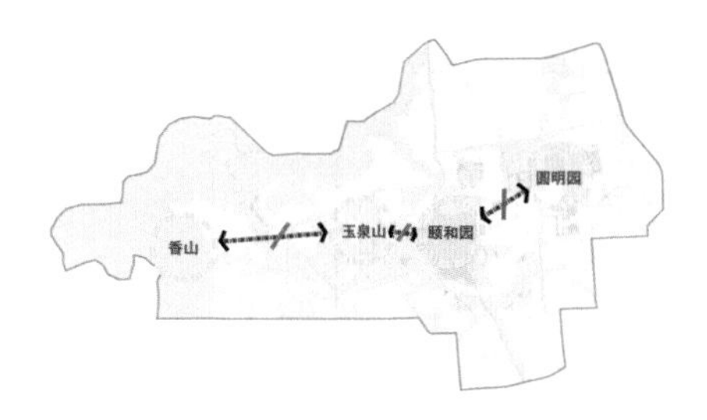

景区连通性分析
Connectivity of Scenic Spots

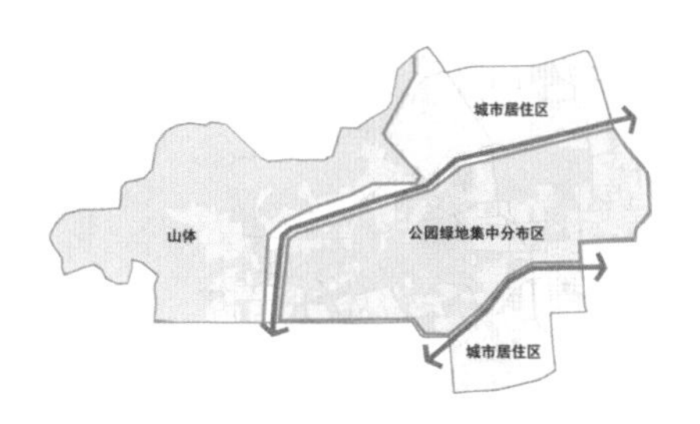

公园绿地分布分析
Distribution of Parks

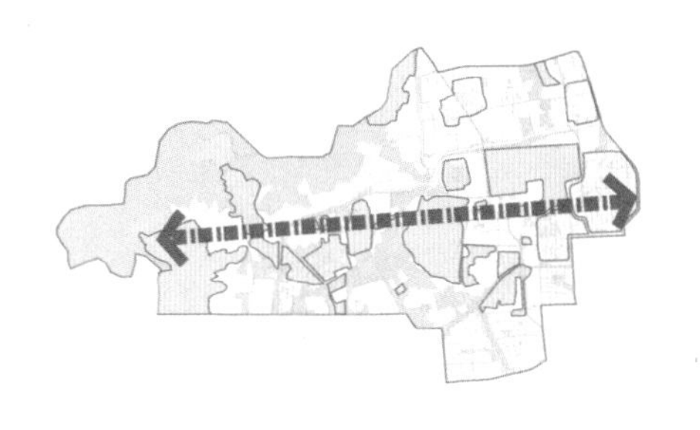

绿地封闭性分析
Closure of Green Land

河道生态效益分析
Eco-efficiency of Rivers

策略：增加游憩功能设施，完善三山五园地区的游憩设施系统。

问题四：居住区范围内的公园绿地较少，且与中西部联通性较弱。东北部及南部以居住用地为主，该区域内公园绿地面积相对较少，且因河道或城市快速路的阻隔，减弱了其与中西部的连通性。

策略：社区内建立线形的生态廊道，同时满足生态和游憩的需求，活跃该片区生态社会效益。

问题五：整个片区内，封闭性绿地比例占总绿地的一半，且大都位于东西中轴线内，形成了绿地游憩空间的阻隔，也是除北五环外另一个阻隔山体与城区联系的因素。

策略：将部分郊野公园或防护绿地开放，加强片区南北部的绿地结构上的沟通。

问题六：河道生态效益较差。线性绿地在建立绿地生态系统中作用重要，三山五园的河道水系水质较差，河道周边绿地狭窄，并未起到生态防护功能，游憩功能缺失。

策略：扩宽相应的河道绿地宽度，增加生态廊道连接性。

历史遗产现状专题研究

Specialized Studies on Existing Heritage

李婉仪、葛韵宇、张芬、胡盛劼、李雅祺、赵欢、邓力文
Li Wanyi/Ge Yunyu/Zhang Fen/Hu Shengjie/Li Yaqi/Zhao Huan/Deng Liwen

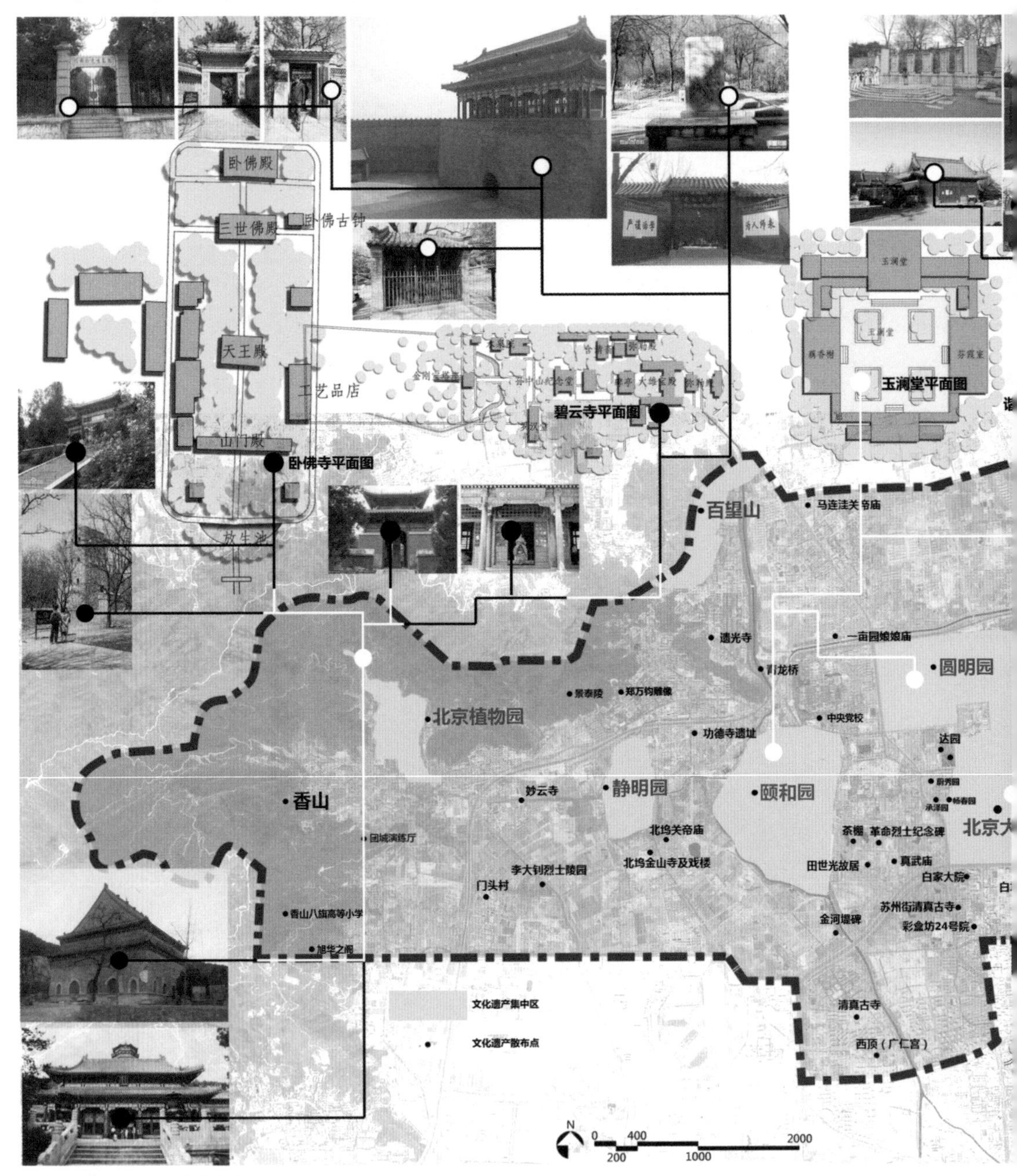

三山五园地区文化遗产总体分布图
Existing Heritages in "The Three Mountains and Five Gardens "

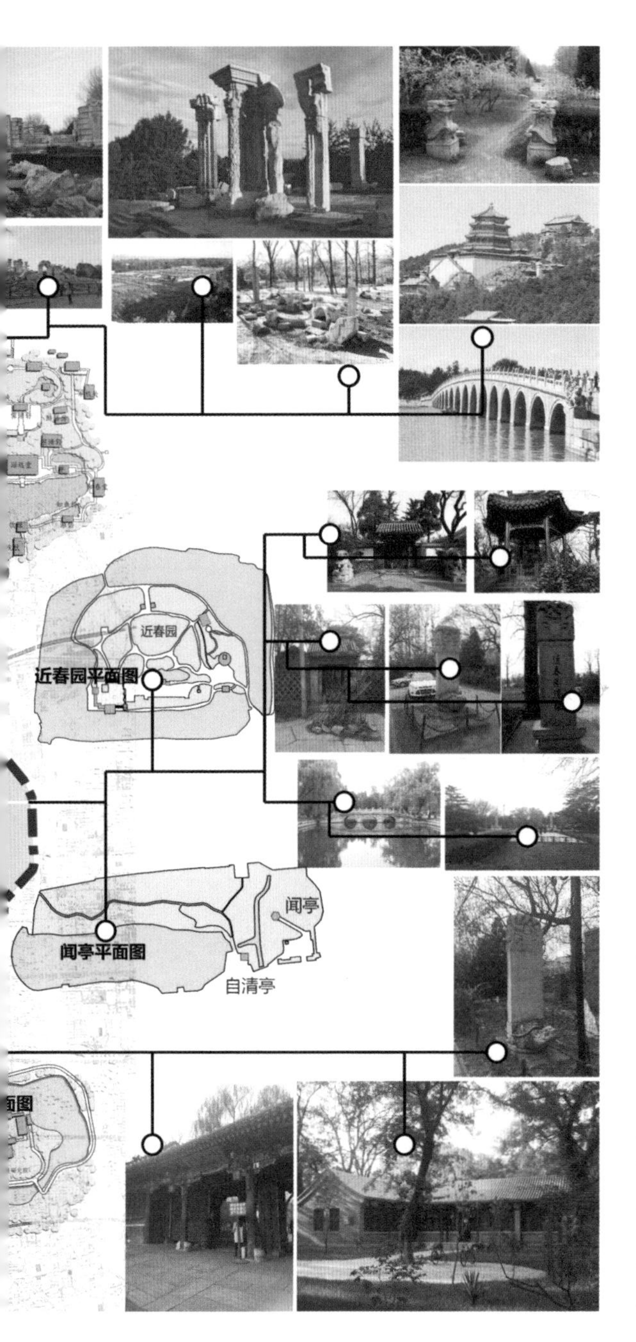

三山五园地区历史文化遗产的知名度普遍比较高，中国传统皇家园林享誉世界。其中，颐和园单日人流量可达 40 万，国外游客占游客总数的 9.3%。清华大学与北京大学两大区域，以及近现代建筑文物，主要凭借两大学校的影响力，也有着比较高的知名度。其他的文物主要分布在西山，集中在植物园与香山。此外还有大量市级、区级重要历史遗产。

针对三山五园地区的文物等级进行调查，分为全国重点文物保护单位、北京市级文物保护单位、海淀区级文物保护单位、文物普查登记项目四类。全国重点文物保护单位有 18 处，占总数的 38%；北京市级文物保护单位有 1 处，占总数的 2%；海淀区级文物保护单位 7 处，占总数的 15%；文物普查登记项目 21 项，占总数的 45%。文物类型分为综合类、建筑类、墓园类、纪念碑或雕像类四类。综合类有 20 处，占总数的 60%；建筑类有 26 处，占总数的 43%；墓园类有 9 处，占总数的 15%；纪念碑或雕像类有 5 处，占总数的 8%。文物现状类型分为军事用地及科教用地、参观游览景点、未开放的遗址保护景点以及村落及其他四类。军事用地及科教用地 19 处，占总数的 15%；参观游览景点有 76 处，占总数的 60%；未开放的遗址保护景点有 3 处，占总数的 2%；村落及其他有 2 处，占总数的 2%。

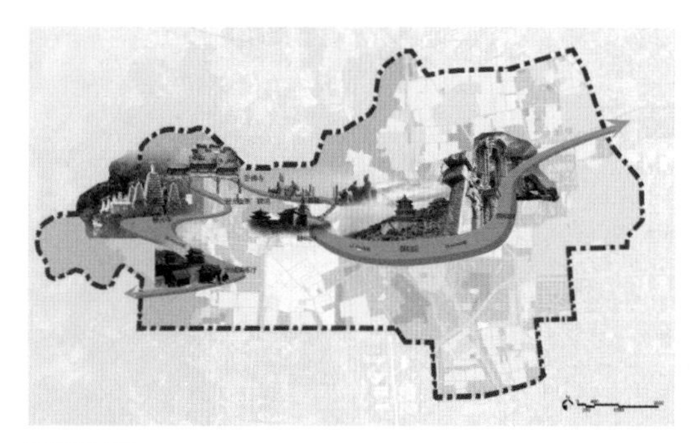

国家级文物
National-level Heritages

北京市级文物
City-level Heritages

三山五园地区文化遗产现状调查统计
List of Heritages

序号	所在位置	文化遗产名称	遗产时间	规模及类型	历史意义	现状保护利用情况	文物等级
1	北坞村	北坞关帝庙	建于清早期	历史纪念物	北坞关帝庙建于清早期，坐北朝南，有很高的历史价值	处于北坞公园内，保护完好	海淀区级文物保护单位
2	北坞村	北坞金山寺、戏台	明天顺五年建	历史纪念物	寺内佛道共存。该寺三进院落，主要建筑有灵官殿，正殿，还有大戏台	处于北坞公园内，保护完好	海淀区级文物保护单位
3	海淀区海淀镇北1公里	达园	始建于民国初年	历史纪念物	整个庭院融江南园林与北方建筑于一体，湖水山石叠映，亭榭长廊相连，总体建筑	京郊私家园林中保存最为完整的一座	北京市级文物保护单位
4	香山地区	妙云寺	建于乾隆年间	历史纪念物	有山门、前殿、配殿及后殿，具有历史价值	修复后保护较好	海淀区级文物保护单位
5	颐和园西侧，玉泉山上	静明园	清康熙十九年建行宫	历史纪念物	玉泉山，金、元以来的"燕京八景"之一，名曰"玉泉垂虹"		全国重点文物保护单位
6	绮春园宫门内西侧湖中	鉴碧亭	建于嘉庆十六年（1811年）前	历史纪念物	重檐方亭四面各显三间，具有很高的历史价值	复建后保护状况良好	全国重点文物保护单位
7	绮春园中心位置	涵秋馆	建于嘉庆年间	历史纪念物	涵秋馆是绮春园春（春泽斋）、夏（清夏斋）、秋（涵秋馆）、冬（生冬室）四季	立石刻图以志，古建基址今保存，建筑已不存在	全国重点文物保护单位
8	凤麟洲湖西岸山凹	仙人承露台	建于嘉庆年间	历史纪念物	露水神台作为点景之物，价值极高	1989年整理台基，重制墨玉石雕承露仙人，成为遗址公园一景	全国重点文物保护单位
9	绮春园东湖之中	凤麟洲	建于嘉庆年间	历史纪念物	嘉庆帝誉之为绮春园避暑最佳处	现存仅有废墟和石块，立石刻图	全国重点文物保护单位
10	敷春堂西宫门外	石残桥	建于嘉庆年间	历史纪念物	圆明园罹劫并经百年风雨后，仅残存这座单孔石拱桥	仅剩部分石块，未经修整	全国重点文物保护单位
11	绮春园南向正宫门内	迎晖殿	嘉庆十四年（1809年）建成	历史纪念物	本景即绮春园宫门区，位置独特	迎晖殿等遗址建筑台基仍在，今为一小型花园	全国重点文物保护单位
12	绮春园南部中间	正觉寺	乾隆三十八年(1773年)建成	历史纪念物	清帝御园圆明园附属的一座佛寺	寺里原存990平方米古建筑现已全面修缮	全国重点文物保护单位
13	居后湖南岸，东邻涵秋馆	展诗应律	嘉庆六年(1801年)已建成	历史纪念物	是一座看戏殿，具有历史研究意义	现古建基址已不见踪迹。1986年后绿化美化，1992年竖石刻图以志	全国重点文物保护单位
14	绮春园宫门内	敷春堂	—	历史纪念物	绮春园宫门内的中心景观	敷春堂一景北半部遗址今为圆明园管理处所在地	全国重点文物保护单位
15	福海中央	蓬岛瑶台	建自雍正初年	历史纪念物	圆明园四十景之一，在整个福海景区的点景与观景功能显著	1985年清整补砌、修复，现保存比较完好	全国重点文物保护单位
16	福海东岸北半部	涵虚朗鉴	建于乾隆三年(1738年)前后	历史纪念物	亦总称雷峰夕照，圆明园四十景之一	现仅存遗迹	全国重点文物保护单位
17	福海东岸南半部	接秀山房	雍正九年(1731年)前后	历史纪念物	圆明园四十景之一，欣赏风景最佳处	挖掘廓清观澜堂高台殿基及部分甬路，并略施修补	全国重点文物保护单位
18	福海东北海湾内之北岸	方壶胜境	建自乾隆三年前后	历史纪念物	园中最为壮观的建筑群之一	1985年来经过发掘清理，出土了大量建筑残件	全国重点文物保护单位
19	福海北岸	平湖秋月	建自雍正朝	历史纪念物	为秋夜赏月佳处，由一组散布的临水建筑组成。	平湖秋月基址现今发掘出土，两峰插云五孔石桥也加以修复	全国重点文物保护单位
20	福海正南岸中部	夹镜鸣琴	清代	历史纪念物	隐含了追求高洁人生的人格憧憬，表达出卓尔不凡的园林意境	古建基址全面清整	全国重点文物保护单位
21	后湖西岸	坦坦荡荡	建自康熙后叶	历史纪念物	圆明园四十景之一	2004年清理发掘	全国重点文物保护单位
22	东邻慈云普护，南俯后湖	上下天光	建于雍正初年	历史纪念物	上下天光就是模拟洞庭湖景色而建	现仅存遗迹	全国重点文物保护单位
23	后湖正北，濒临后湖	慈云普护	康熙朝后叶	历史纪念物	该处寺庙建筑充分体现了雍正崇教的复杂性	现仅存遗迹	全国重点文物保护单位
24	京西香山东南的万安公墓内	李大钊烈士陵园	1983年10月落成	历史纪念物	李大钊烈士陵园已成为爱国主义教育和革命传统教育的重要基地	李大钊烈士陵园建于1983年，当时利用万安公墓的老建筑，年代久远，破损严重，因而进行了大	海淀区级文物保护单位
25	海淀区西部	门头村	明嘉靖年间	历史纪念物	门头村历史悠久，自然环境优美，地理位置优越	保护较好，但是很多历史遗存风化严重	海淀区级文物保护单位
26	颐和园昆明湖西北岸边	功德寺	建于元朝天历二年	历史纪念物	历史悠久，有很高的研究价值	局部保护完整	海淀区级文物保护单位
27	圆明园遗址公园正大光明门东南侧	一亩园娘娘庙	建于光绪年间	历史纪念物	历史悠久，有很高的研究价值	保存一亩园昔日演耕譛耕田的原址	海淀区级文物保护单位

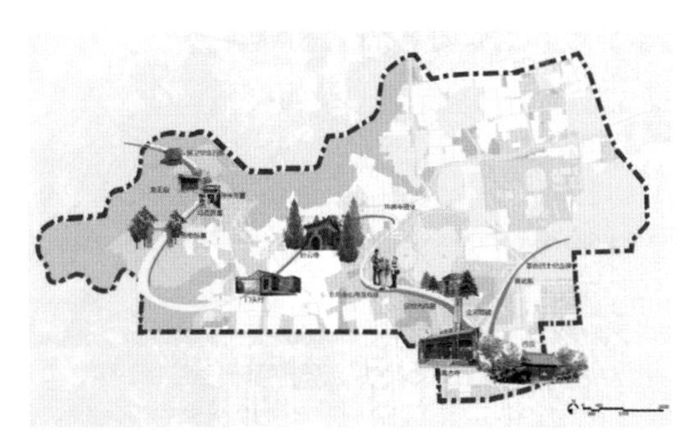

海淀区级文物
District-level Heritages

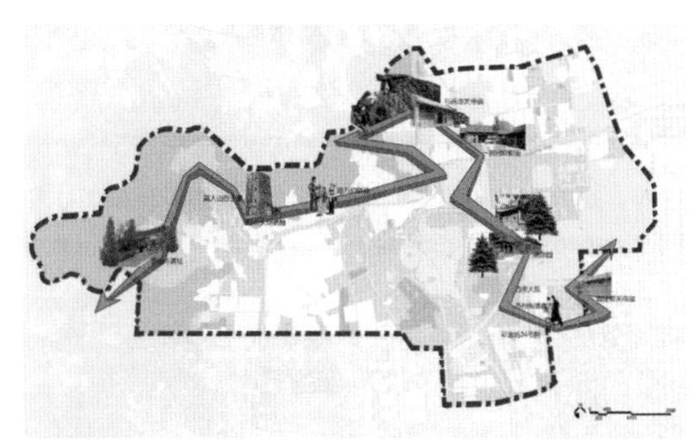

普查登记项目
Other Registered Heritages

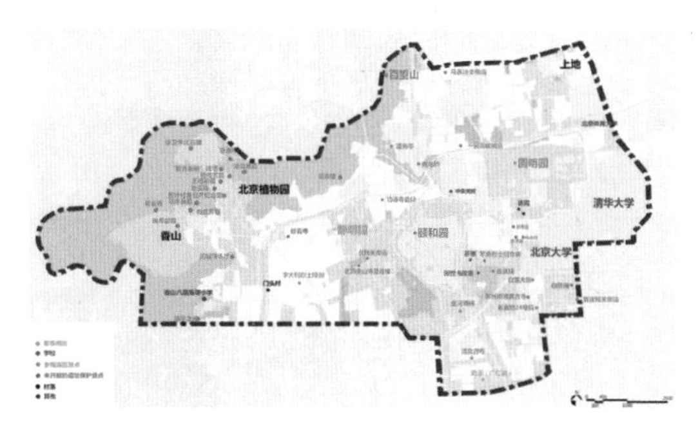

文物现状分类图
Existing Relics

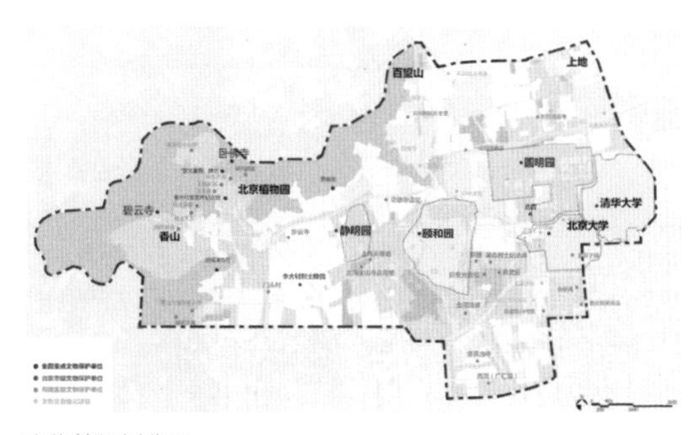

文物等级划分图
Levels of Relics

文化遗产利用及保护分析图
Conservation and Utilization of Heritages

文化遗产现状规模分析图
Aggregation of Heritages

非物质文化遗产及历史园林专题研究

Specialized Studies on History of Garden and Intangible Heritages

于静、洪卫静、乔丽霞、乔菁菁、伊琳娜、朱青、丁宁

Yu Jing/Hong Weijing/Qiao Lixia/Qiao Jingjing/Yi Linna/Zhu Qing/Ding Ning

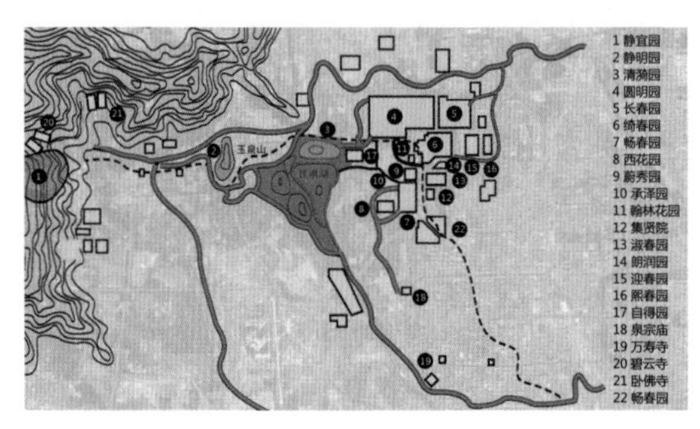

乾隆时期三山五园地区园林分布图
Historic Gardens during the Reign of Emperor QianlongYears

北京西北郊泉水丰富，风景秀丽，早在金朝便已在西山地区建立了名为“八大水院”的八处离宫。元大都的兴建当属元代北京最重要的建设工程，而西山一带的离宫别园则成为皇帝政治生活以外消遣散闷的去所。 元世祖忽必烈常到香山游乐。元代香山风景建设最重要的事件莫过于碧云寺的前身碧云庵的兴建。元文宗图帖睦尔至顺三年（1332 年），大承天护圣寺落成，寺址在玉泉山脚下。

清朝入关后，康熙开始经营西山园林，开启了在西山大规模建园的序幕，康熙十九年 (1680 年) 将玉泉山南麓改为行宫，初名澄心园，在香山寺旁建行宫。康熙二十三年 (1684 年)，在清华园废址上修建畅春园，成为北京西郊第一处常年居住的离宫。继康熙在西郊兴建园林之后，雍正三年 (1725 年)，将圆明园升为离宫，开始大规模扩建，将其面积由 300 亩扩大至约 3000 亩，并命名了“圆明园二十八景”。乾隆帝即位后，开始了大规模的园林兴建。首先将圆明园二十八景扩建为四十景，随后乾隆十年 (1745 年) 在其东边修建长春园。同年在香山修建静宜园,建成二十八景。乾隆十四年 (1749 年)，十五年 (1750 年) 扩建玉泉山静明园 (1692 年由澄心园改名)，将玉泉山全部圈占，并修建了静明园十六景。到乾隆三十四年（1769 年），“三山五园”工程基本完成。

清咸丰十年（1860 年），英法联军侵入北京，火烧圆明园、静宜园、静明园，对“三山五园”造成了严重的破坏。1947 年，古典园林面积进一步缩小，附属园林逐渐破败。由于屡经战乱，大量难民涌入，颐和园周围农业生产扩张无序，稻田密布，古典园林风貌几近消逝。该地区也逐渐遭到城市入侵，成为驻军、学校的选址处。

至 2015 年，该地区古典园林总体规模较为可观。建筑、道路合计已占总面积的一半以上，大型交通设施渗入“一山两园”(万寿山、颐和园、圆明园) 周边。科技园与住宅区的建设渗入该地区，“三山五园”整体表现为兼具自然风貌和高新产业的矛盾体。水体面积减小，水田稻作面临消失的危机。公园与绿道系统开始启动，该地区的整体风貌得到极大改观。

根据联合国教科文组织《保护非物质文化遗产公约》定义：非物质文化遗产 (intangible cultural heritage) 指被各群体、团体、有时为个人所视为其文化遗产的各种实践、表演、表现形式、知识体系和技能及其有关的工具、实物、工艺品和文化场所。海淀区非物质文化遗产 68 项，三山五园区域内有 12 项，国家级别 1 项，市级别 5 项，区级别的 6 项。其中包括的类别有民间文学、传统音乐、传统舞蹈、传统戏剧、曲艺、传统杂技竞技、传统美术等。例如和三山五园有关的传说：香山传说、圆明园传说、颐和园传说、曹雪芹传说（流传于香山一带）；还有六郎庄五虎棍、蓝靛厂少林棍、宋氏三皇炮捶拳等当地的传统杂技竞技；以及充满宫廷特色的京西皮影戏等。这些宝贵的民间艺术瑰宝在社会上知名度不高，处于濒危状态，亟需加强整理和保护。

自康熙兴畅春园开清代皇帝园居之风，北京的政治中心实际上变成了两个：紫禁城与“三山五园”，这便是所谓的双城体制。“三山五园”通过水（直接连接皇城三海与颐和园昆明湖的长河，此路更大程度上供皇家游玩，沿线分布了一系列的寺庙、戏楼、茶馆、街市等供皇室游娱休息的设施）、陆（始于西直门，沿长河经海淀镇达“三山五园”，该御道用专门的石块铺砌而成）两条御道与北京城、紫禁城联系在一起，形成所谓的双城体制。如此，长河御道以及“三山五园”便形成清代北京城的政治中轴，这条中轴是清廷政治生活的主要载体。

清漪园建成、昆明湖开拓后，构成了万寿山和里湖的南北中轴线。静宜园的宫廷区、玉泉山主峰、

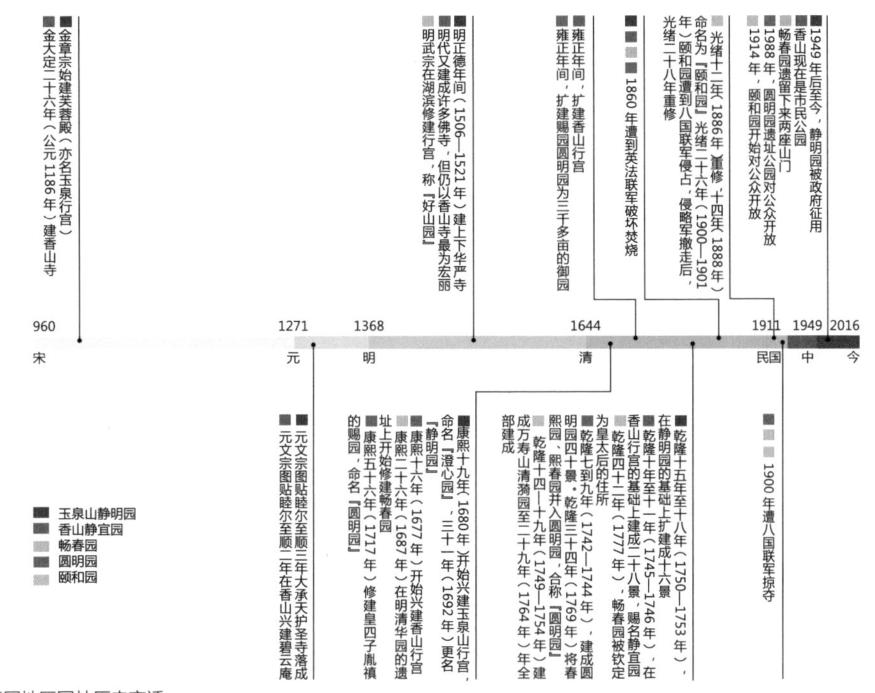

三山五园地区园林历史变迁
Evolution of Gardens in The Three Mountains and Five Gardens

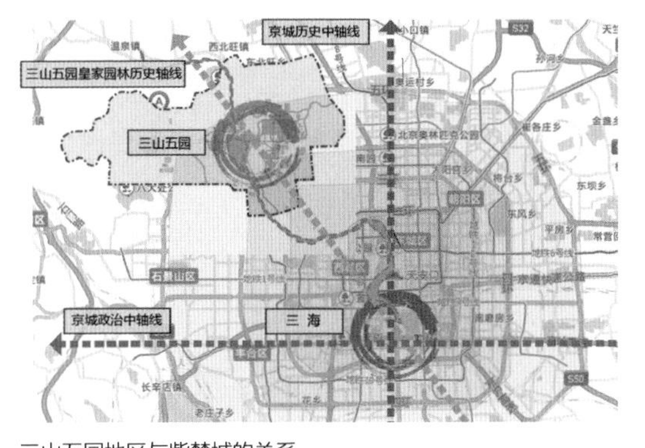

三山五园地区与紫禁城的关系
The Three Mountains and Five Gardens and The Forbidden City

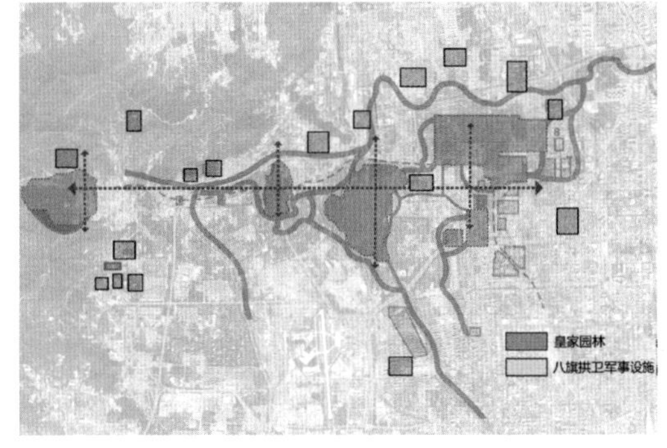

三山五园边界及八旗军事防卫体系
The Three Mountains and Five Gardens and the Defense System

清漪园的宫廷区此三者又构成一条东西向的中轴线，再往东延伸交汇于圆明园与畅春园的南北轴线的中心点。这个轴线系统把三山五园串缀成为整体的园林集群；在这个集群中，清漪园的建成无疑起了关键性作用。概括起来说“一园建成，全盘皆活。”

三山五园的构成可以分为三个层次，位于核心区的皇家园林（包括颐和园、静宜园、静明园、畅春园和圆明园五个皇家园林核心区），各园林之间的过渡地带（根据条件用作农业、商业、服务等各种不同的功能）以及外围的八旗拱卫军事设施圈，形成一个非常清晰的功能结构。三山五园不仅是互相独立的三山和五园之和，更是一个不可分割的整体。它的总体规划、系统管理以及封建体系的历史文脉，都说明我们应该用一种整体的眼光去认识和看待三山五园。它的自然背景、中国古典园林的设计手法和传统的天人和谐的思想都使得三山五园的景观交融成一个整体。

畅春园：畅春园是位于北京西北郊的第一座皇家园林，打开大型皇家园林群“三山五园”之先河。

静明园：康熙十九年（1680年）就在玉泉山的南坡裂帛湖一带改建为行宫，命名“澄心园”，三十一年(1692年)易名静明园。乾隆十八年(1753年）置总理大臣兼领清漪、静宜、静明三园事务，并命名“静明园十六景”。

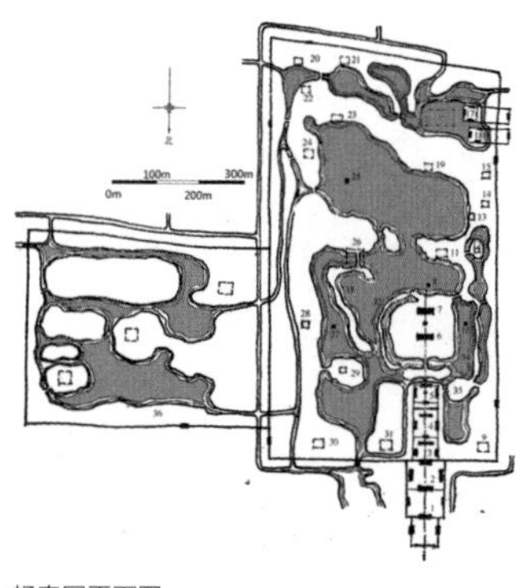

畅春园平面图
Plan of Changchun Garden

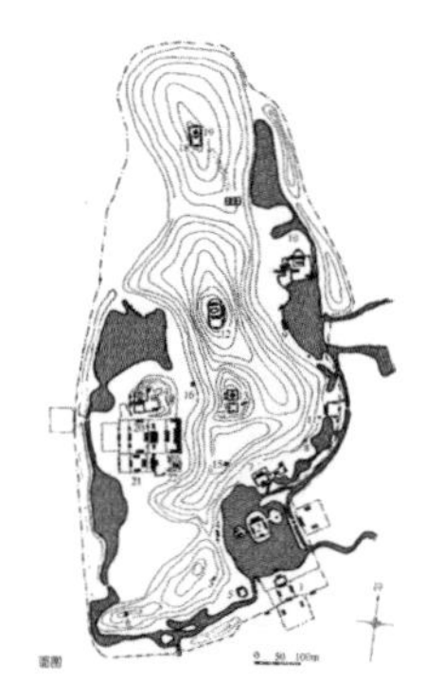

玉泉山静明园平面图
Plan of Jingming Garden

香山静宜园平面图
Plan of Jingyi Garden

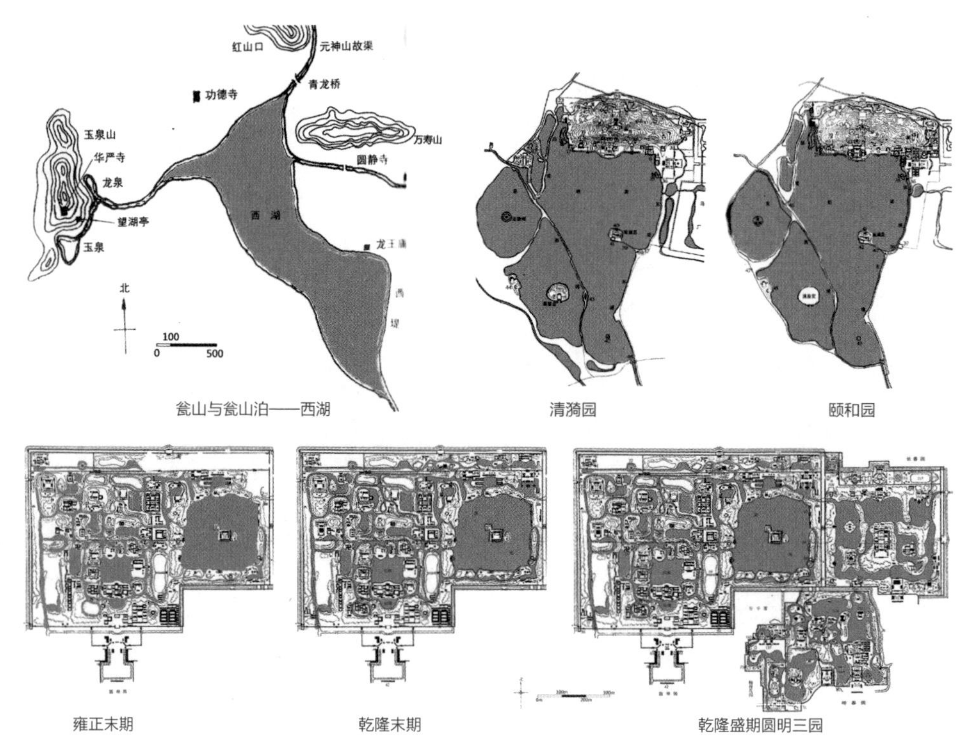

瓮山与瓮山泊——西湖　　清漪园　　颐和园

雍正末期　　乾隆末期　　乾隆盛期圆明三园

清漪园变迁示意图
Evolution of the Summer Palace

静宜园：康熙十六年（1677 年）重修香山寺并扩建香山行宫。乾隆十年（1745 年）开始扩建香山行宫，十一年（1746 年）完工，面积为 140hm^2。乾隆亲自命名香山静宜园二十八景。

清漪园：其所在地区早在金元代时代就已成为郊野风景名胜区，有行宫别苑的建置。元朝以后称为瓮山。翁山的南面一带，地势低洼，两边群山和玉泉山诸山，潴而成一片湖水名曰翁山泊。清漪园从乾隆十四年到乾隆二十九年全园建成，共花费 15 年的时间。清漪园的落成，北京西郊就出现了“三山五园”的说法。咸丰十年（1860 年）九月，英法联军进犯北京，对清漪园进行了劫略和焚烧，直到光绪十二年（1886 年）开始修复，光绪十四年（1888 年）改名为颐和园。

圆明园：历史上圆明园是皇四子胤禛“藩邸所居赐园”也。雍正元年 (1723 年) 圆明园升格为御园，经雍正朝 13 年大规模拓建和乾隆初年增建，乾隆九年 (1744 年) 最终形成著名的“圆明园四十景”。乾隆二年到九年 (1737—1744 年)，圆明园的格局已基本成型，通过扩建、改建和添建，园林格局更丰满。在乾隆三十五年 (1770 年) 基本形成圆明三园格局。

海淀区非物质文化遗产统计表
The List of Intangible Heritages in Haidian district

级别	项目	类别	位置	传承人
国家级非物质文化遗产	曹雪芹传说	民间文学	西山	
	口技	传统杂技与竞技		
	面人汤面塑	传统美术		
	传统插花			
	北京面人郎			
	北京风筝哈制作技艺			
	宏音斋笙管制作技艺	传统技艺		
	曹氏风筝技艺			
	葛氏捏筋拍打疗法	传统医药		
市级非物质文化遗产	香山传说	民间文学	静宜园	
	圆明园传说		圆明园	
	颐和园传说		颐和园	
	凤凰岭传说			
	六郎庄五虎棍	传统舞蹈	六郎庄	葛金山
	西北旺高跷秧歌			
	苏家坨太平鼓			
	西北旺少林五虎棍			
	海淀扑蝴蝶			
	太平歌词	曲艺		王双福
	吴氏太极拳	传统杂技与竞技		
	孙氏太极拳			
	珍珠球			
	踢石球（蹴球）			
	传统弹弓术			
	京剧盔头制作技艺	传统美术		
	绣花鞋制作技艺			王冠琴
	彩塑京剧脸谱			
	颐和园听鹂馆寿膳制作技艺	传统技艺	颐和园	
	山石韩叠山技艺			
	程氏针灸	传统医药		
区级非物质文化遗产	民间气象谚语	民间文学		
	怯音乐	传统音乐		张贵
	南安河武松打店棍会	传统舞蹈		
	蓝靛厂少林棍			索凤才
	太少狮			
	京西皮影戏	传统戏剧		王丽娟
	西路评剧			
	宋氏三皇炮捶拳	传统杂技与竞技		张成仁
	屯佃中幡			
	纪氏太极拳法			宋荣宇
	花样空竹表演技法			
	白猿通背拳			
	飞叉			
	临清潭腿			
	祁家通背拳			
	彩灯工艺	传统美术		小灯张
	平刻微雕			
	北京绢人			齐聪颖
	面塑			潘大洪
	金属锻錾			
	团花剪纸			
	颖拓艺术			
	京绣			
	齐派篆刻			
	绣花鞋制作技艺	传统技艺		蒋丽娟
	京西水稻种植技术			
	惠丰堂鲁菜制作技术			
	京派内画鼻烟壶			
	御膳制作技艺			
	蒙镶			
	山石韩叠山技艺			
	中式盘扣技艺			
	蔡氏脉象	传统医药		
	正体复本术			
	苏家坨立夏习俗	民俗		
	喜轿习俗			

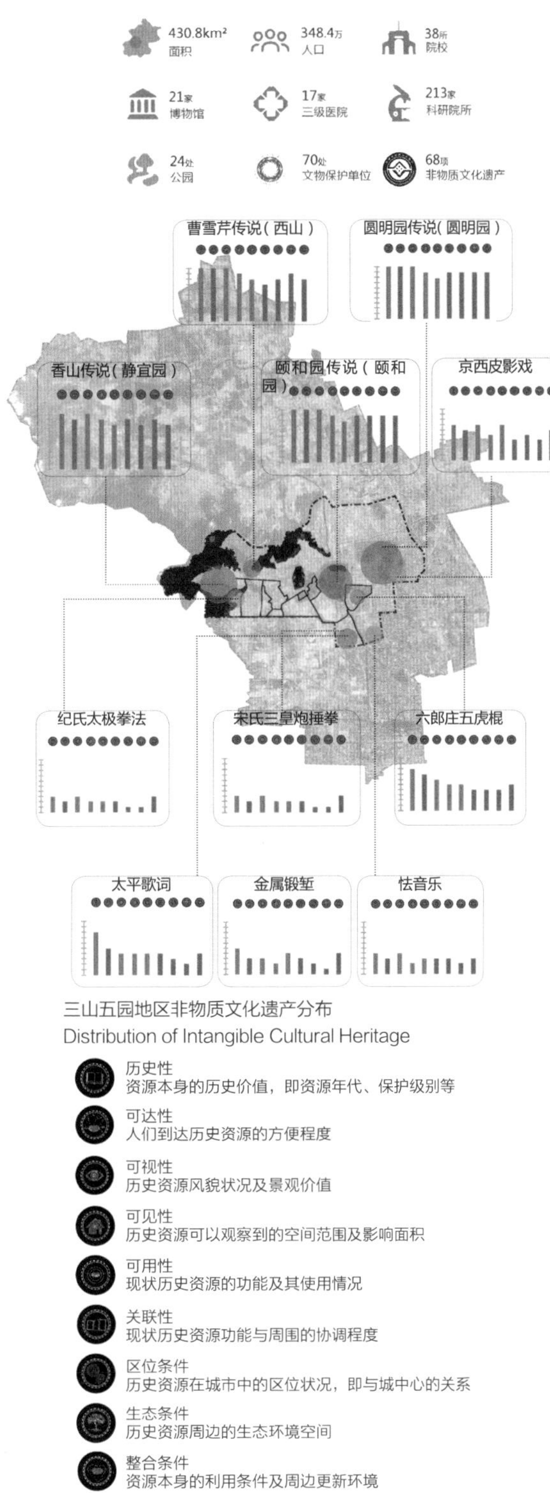

三山五园地区非物质文化遗产分布
Distribution of Intangible Cultural Heritage

历史资源评价体系
Evaluation of Historic Assets

历史水系专题研究

Specialized Studies on History of Water Systems

佟思明、崔滋辰、李璇、王晞月、李媛、徐慧、莫日根吉、Mosita

Tong Siming/Cui Zichen/Li Xuan/Wang Xiyue/Li Yuan/Xu Hui/MorigenJi/Mosita

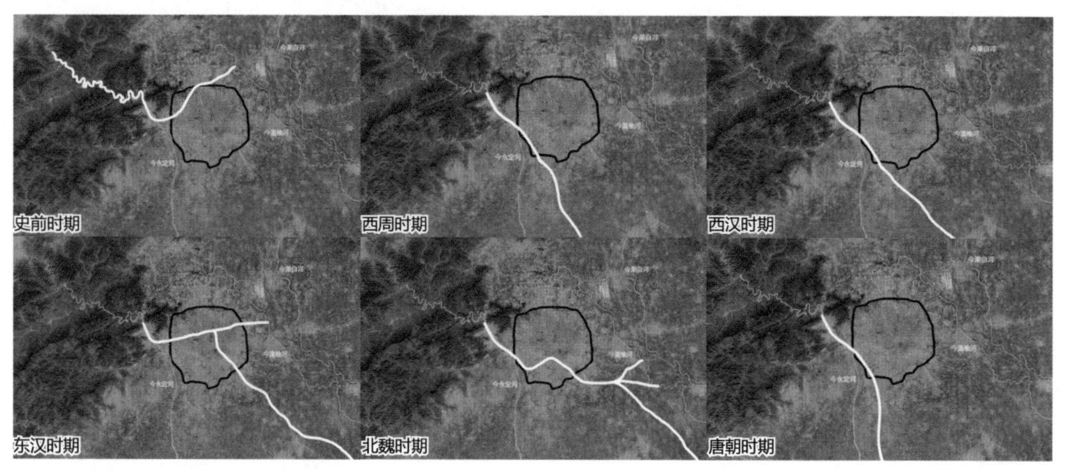

永定河变迁图

The Transition of Yongding River

北京处于华北平原西北边缘，西部—太行山脉，北部—燕山山脉，东南—海拔 100m 以下的北京小平原。市域内有五大水系：永定河水系、潮白河水系、北运河水系、泃河水系、拒马河水系。五大水系均属海河流域。

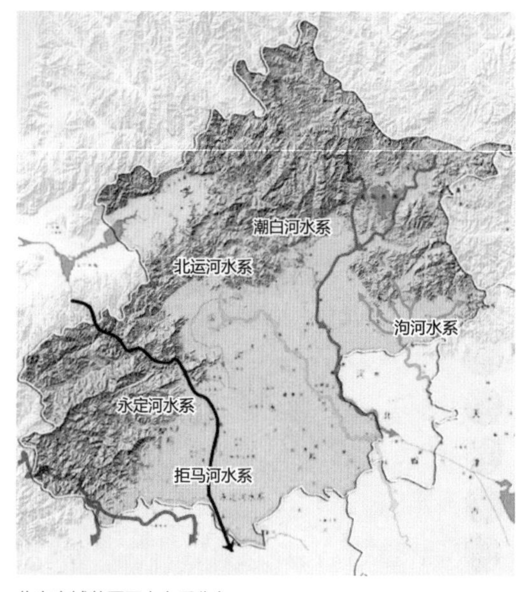

北京市域范围五大水系分布

Five Main Rivers in Beijing

三山五园位于北京城区西北郊，7000 多年前，古永定河自西南向东北经三山五园区域汇入温榆河。经过两千年的历史变迁，古永定河道向南摆动，而曾为永定河故道的三山五园地区成为一片湖泊溪流遍布的肥沃低地，万寿山和玉泉山则鹤立于这片低地中央，与西北的西山遥相呼应。

史前时期、西周时期、西汉时期、东汉时期、北魏时期、唐朝时期的永定河位置。史前时期永定河从西北部山脉流入平原后向北发展汇入今温榆河流域。西周时期摆动到平原南部。西汉时期微微向北侧摆动，依然流过平原南部。东汉时期永定河剧烈变化，分为两支流，一支流东西走向与温榆河交汇，另一支流继续流向东南方向。北魏时期永定河与温榆河水系相接的支流逐渐消退，在下游与温榆河交汇。唐朝时期对永定河河道进行疏通整治，稳定河道位置，流经城南向南流去，与今永定河道形态基本重合。

三山五园地区水系从金代到近现代经历了多次历史变迁，形成了现在的水系格局。从整体来看，其水系变化主要为以下几方面：

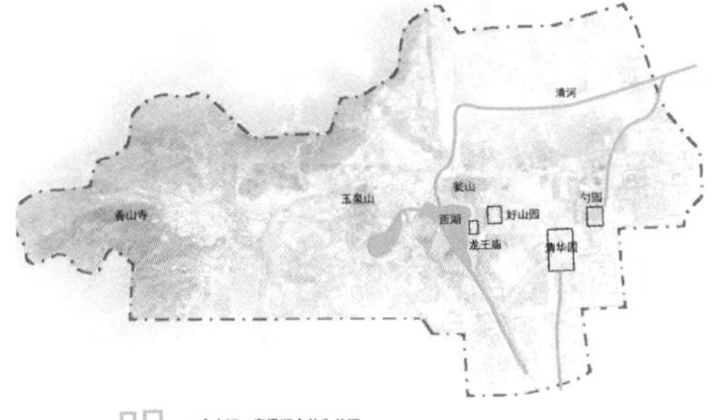

明
1金水河、高梁河合并为长河
2西湖（瓮山泊）水门缩小，明初进行修治，水形变化
3西北郊燕山天寿寺建明十三陵，为保护皇陵地脉而废弃白浮河

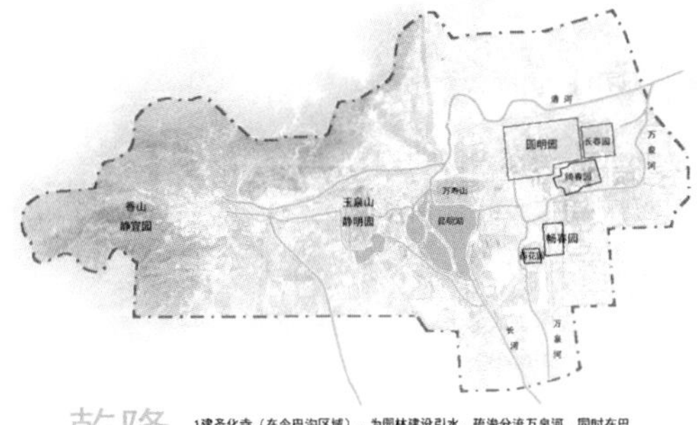

乾隆年间
1建圣化寺（在今巴沟区域），为园林建设引水，疏浚分流万泉河，同时在巴沟一代开垦水田，宛然江乡风景。

2 扩建圆明园，加上瓮山泊淤积，蓄水量减少，不能满足用水需求。于是对瓮山泊进行大规模开挖、清淤和疏浚，并引水入圆明园后汇入清河。同时建石槽引西山碧云寺和卧佛寺附近的泉水，经玉泉山入昆明湖，扩充水源。

3 扩建昆明湖同时，疏浚玉泉水系，开辟高水湖、养水湖，使水次第节蓄，灌溉京西农田。

4建北旱河、南旱河，作为泄洪水道。
在此阶段河流水系的生态、景观、调蓄等价值均达到最高峰

从金代到咸丰年间水系变迁
Changes of Waterways from Jin Dynasty to Mid-Qing Dynasty

历史水系价值评估表
Evaluation of Historic Waterways

年代	景观价值	生态价值	调蓄价值	文化价值	交通价值
金元			昆明湖		白浮河、金河
明		万泉河	昆明湖、清河		长河
康熙	万泉河		昆明湖、长河		清河、长河
乾隆	万泉河、昆明湖		高水湖、养水湖		长河
新中国成立前	万泉河		南旱河、北旱河清河		长河
现状	长河、京密引水渠、昆明湖、清河	长河、京密引水渠	万泉河、清河	昆明湖	京密引水渠昆玉段

康熙年间
1 清康熙中期，首先在清华园遗址上兴建畅春园，伴随园林建设扩展万泉河水系支流

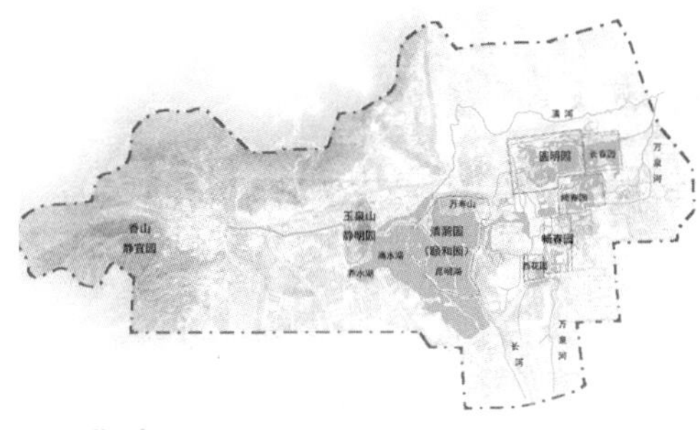

咸丰年间
1扩大高水湖、养水湖水域面积

1. 水量逐渐减少，水质逐渐恶化

随着城市规模扩大，西郊大范围超量开采地下水，西山诸泉泉流逐年减少至枯竭断流。同时，伴随城市建设和人口增长，每年排入河道的工业废水和生活污水增加。水源不足，河水滞留也导致水质恶化。

2. 水利工程逐渐完善，滨水开放空间逐渐减少

城市发展修建水闸、水渠，保障了首都防汛的安全，但结果是绿化隔离带和护河围栏使本来京城纳凉消夏、休闲游玩的滨水空间与城市相互隔离，牺牲了河道的生态环境。

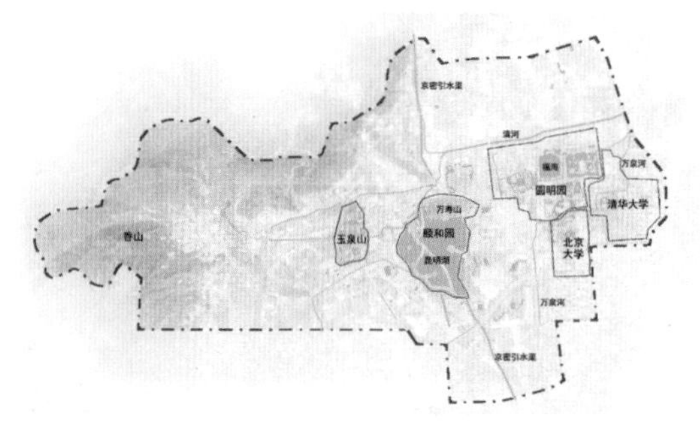

三山五园地区现状水系图
Existing Water System

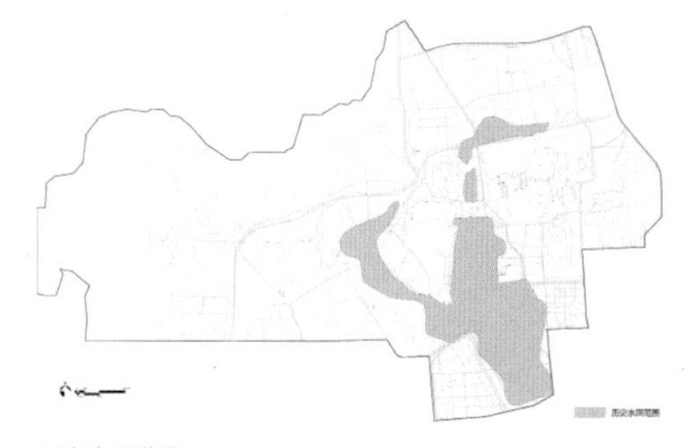

历史水田分布
The Distribution of Historical Paddy Fields

清代皇家宫苑建设
Timetable of the Construction of Royal Gardens in Qing Dynasty

畅春园 1684—1687年 1650 1700 1750 1800 1850	1684年清朝康熙皇帝南巡归来启建。康熙六十一年（1722年）去世于园内清溪书屋。咸丰十年(1860年)，英法联军攻入北京焚烧圆明园时将其一并烧毁
静宜园 1745—1746年 1650 1700 1750 1800 1850	金大定二十六年(1186年)建香山寺，明代又有许多佛寺建成，但仍以香山寺最为宏丽。清康熙年间(662—1722年)，就香山寺及其附近建成“香山行宫”。乾隆十年(1745年)加以扩建，翌成竣工，改名“静宜园”
静明园 1750—1759年 1650 1700 1750 1800 1850	清康熙十九年(1680年)将玉泉山辟为行宫，名“澄心园”。康熙三十一年(1692年)改名为“静明园”。乾隆十五年(1750年)再次修葺，增建玉峰塔等景观并命名了“静明园十六景”。静明园1860年遭英法联军、八国联军两次焚毁
清漪园 1753—1764年 1650 1700 1750 1800 1850	乾隆十五年（1750年），乾隆皇帝在这里改建为清漪园，瓮山改称为万寿山，瓮山泊改称为昆明湖。咸丰十年（1860年），清漪园被英法联军焚毁。 光绪十四年（1888年）重建，改称颐和园，作消夏游乐地。 光绪二十六年（1900年），颐和园又遭“八国联军”的破坏，珍宝被劫掠一空
圆明园 1709—1782年 长春园 1745—1759年 绮春园 1769—1801年 1650 1700 1750 1800 1850	兴建于康熙末年和雍正朝，康熙四十八年（1709年）。雍正二年（1724年），圆明园的扩建工程正式开始。乾隆三十四年(1769年)春和园归入圆明园，正式定名为“绮春园”。绮春园宫门，建成于嘉庆十四年（1809年）。咸丰十年（1860年），圆明园被英法联军焚毁

3. 泄洪问题逐渐严重，旱河增加

北京城市水源的供应一直呈现下降的趋势，许多河道成为旱河。同时，城市防汛压力增加，这些旱河有存在的必要。然而北京雨洪季节很少，很多河道在 90% 的时间处于枯水期。城市用地的紧缺和旱河使用效率的低下，成了一对很难调和的矛盾。

4. 河道周围城市用地的矛盾日益凸显

滨河区域是城市最为复杂、活跃的区域，是城市连续、开敞的空间，北京城市滨水区域的规划建设，城市各种用地与河道间相互侵占，建筑、道路与河道缺乏联系，不能形成具有京城特色的滨水景观区域。三山五园位于北京湾的西北部，大约在七千到五千年以前，永定河古道流经此地，地下水源丰富，西山一带多石灰岩，溶洞较多，透水性强，形成玉泉山和万泉河两大水系。

这两大水系不仅滋养了世世代代在这里的人们，而且也给园林艺术家们带来丰富的想象和无尽的灵感。他们巧妙地利用这两条水系中的河、溪、湖、泽，取自然之势而创造出风格各异、特色纷呈的皇家苑囿别墅，达到“虽有人造，宛自天开”的效果，成就世界园林艺术。

除了园林艺术上的造诣，三山五园地区水系在城市防洪防涝方面也发挥了巨大缓冲作用。

在历史发展过程中，清代皇家宫苑的建设对三山五园水系的形成有很大的影响和作用。

山林、水田、宫苑、村庄、营房构成了历史时期三山五园的景观风貌。明清时期，三山五园地区形成了东起海淀、西抵玉泉山、南达长春桥、北抵青龙桥的万亩稻作景观。乾隆年间，巴沟低地中的水田累计已逾万亩。根据岳升阳所绘《京西稻作景观分布图》，清河南北两侧、静明园与清漪园之间等区域也分布着成片的田野。1783 年，农田面积达到了 8420hm^2，仅巴沟低地中的稻田就超过了万亩，功德寺、高水湖和养水湖一带也分布大片农田，“三山五园”大部分都在农田水系的围绕之中，展现了北方罕见的水乡景观。新中国成立后农田面积持续下降，反映出水体、稻作景观面临消失的危机。

海淀至玉泉山一带，康乾年间京西御稻田达一万余亩。乾隆帝的《万泉郊行即事》诗写道：

疑是山村是水乡，禾苗低亚稻苗黄。

绿杨十里蝉声沸，飒爽风中饘粥香。

河道现状照片
Photos of Current Rivers

历史水系现状一览表
Profiles of Historic Waterways

河流名称	河段	三山五园境内流域面积（㎡）	河道平均宽度（m）	河流平均深度（m）	水流方向	活水/死水/无水	水质	驳岸类型（硬质河道/软质）	周边用地-与河道相接的用地类型		周边用地-河道两侧是否有道路		周围植被-河道防护绿地宽度		周围植被-河道防护绿地植物状况		是否为周边湖泊供水	是否有文化遗址	视线上是否有能看到远山的视点	功能
									南侧（东侧）	北侧（西侧）	南侧（东侧）	北侧（西侧）	南侧（东侧）	北侧（西侧）	南侧（东侧）	北侧（西侧）				
京密引水渠	京密引水渠-火器营桥北段	2300.4	40	2.4~2.8	由北向南	活水	II类	硬质	绿地+公共服务设施	绿地+商业	昆明湖东路	蓝靛厂北路	20m绿化	20m绿化	乔灌草	乔灌草	是	金河堤碑	可见远山	引水
	京密引水渠-万柳南段	3484.9	35	2.4~2.8	由北向南	活水	II类	硬质	绿地	居住+公共	蓝靛厂北路	蓝靛厂北路	15m绿化	15m绿化	乔灌草	乔灌草	是	无	可见远山	引水
	京密引水渠-万柳北段	3228.4	35	2.4~2.8	由北向南	活水	II类	硬质	绿地	绿地	蓝靛厂北路	蓝靛厂北路	15m绿化	16m绿化	乔灌草	乔灌草	是	佛香阁、巴沟山水园	可见远山	引水
	京密引水渠上段	2300.4	30	—	北向南	活水	II类	底部硬质，两岸软质	道路	道路	主干道	主干道	6m绿化带	6m绿化带	乔草	乔草	是	无	否	引水
	京密引水渠-中段	3484.9	30	—	北向南	活水	II类	底部硬质，两岸软质	道路	道路	主干道	主干道	6m绿化带	6m绿化带	乔草	乔草	是	青龙桥	可，护林防火山	引水
清河	清河-圆明园段	761.27	35	—	温榆河	活水	V	硬质	公园	居住区	清河路	主干道	5m绿化	5m绿化	灌+草	无	否	无	西山	泄水
	清河-北体段	287.95	38	—	温榆河	活水	V	硬质	公共	公共	无	主干道	4m绿化	4m绿化	柳+草	无	否	无	西山	泄水
	清河-安河桥段	235	12	—	南向北	死水	V	软质，施工中	围墙	道路	无	次干道	无	无	乔草	无	否	无	百望山	泄水
旱河	北旱河东段	0	15	4	—	无水	—	硬质，有人行路	复杂	复杂	无	主干道	无	无	乔草	无	否	无	可见护林防火山	泄水
	北旱河玉泉山西段	0	8	5	—	无水	—	硬质	公园	公园	无	主干道	10m营城	3m绿化带	沿河柳树	沿河柳树	否	无	否	泄水
	北旱河健康步道段	0	1	1	—	无水	—	软质	公园	公园	无	主干道	0.5m绿化	0.5m绿化	无	乔草	否	养水池	否	泄水
	北旱河香山路段	0	4（中间1.5）	2	—	无水	—	硬质	公园	公园	无	主干道	0.5m绿化	0.5m绿化	无	乔草	否	无	否	泄水
	北旱河-植物园引水段	0	4	2	—	无水	—	硬质	道路	公园	支路	无	0.5m绿化	0.5m绿化	无	乔草	否	无	可看香山	泄水
	北旱河-旱河路北段	150	5	3	—	死水	—	北软南硬	道路	公园	主干道	无	3m护坡	3m护坡	柳树	乔草	否	无	否	泄水
	南旱河-旱河路南段	365	12	3	—	死水，一点水	—	部分硬质，部分软质	道路	公园	主干道	无	5m绿化	5m绿化带	无	无	可能是	无	否	泄水
万泉河	万泉河-圆明园段	959.4	9	0.4	清河	活水	V1	硬质	公园	居住区	颐和园路	无	无	无	无	无	否（曾经是）	无	否	泄水
	万泉河-稻香园段	847.3	9	1	由南向北	活水	V1	硬质	居住	道路	无名河堤路	万泉河快速路	无	2m绿化带	硬质+灌	乔草	否（曾经是）	无	否	泄水
	万泉河-西苑段	993.3	9	1	由南向北	活水	V1	硬质	居住	道路	无名河堤路	万泉河快速路	无	无	乔	乔灌	否（曾经是）	畅春新园	可见远山	泄水
	万泉河-海淀公园段	521.3	9	1.5	由南向北	活水	V1	硬质	居住	道路	无名河堤路	万泉河快速路	无	5m绿化	乔灌	乔灌草	否（曾经是）	无	可见远山	泄水
金河	金河-南	600.5	9	1	由北向南	活水	V	软质	绿地	绿地	金河路	无	2m绿化	大片绿地	乔草	乔	否	金河堤碑	否	泄水
	金河-沿路段	380	7		由东向西	活水	V	硬质	绿地	商业	四环	支路	5m绿化	无	乔草	乔	否	无	可见远山	泄水
	金河-北	689	3	3	—	死水，一点水	有味道	软质	公共	绿地	四环	无	2m护坡	2m护坡	无	乔草	否	无	否	泄水
	金河-北坞公园东北段	0	3	1	—	无水	V	硬质	公园	废弃地	四环	无	1.5m绿化带	1.5m绿化带	无	乔草	否	无	否	泄水
场地内其他河道	长河-玉泉山东	582	12	3	—	死水	V	西部软质，东部硬质	公园	公园	主干道	无	2m护坡	2m护坡	沿河柳树	沿河柳树	否	1993年保护水利设施碑亭	可见玉泉山	泄水
	北坞村路西路旁河道	269	0.8	1.5	—	无水	未知	硬质	道路	居住	主干道	无	1.5m绿化带	0.5m绿化带+围栏	草	草	否	无	否	泄水
	闵庄路东段路旁河道	386	1	2	—	死水，一点水	未知	硬质	道路	居住	主干道	无	人行道	2m绿化带	无	草灌	否	无	否	泄水
	闵庄路西段路旁河道	211	1	2	—	死水，一点水	未知	硬质	道路	公园	主干道	无	0.5行道树	0.5绿化带	草灌	无	否	无	否	泄水
	万安东路路旁河道	245	4	3	—	死水，一点水	未知	软	道路	居住	次干道	无	无	无	草	无	否	无	可看玉泉山	水源保护
	南水北调保护区	2885.9	38	—	—	死水	未知	—	—	—	—	—	—	—	—	—	—	—	—	—

03 规划设计

PLANNING & DESIGN

三山五园地区因其得天独厚的地理环境条件，自北京建城以来就成为维持城市生存最重要的水源地和自然生态屏障，并随着皇家园林、行宫等的兴建而逐渐成为北京城最集中的自然、文化遗产聚集区。近年来，随着北京城市建设的扩张，该地段呈现出日益复杂的发展趋势，如何保护其整体自然文化价值，在与城市发展协调的同时提升环境品质，成为一个重要的课题。

在规划研究中，指导教师团队明确提出了“构建绿色网络、激活城市遗产”这一目标。研究团队在详细的场地调研分析的基础上，综合分析场地问题和机遇，探讨了该地段的未来发展方向和内容，通过详细的空间潜力评价和适宜性分析方法，在三山五园地区构建了一个连贯的多功能生态空间网络。该网络将发挥生态环境优化、文化遗产保护、自然休闲游憩、居民生活服务、空间活力重塑等一系列功能。团队在规划绿色网络的基础上，选择了香山环岛、青龙桥、北宫门等 8 个最重要并具有代表性的节点，开展详细设计研究，探索了不同场地多元化发展的可能性，勾画了三山五园地区未来自然与文化共存、绿色与城市共生、旅游与生活共荣的美好蓝图。

Thanks to its geographical conditions richly endowed by nature, the 'Three Hills and Five Gardens' area has become the most important water head site and ecological barrier for living since the establishment of Beijing city. Along with the construction of royal gardens and palaces, this area has gradually become the most concentrated area of natural and cultural heritage in Beijing. In recent years, with the expansion of urban construction scale in Beijing, the area shows an increasing complex development trend. As a result, how to protect its overall natural and cultural value and enhance the environmental quality while coordinating with urban development have become essential topics for us.

In the research, our instructor team clearly advances the goal of setting up the green network and activating the urban heritage. Based on detailed field research, our team analyzes the problems and opportunities comprehensively, and explores the future development direction of the area. Through detailed spatial potential evaluation and suitability analysis, we construct a coherent multi-functional ecological network, which will develop a series of functions, including ecological environment optimization, cultural heritage protection, natural recreation, residential life service and vitality reconstruction, etc. We chose the 8 most important and representative nodes such as Xiang Shan Island, Qing Long Bridge and the north gate of Summer Palace to carry out detailed design and researches. In a word, we have always been exploring the possibility of diversified development in different sites, so as to outlined the blueprint for the coexistence of nature and culture, city and environment, as well as residence and tourism in the 'Three Hills and Five Gardens' area.

规划方案一 及 圆明园大宫门遗址片区概念设计

Plan 1 & The Conceptual Design of Dagongmen Site in Yuanming Yuan

何二洁、罗亚文、朱友强、甘其芳、施晨、孟宇飞、施莹、张俞
He Erjie/Luo Yawen/Zhu Youqiang/Gan Qifang/Shi Chen/Meng Yufei /Shi Ying/Zhang Yu

本规划方案中，绿道体系将山林、公园出入口与公共交通站点便捷连接，同时高度重视视觉廊道设计，在多个视觉景观节点预留景观驿站，构成本方案的最大特色。

大宫门是圆明园的正门，近代以来历史遗迹破坏严重，现状场地被城中村占据。设计中对高度拥挤的城中村进行了清理，恢复了壮丽的轴线景观并建设了遗址纪念馆，成为连接西苑公交枢纽、圆明园遗址公园和北京大学家属区之间的绿色开放空间。

In our planning scheme, we built the greenway system to connect the mountain forests, the park entrance and public transport sites. At the same time, we attached great importance to the visual corridor design, and remain landscape stations in different visual landscape nodes, which makes the most important feature of this scheme.

Dagongmen is the main gate of Yuanming Yuan. Since the modern times, plenty of historical relics here have already been damaged seriously, and now the site is occupied by villages in the city. In the design, we cleaned up the highly crowded villages, restored the magnificent axis landscape and built a site memorial hall, making the region a green open space to communicate the Xiyuan public transport hub, the Yuanming Yuan Ruins Park and the family district of Peking University.S

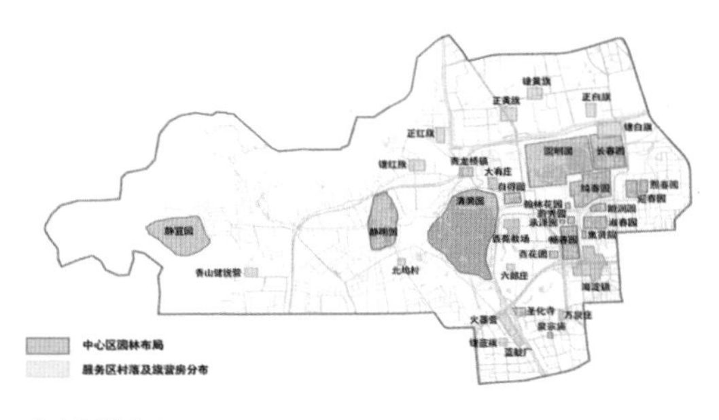

皇家园林分布
Distribution of Royal Gardens

非物质文化遗产
Distribution of Intangible Cultural Heritage

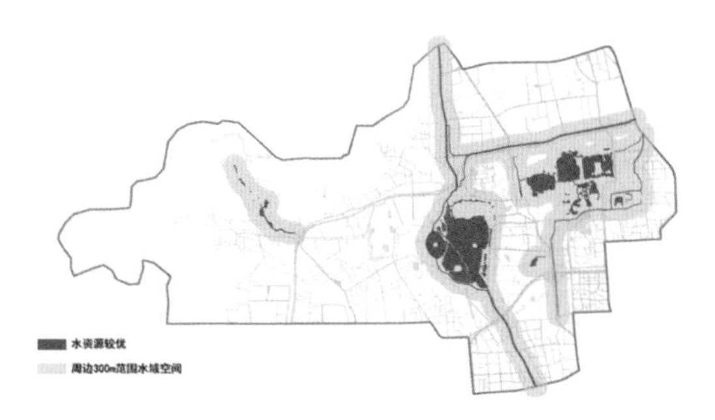

水系评价
Distribution of Water Area

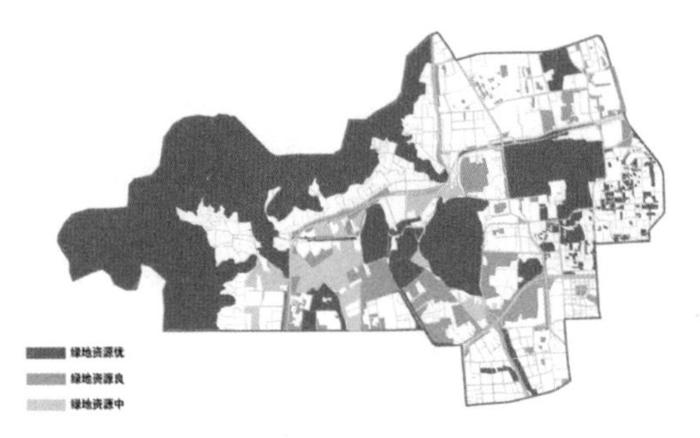

绿地评价
Distribution of Green Land

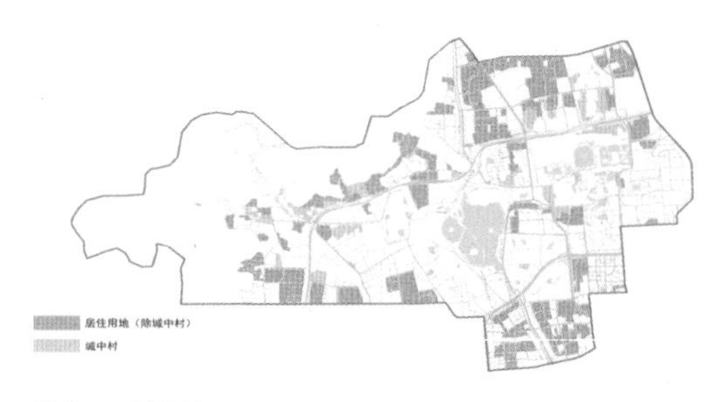

居住区及城中村
Distribution of Residential Area with Urban Villages

商业区及商业用地
Distribution of Commercial Land

1 绿道选线规划

根据上位规划的要求形成三山五园地区的绿地格局总体结构。同时，结合三山五园地区绿地现状，构建绿地评价体系。绿地评价方法的制定首先基于绿道功能的分析，确定绿地评价体系包含生态环境、服务功能、文化资源和经济价值四方面，经过每个地块的打分评级，然后进行评价结果叠加，从而确定基于绿地要素的绿道选线走向。其后，再通过对三山五园的选线现状及自然文化资源的八项分析(皇家园林布局、文物保护单位分布、非物质文化遗产分布、特色历史游线分布及节点、梳理水脉、强化山脉、绿地评价、水系评价) 得出三山五园绿道的基准选线叠加图，在基准选线模型的基础上，综合中小学分布分析、环境设施分析、商业用地及主要商业区分析、公共交通接驳点分析、居住区及城中村分析、快速路网分析、空间限制分析、旅游发展分析，对基准选线进行调整和优化，得出三山五园绿道的选线深化叠加图。

公共交通接驳点
Distribution of Public Transport Stops

中小学分布
Distribution of Primary and Secondary Schools

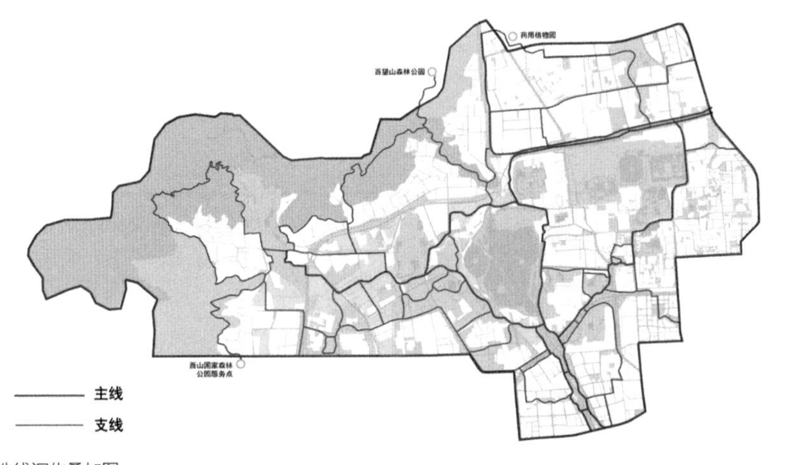

选线深化叠加图
Overlay of Selected Greenways

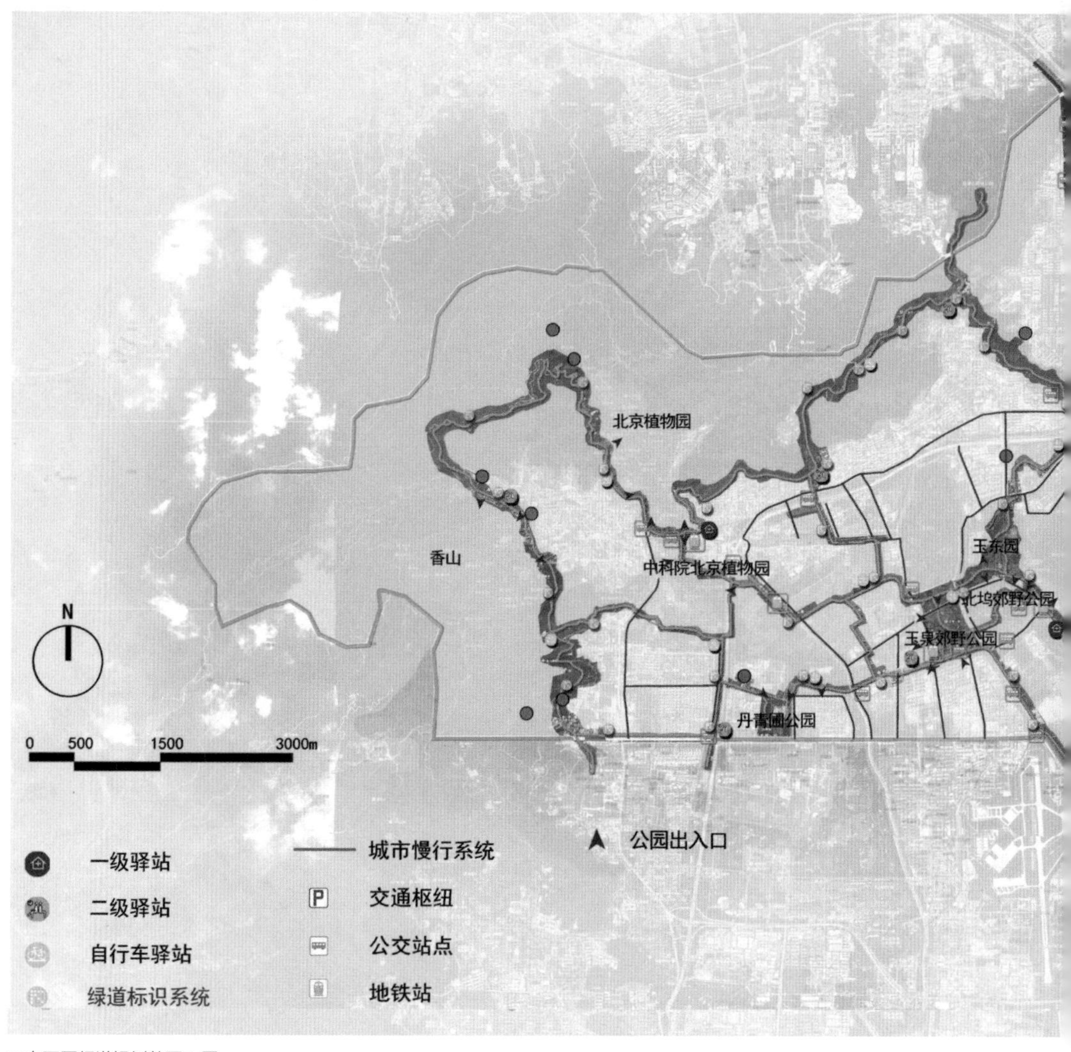

三山五园绿道规划总平面图
Master Plan of Greenways

视线分析
Visual Analysis

文化资源分析
Natural Resource Analysis

规划绿道总长度 83.5km，配套设置了驿站、自行车驿站、绿道标识系统。慢行系统连接各资源点，为游客提供丰富的游憩体验和选择。

规划绿道综合多处视线焦点，为游客设置了多条远眺视线，可观赏香山、西山、玉泉山、百望山、平顶山、万寿山等主要山系，除了临近山系间互相通视外，视点的设置给游客和摄影爱好者提供了远眺的机会。预留的视线给游客的旅行提供了远近变化的视觉体验，步移景异。让游客回归自然，充分体验三山五园地区内的自然山水。

绿道标识系统包括立体标识和地面标识。立体标识包括：空间引导标识、文字标识、距离指示标识、导向指示标识、警示禁止标识以及服务设施命名标识六种类型。地面标识包括慢行道路标识线、北京三山五园绿道 logo、骑行标识以及方向指示。

绿道服务设施由游客服务站、休憩设施、停车设施、卫生设施以及基础配套设施组成。游客服务站分为游客服务中心和驿站两个级别。

绿道慢行道路系统保持连贯性和通畅性。慢行道与城市道路交叉时，平面交叉和立体交叉两种形式并用。绿道与河流水系相交时，结合现有的桥梁或新建桥梁跨过河流水面，或利用水上交通的方式通过水面，实现水上交通与绿道的无缝衔接。绿道与高速公路、城市主干道交叉时，宜优先采取安全的立交方式。

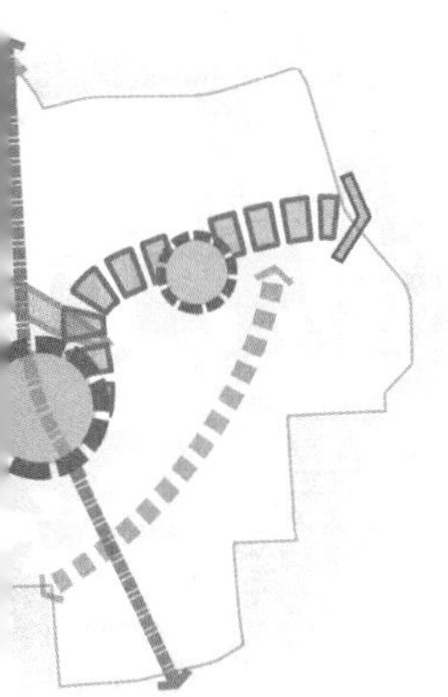

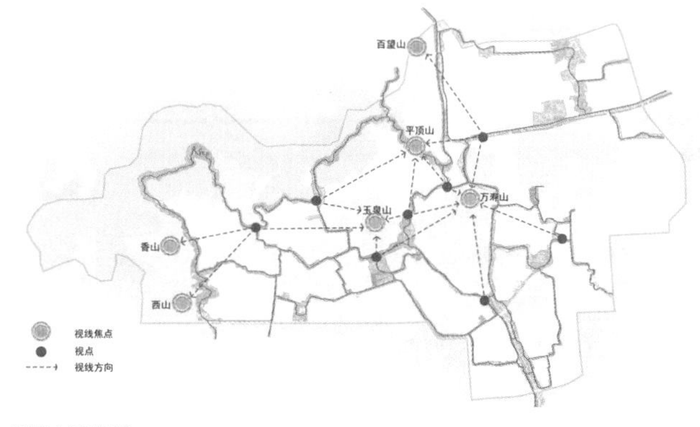

绿道主题分析
Structural Analysis

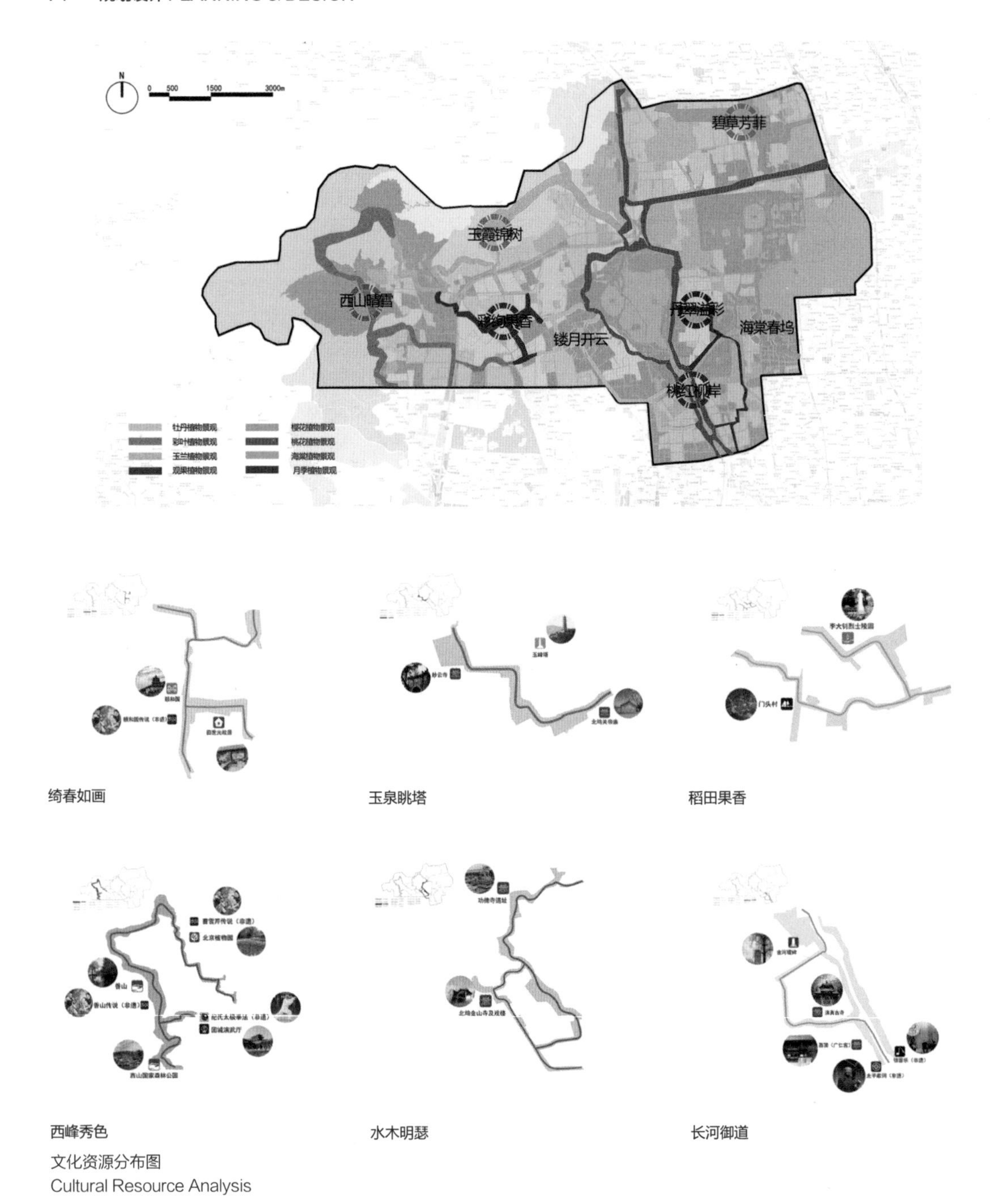

绮春如画　玉泉眺塔　稻田果香

西峰秀色　水木明瑟　长河御道

文化资源分布图
Cultural Resource Analysis

2 绿道文化资源点分析

规划的绿道总共分为十一个主题，分别为：西峰秀色、稻田果香、水木明瑟、玉泉眺塔、芹溪绿道、盛景长春、山水之道、畅春芳菲、绮春如画、长河御道、清河柳荫。结合了绿地片段的历史、文化和地标符号，形成了特色鲜明的旅游景观。在游览的同时，能够感受到三山五园几百年来形成的独特历史气息。

三山五园有着悠久的历史，留下了大量的物质和非物质文化遗产。除了闻名中外的颐和园、圆明园等大型皇家园林外，还有许多小型的历史遗迹和文物保护单位。绿道起到了串联这些分散的文化资源的作用，给游客接近这些文化资源点提供了便利。

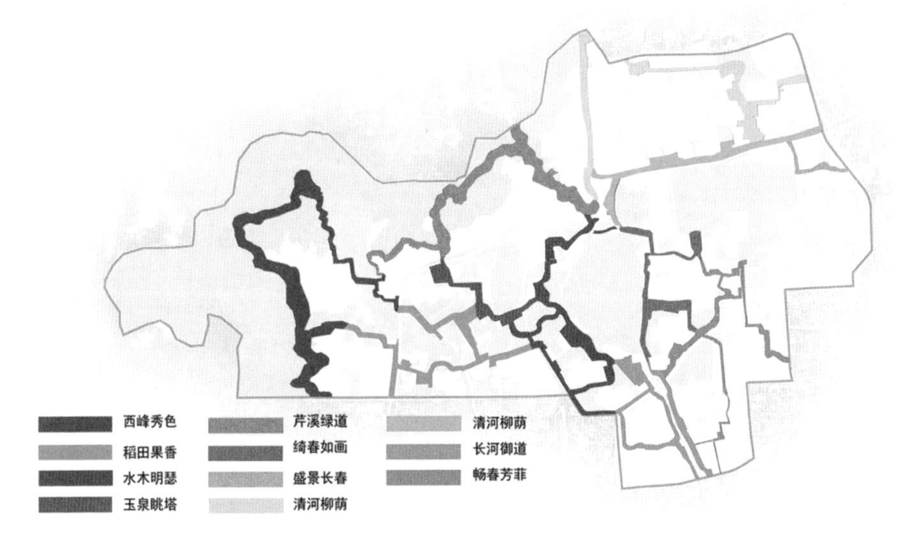

绿道主题规划图
Themed Analysis

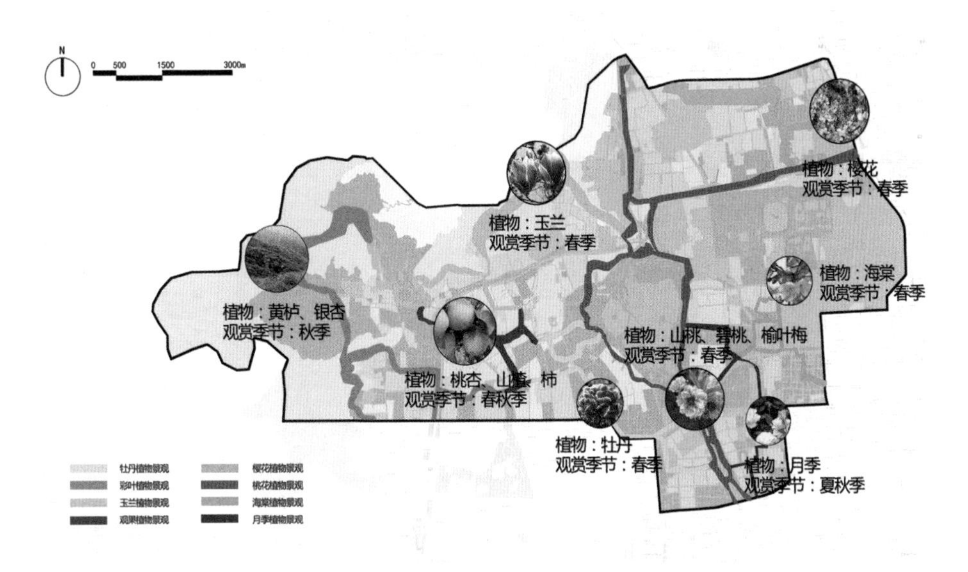

植物景观规划图
Plants Landscape Analysis

3 绿道植物规划

重点运用皇家园林植物材料牡丹、桃花、梅花、玉兰和海棠，形成五种各具特色的植物景观。分别是：蔷薇植物景观、牡丹植物景观、彩叶植物景观、玉兰植物景观及观果植物景观。在三山五园绿道不同的区段，观赏不同品种的花卉。 使之成为三山五园绿道规划中有特色、分区域、有层次的植物景观。此外，还有八组色彩绚丽的植物主题分别是，西山晴雪、玉霞锦树、碧草芳菲、彩绚果香、镂月开云、丹翠溢彩、桃红柳岸、海棠春坞，对应应用的植物素材有：牡丹、玉兰、樱花、桃花、海棠、月季、彩叶植物和观果植物。做到三山五园绿道在四季中均有植物景观可赏可观。所谓，朝而往，四时之景不同，而乐亦无穷也。

设计总平面图
Site Plan

功能分区
Functional Zones

景观结构
Landscape Structure

4 重点地段规划

地段毗邻颐和园、圆明园和北京大学的直接联系作用，也临近清华大学和海淀公园。由于该区域的混乱无序发展导致颐和园和圆明园的联系被切断，加之西苑交通枢纽的存在及慢行系统的不完善，进一步阻隔两园之间的步行连接，这些情况极大地影响三山五园的整体性，严重地制约圆明园遗址的保护与利用。

该节点地段设计的三大目标是：第一，构筑连接颐和园、圆明园、北京大学的绿色廊道；第二，加强历史文化遗迹的保护和保护性利用；第三，建立城中村社区主客共享的开放空间体系。同时以目标为导向，提出相应的设计策略：连接历史文化资源点；延续两条历史轴线；优化遗址的展示；增加综合性绿地；线性绿地连接；完善区域绿网；复合功能与产业植入；通过场地的资源连接最终提升区域资源的连接性。

详细设计地段总面积 75hm^2，北部地块为圆明园大宫门遗址区域，西邻颐和园路，东临达园宾馆，南接颐和园东路，内部包括西苑交通枢纽与一亩园。南部地块北临颐和园东路，西临万泉河快速路，东临北京大学，设计范围内包括挂甲屯社区、承泽园及蔚秀园。

地理位置优越：位于香山静宜园、玉泉山静明园、万寿山清漪园的东西轴线上，北邻圆明园，西邻颐和园，东邻北京大学。历史文化资源丰富：北部正在考古的圆明园大宫门遗址，承泽园、蔚秀园为清代历史名园，除此之外，场地内还有娘娘庙遗址、虎城遗址、挂甲屯、一亩园、御道等遗迹。

现状绿地零散分布缺少联系，场地缺乏城市开放空间。历史文化资源保护力度不够，多数遗迹被破坏。西苑交通枢纽影响圆明园与周边景区的连通性。还有三山五园绿道规划区域内最亟待解决的城中村问题。

整体方案功能分为七区，强调与三山五园绿道的整体性，优化与周边公共交通的衔接，完善与西山及万寿山的视线连接，并增加相应的服务设施，如游客中心、绿道驿站等。

N　0　20　60　100　200m

1 大宫门遗址展示区
2 眺望台
3 地下参观区
4 影壁水景
5 圆明园遗址纪念馆
6 娘娘庙
7 节庆观演台
8 纪念广场
9 野趣草坪区
10 漫步幽径
11 景观天桥
12 交通岛公园
13 挂甲屯特色商业街
14 文化创意区
15 门楼遗址绿地
16 承泽园
17 绿道驿站
18 挂甲屯住区
19 蔚秀园
20 蔚秀园住区
21 两园休憩道

清华西路　颐和园路　北京大学　春园

公共交通系统
Plan of Public Transit

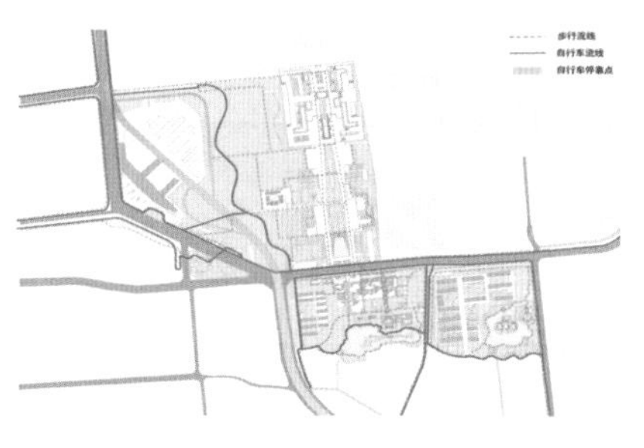

慢行交通系统
Plan of Non-motorized traffic

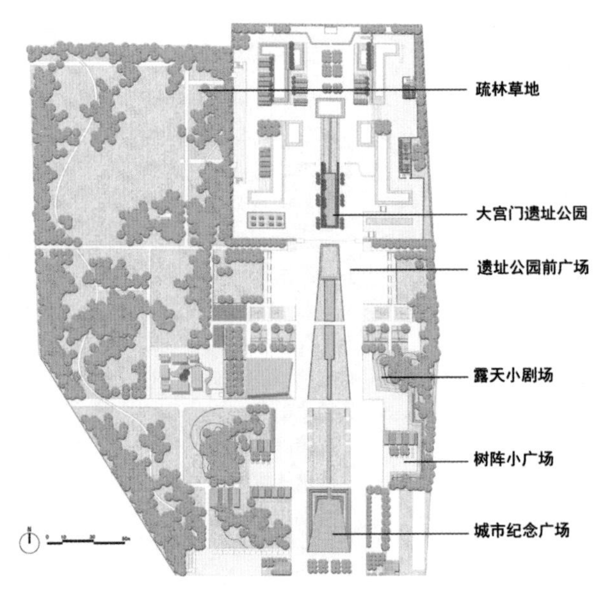

大宫门节点设计总平面图
Detailed Plan of Dagongmen Area

遗址公园设计分析图
Analysis of the Ruin Park

大宫门节点鸟瞰图
Aerial View of Dagongmen Area

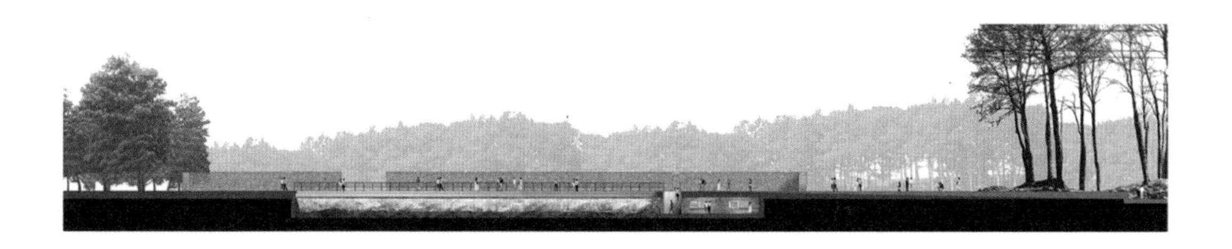

遗址公园保护游览剖面详图
Detailed Section of the Ruin Park

5 大宫门节点设计

圆明园大宫门遗址节点的保护性规划设计充分体现了前期绿道规划的目标，建立文化资源保护的核心绿色保护廊道。

圆明园大宫门节点的设计主要包含遗址公园、前广场和城市开放纪念性广场及城市开放绿地。设计延续圆明园大宫门轴线，控制整个区域。东部以广场为主，西部逐渐过渡到绿地。整个氛围肃重而宁静。

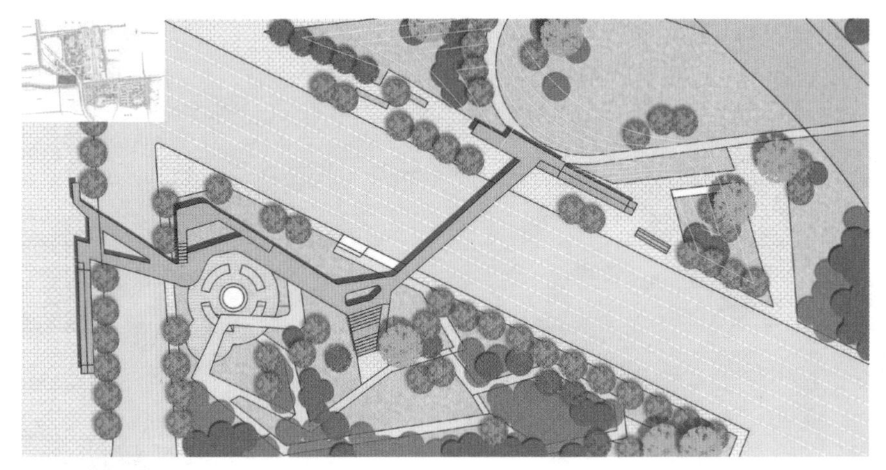

人行天桥平面图
Detailed Design of the Footbridge

人行天桥效果图
Footbridge Rendering

6 场地内及其周边连接性分析及策略

策略一：增加人行天桥、地下通道；

策略二：利用绿道建立慢行系统，加以完善；

策略三：土地置换，合理增加绿地；

策略四：置入绿道，增强连接性。

7 承泽园、蔚秀园改造措施

将绿道功能置入承泽园、蔚秀园，彰显文化价值，提升游览性；保护现状古树名木，在现状的基础上丰富植物群落，并结合历史文化特色设置景观构筑物；恢复承泽园中轴线；蔚秀园在现状的基础上恢复部分建筑群的空间格局；在承泽园基址上建造与其建筑风格相协调的仿古建筑，作为绿道的服务驿站。

8 城中村改造专题（挂甲屯为重点）

（1）策略分析

策略一：整合各类用地，商住分离（挂甲屯）；

策略二：拆除城中村，调整为公园绿地（一亩园）；

策略三：重建建筑街区，梳理内部交通。

（2）设计理念：肌理、街坊、传承

从挂甲屯街区肌理提取适宜的建筑空间结构，并从宫城建设肌理中演变出方格街坊制，从而形成改造后挂甲屯的结构。

9 建筑设计

圆明园遗址纪念馆位于圆明园大宫门遗址内，

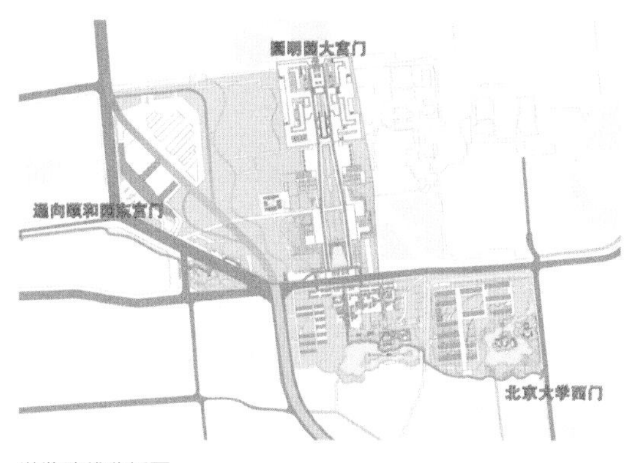

游览路线分析图
Plan of Tour Route

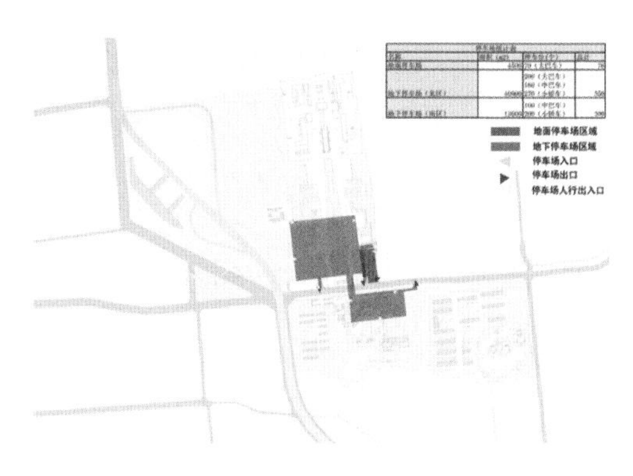

停车场示意图
Plan of Parking Areas

挂甲屯鸟瞰图
Aerial View of Guajiatun Area

位于纪念性中轴西侧，建筑面积 4540.5m^2，占地面积 2503m^2。

在建筑定位上，从认知、情感、发展三方面诉求分析，将建筑定位为圆明园遗址纪念馆，以遗址为纪念起点，以文化遗产为纪念终点。

在形式推演上，以“掀开历史的一角”作为构思起点，以形式呼应主题，建筑致力于寻求与环境的融合，将大部分建筑置于地下或半地下，使建筑与周围开放式的公园景观和谐地融合，呼应中轴，简洁的雕塑感立面呈现纪念碑式的情感特质，镜面水池隐喻扇面河的历史，营造纪念性的场所氛围。在参观动线上，通过单线串联式的流线引导游客依次进入万园之园、毁灭之劫、复兴之路与文化之思四个主要展览空间。

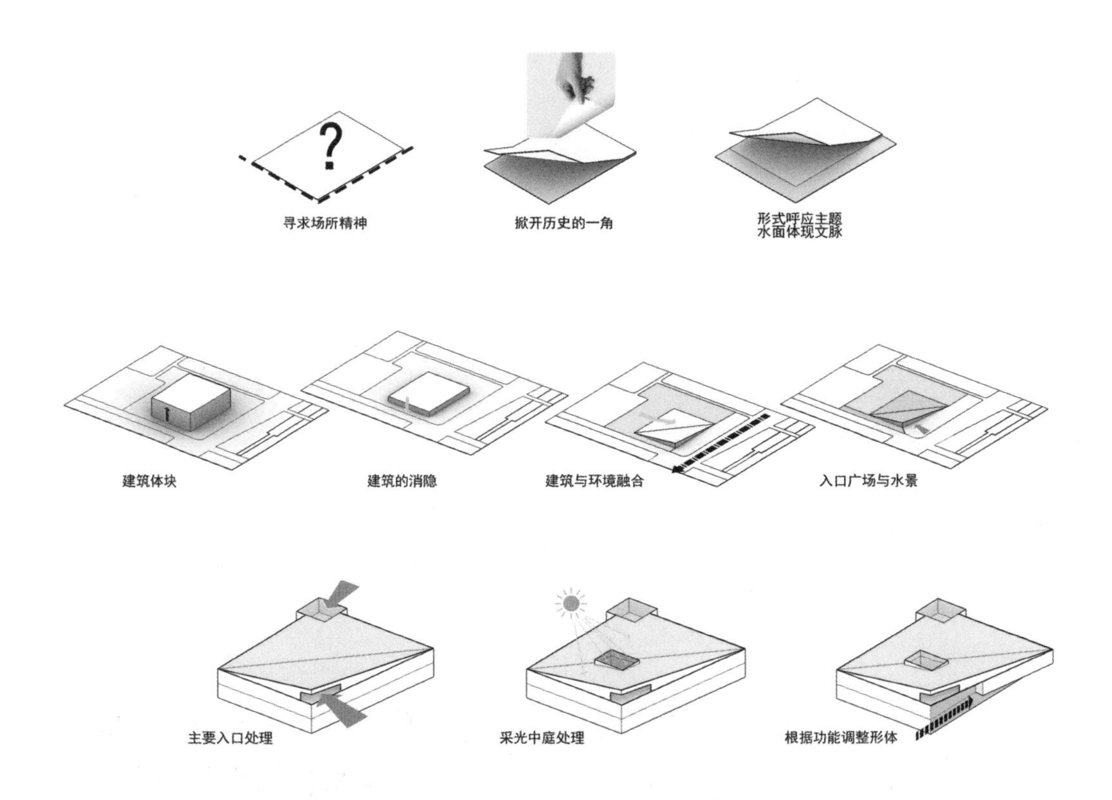

建筑设计构想
Architectural Design Idea

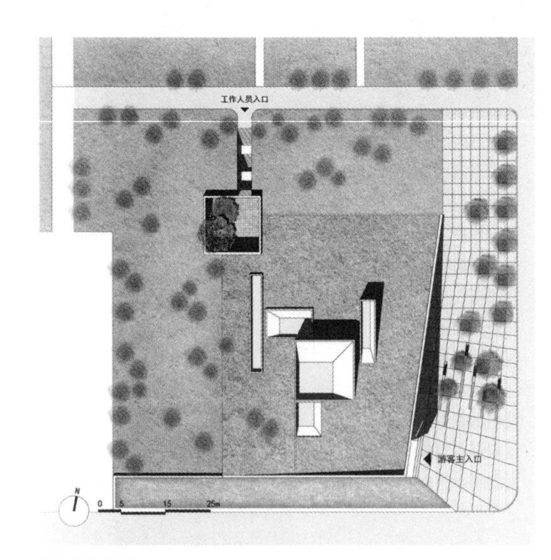

建筑总平面图
Site Plan of the Museum

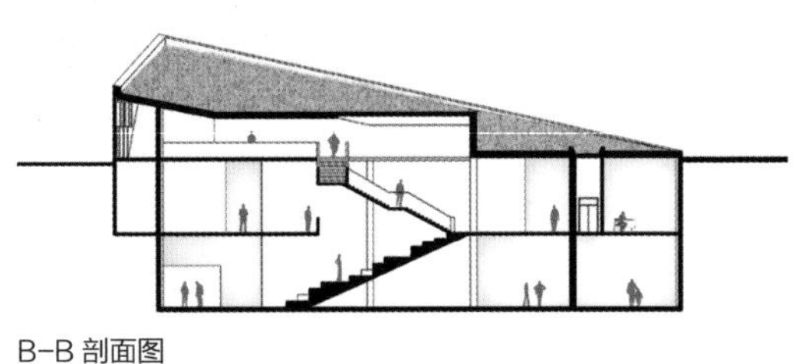

B–B 剖面图
B–B Section

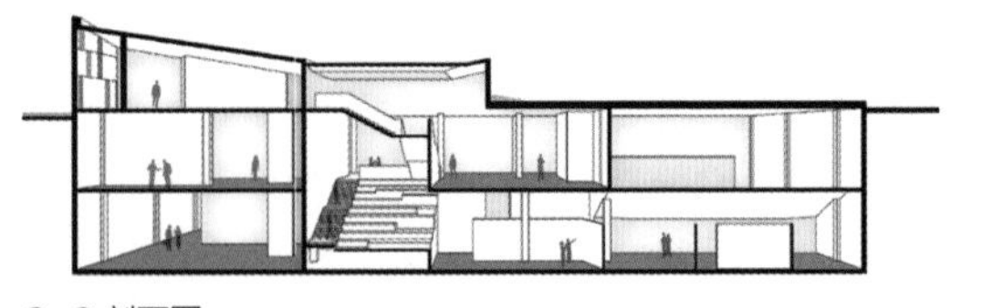

C–C 剖面图
C–C Section

建筑效果图
Building Rendering

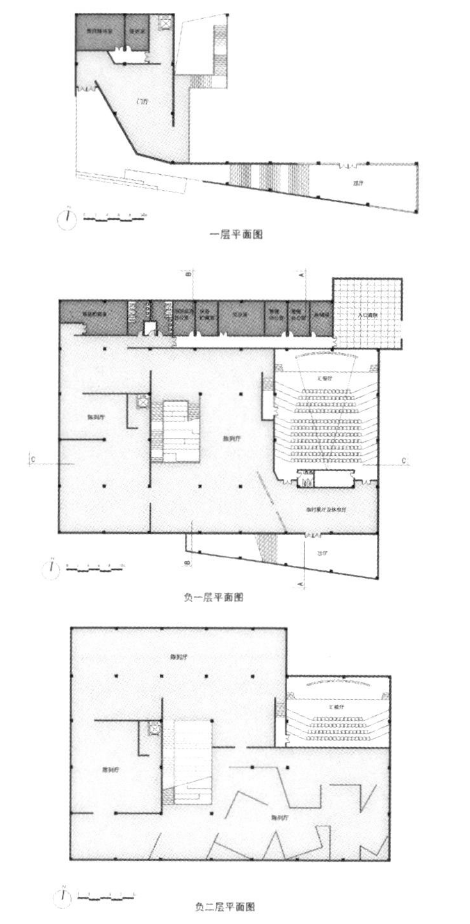

各层平面图
Floor Plans

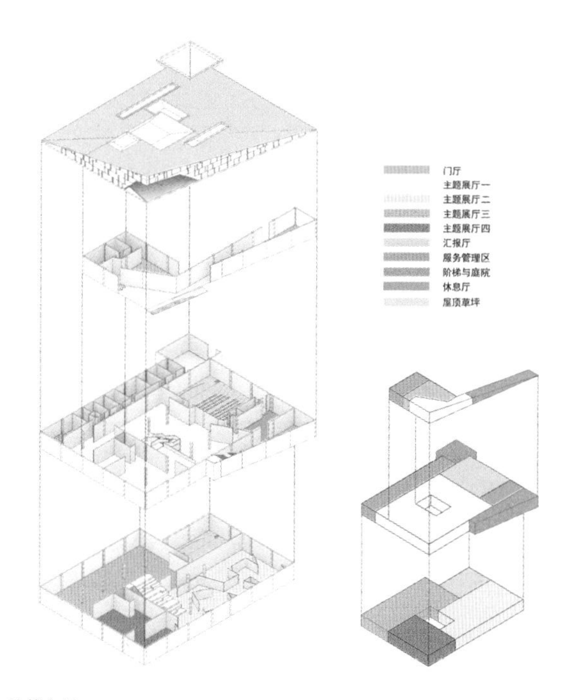

功能布局
Function Layout

规划方案二及香山街道片区概念设计

Plan 2 & The Conceptual Design of Fragrance Hill Street Section

尚尔基、李娜亭、周珏琳、王训迪、孙津、高琪、吴晓彤
Shang Erji/Li Nating/Zhou Juelin/Wang Xundi/Sun Jin/Gao Ji/Wu Xiaotong

本规划方案中，充分考虑住民、游客密度的空间分布，着重在“需求”较高的地段加大绿道建设密度，以实现绿色开放空间更理性地“按需设置”。

香山街道片区位于香山公园大门与公交枢纽之间，现状建设混乱，游客与居民混杂，地段坡降较大。设计中重新梳理了步行游览开放空间，完善了旅游设施，并重新安排了顺应地势、完整连续且景观效果良好的雨洪自然排涝系统。

In our planning scheme, we took full account of the spatial distribution of the density of the residents and tourists, and focused on increasing the density of greenway construction in the areas where demands are high, so as to achieve a more rational 'on-demand setup' of green open space.

Fragrance Hill Street section is located between the main gate of Fragrance Hill Park and the public transport hub. The current situation of the construction here is in confusion, with mixed tourists and residents, and a large gradient in the site. In the design, we reorganized the open space of walking, improved the tourism facilities, and also reordered the raindrop water natural exhaust system conforming to the terrain and of great landscape effect with integrity and continuity.

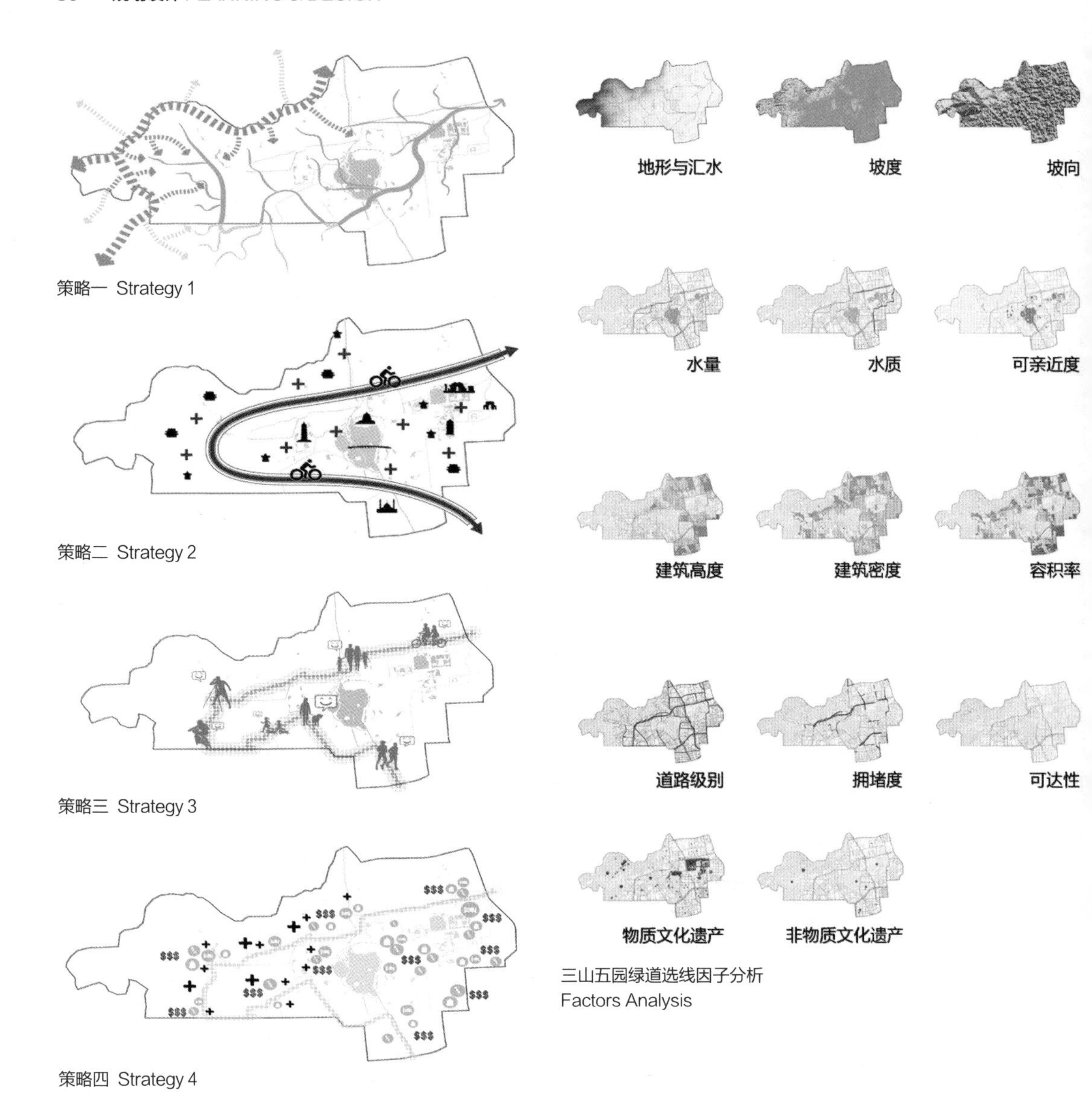

策略一 Strategy 1

策略二 Strategy 2

策略三 Strategy 3

策略四 Strategy 4

三山五园绿道选线因子分析
Factors Analysis

三山五园地区绿道规划主要有以下四个具体策略：

策略一：调整山水关系，改善生态格局，创造自然绿色基底。根据自然汇水线梳理全区水系，形成区域水网，北部加强汇水线和旱河的连接，补充旱河水，一定程度上缓解旱河干旱度；以地表植被覆盖分析为基础，在潜力区增加绿地面积，完善区域绿地组团。以西山山脉为骨架，各组团形成生态斑块，根据斑块的植物群落类型不同，产生的生态效益度不同，形成相互影响格局，网状绿带成为整体格局的沟通纽带，产生由山地乡村向城市中心发展的生态格局，融入城市内部。

策略二：优化游览路径，串联公园体系，丰富观光体验，塑造品质生活，展示三山五园历史文化特征。海淀旅游的“亮点旅游区”，要以生态科技旅游为主题，发展湿地观光、农业科技游览、高档会议度假、国际教育观摩、生态科技主题园等旅游内容。三大旅游区分别侧重皇家园林观光旅游、山地休闲度假旅游和生态科技休闲旅游三个方面发展，共同塑造未来海淀区的三大旅游品牌。

策略三：连通绿色廊道，构建生态宜人的绿道体系。根据《北京城市总体规划（2004-2020年）》和《海淀区空间发展战略》，在生态导向下，形成全市中心辐射型绿道格局，由北京城市中心连通到其

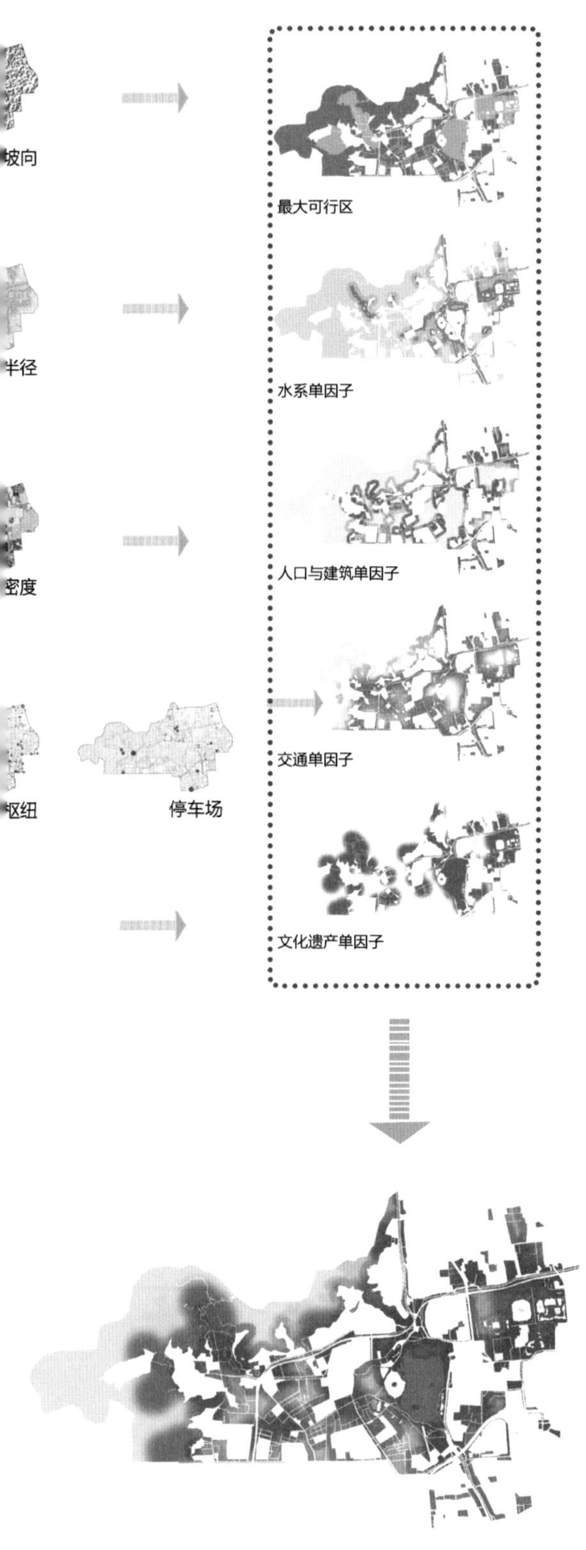

三山五园绿道选线因子叠加
Superposition of the Factors

他各个区县，形成全市性绿道体系预想。规划区的两个对外接口：八家郊野公园和长春健身园，它们作为全市绿带系统构建的预留衔接点。

策略四：以绿道景观为媒介，发展第三产业、绿色经济，推动西北区域发展。西部区域是城市未来的重要发展区域，在维护生态环境的前提下，积极引导旅游休闲、商业物流和教育等生态友好型产业向该地区聚集。以优化、整合和完善现有的发展空间为主，改善生态环境，引导高品质、组团式集约发展，防止高密度连片开发。

三山五园地区的绿道规划以现状绿地分布为基础，同时根据上位规划，移除少量三类居住用地和商业用地，变为绿地与广场用地以及公共设施用地。三山五园绿道的选线通过层次分析法（AHP）将地块肌理、水系、建筑、道路以及文化遗产等要素叠加，最后得出合理的绿道选线范围。

地块肌理：地块肌理是绿道规划的基础。根据现状山水格局和现状用地，分析地形、汇水、坡度和坡向等要素并进行叠加，提炼出绿道规划的最大可行区域范围，其中包括西山、现状绿地和可变成绿地的用地，如废弃地、村庄和三类居住用地等，得到的可行区域是绿道规划的最大潜力范围。

水系：人具有亲水性且水系具有较大的生态潜力和连通性。因此根据规划地区现状，分析水系的水量、水质和可亲近度，并进行适当评价，计算水系缓冲区，与绿地可行区叠加，得出以水系为单因子的绿道规划适宜范围。

建筑：建筑的层高、密度和容积率在一定程度上反映了区域的人口密度。人口密度较高的区域对公共绿地和绿道的需求也更为强烈，因此分析以上建筑各因素可以得到人口密度，将其与绿地可行区域进行叠加，得到以建筑为单因子的绿道规划适宜范围。

交通：道路级别、拥堵程度、可达性、交通枢纽分布、停车场分布是与交通相关的重要因素，将以上各因素与绿地可行区域进行叠加，得到以交通为单因子的绿道规划适宜范围。

文化遗产：三山五园地区的文化遗产十分丰富，因此，绿道规划期望绿道能够串联尽可能多的文化遗产区域，将物质文化遗产与非物质文化遗产按照级别不同，设定不同的辐射范围，并将其与绿地可行区叠加，得到以文化遗产为单因子的绿道规划适宜范围。

将水系、建筑、交通和文化遗产单因子与绿地可行区叠加的结果再次进行叠加，得到绿道规划选线建议范围。颜色越深代表绿道建设的需求与可行性越高。删除选线范围内的封闭型单位，将颜色最深的区域进行串联，得到最终的绿道选线方案。

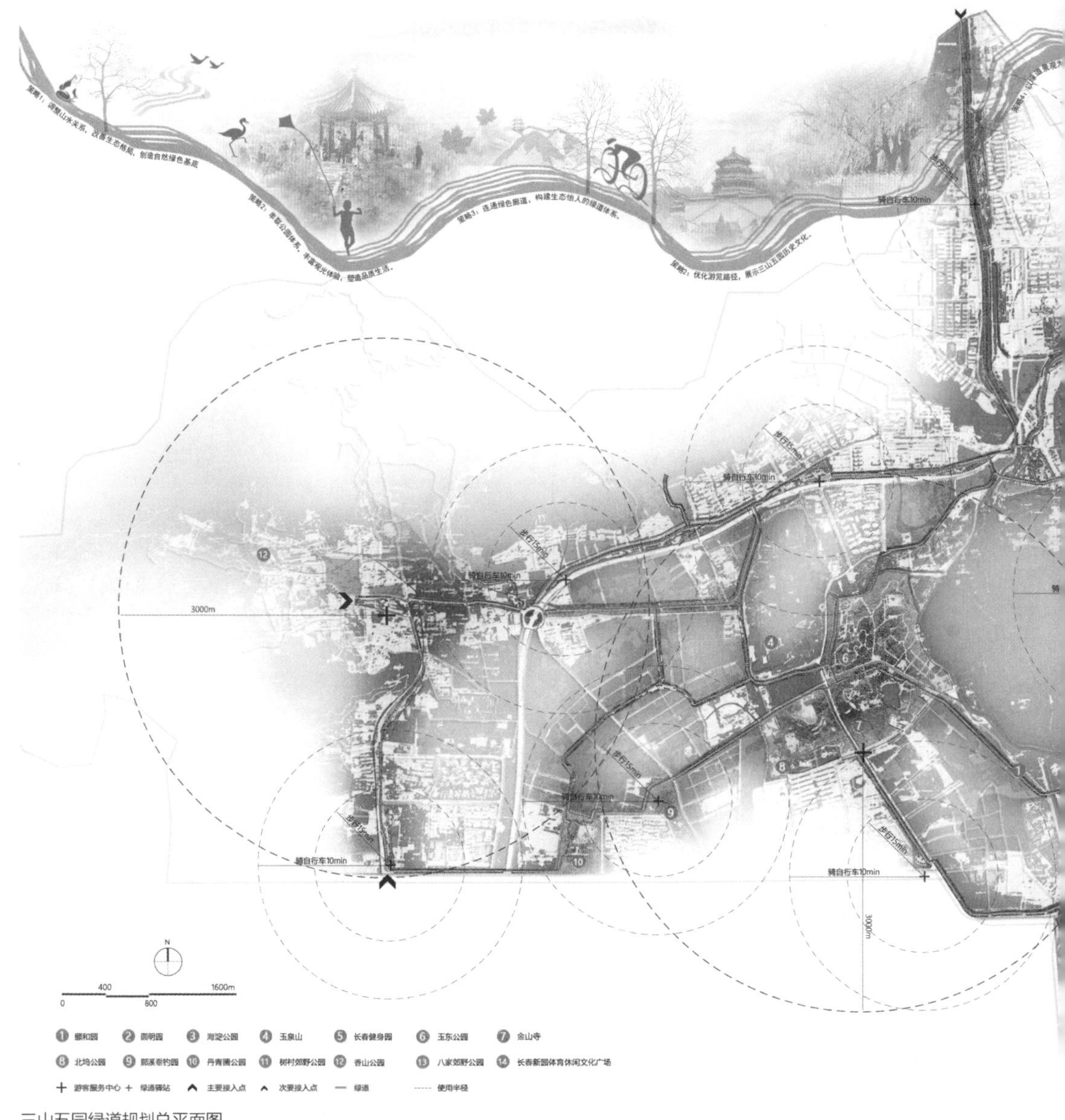

三山五园绿道规划总平面图
Master Plan

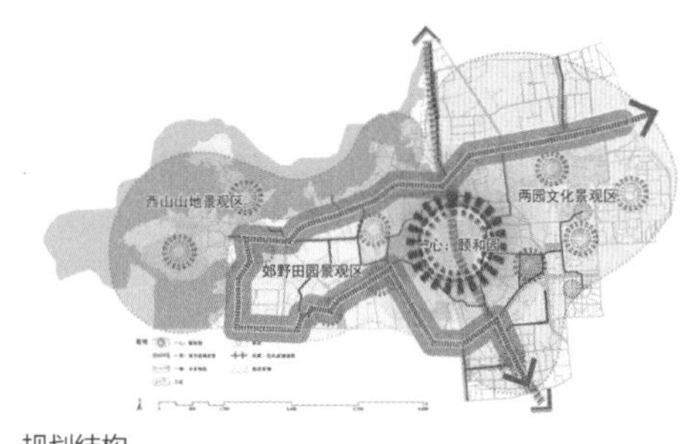

规划结构
Planning structure

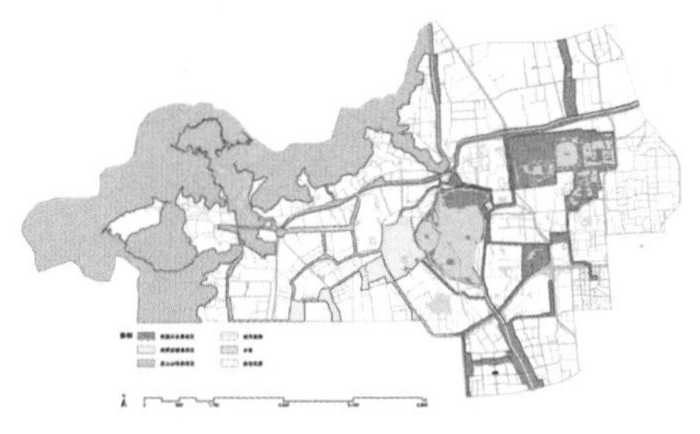

规划分区
Planning Zoning

规划用地调整
Planning Land U

用地平衡表

名称	面积（m²）	百分比
绿地面积	3210740	56.7529%
现状绿地面积	2622247	46.3507%
规划新增绿地面积	588493	10.4022%
水系面积	2020116	35.7075%
现状水系面积	1933789	34.1815%
规划新增水系面积	86327	1.5259%
道路面积	421880	704571%
慢行系统面积	400654	7.0820%
借用城市道路面积	21223	0.3751%
建筑面积	1270	0.0224%
游客服务中心	600	0.0106%
驿站售卖亭	180	0.0032%
驿站卫生间	390	0.0069%
广场面积	2200	0.0389%
停车场面积	1200	0.0212%
规划绿道系统总面积	5657403	100.0000%

三山五园地区规划绿道全长6.6km，总面积约566hm^2，串联地区内的各类绿色开放空间和重要的自然与人文景点。绿道规划充分考虑现状条件，在维持现有绿地格局的基础上进行连接，并给城市发展留有足够的用地。作为“休闲”、“生态”和“人文”兼顾的绿色廊道，规划设计坚持人与自然和谐共生的理念，尊重北京地区的自然山水格局，尽量减少对自然环境的干扰和破坏。提升绿道两侧绿化景观，拆除改造城中村发展为公共绿地，把废弃地和闲置地发展为绿地，在增加绿量的同时，强化廊道的生态与历史人文要素的连通功能。

三山五园地区绿道规划结构以颐和园为核心，城市级绿道带和南北向的水系为轴线，划分为两园文化景观区、郊野田园景观区和西山山地景观区三个片区，各级公园绿地呈多点状分布，社区级绿道网络呈网状脉络分布，其中，形成“一心”、“一带”、“一轴”、“三区”、“多点”与“织网”的结构。

三山五园地区绿道规划从空间结构和功能两个角度出发，将绿道分为城市级绿道和社区级绿道两个层级，二者相互连通形成有机整体。其中城市级绿道总长2.4km，社区级绿道总长4.2km。场内的物质文化遗产与非物质文化遗产总体分布在城市级绿道和社区级绿道周边。此外，依据所连接的自然和人工景观资源特征，绿道分为公园型绿道、道路型绿道、滨水型绿道、郊野型绿道、防护绿地型绿道和山林型绿道六类。绿道规划体系中设置了4个主要接入点6个次要接入点以及多个自行车驿站和停车场，与已有的及规划的地铁站点和公交站点接驳。在特色活动路径方面，根据绿道主题，划分为都市活力绿道、双园文化绿道、田园风情绿道、傍山风光绿道和滨水休闲绿道共五类。绿道不仅仅成为出行的便捷通路，更是北京三山五园地区文化旅游的纽带和名片。场地内具备8条视线通廊和多个观景点，具有较好的观景体验。

绿道标识系统包括立体标识和地面标识。立体标识包括：空间引导标识、文字标识、距离指示标识、导向指示标识、警示禁止标识以及服务设施命名标识六种类型。地面标识包括慢行道路标识线、北京三山五园绿道logo、骑行标识以及方向指示。

绿道服务设施由游客服务站、休憩设施、停车设施、卫生设施以及基础配套设施组成。游客服务站分为游客服务中

绿道类型
Types of Green Corridors

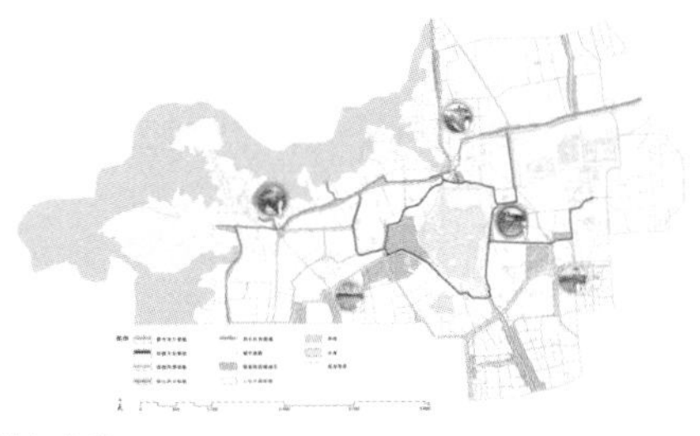

特色主题
Featured Themes Analysis

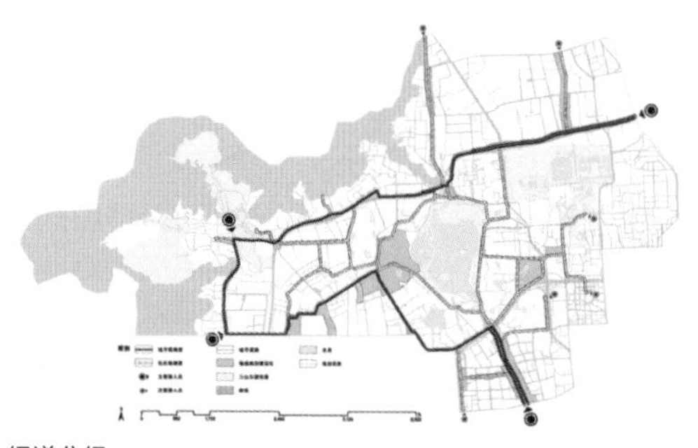
绿道分级
Green Corridors Classification Analysis

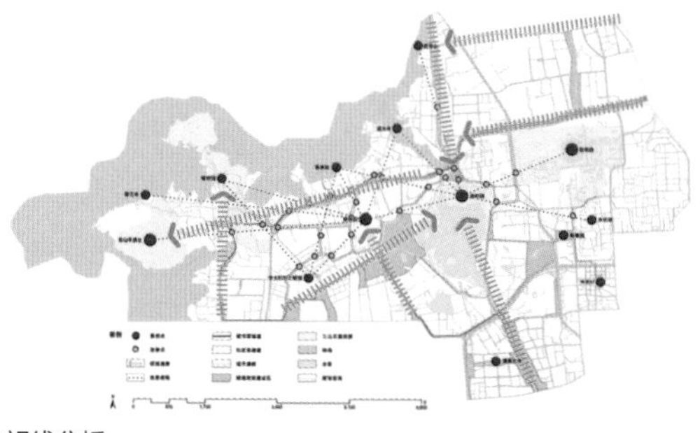
视线分析
Landscape View Analysis

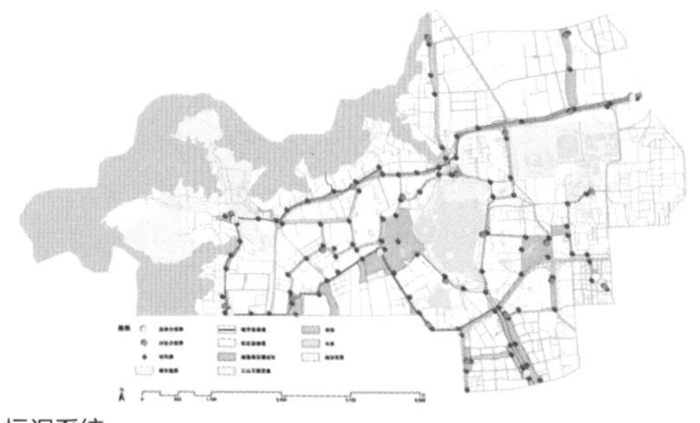
标识系统
Identification System Analysis

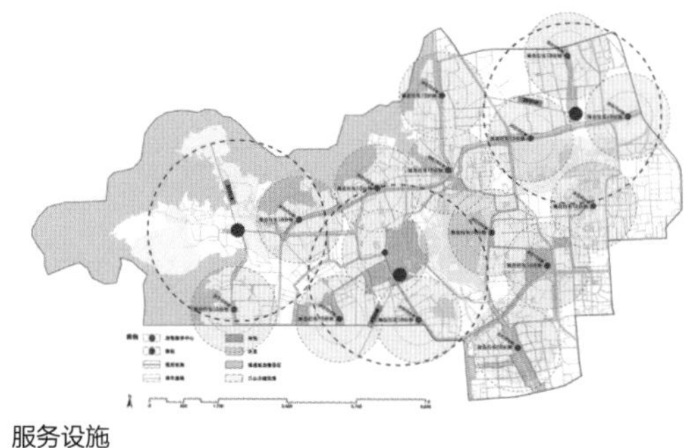
服务设施
Service Facilities Analysis

心和驿站两个级别。游客服务中心承担综合服务功能，驿站承担基本服务功能。绿道休憩设施主要包括亭、廊、花架、坐凳和健身器械等游憩设施和休闲活动场地两类。停车设施包括机动车停车场和自行车停车场。卫生设施中卫生间服务半径为 1km，垃圾箱服务半径为 100m。此外结合绿道的实际使用需求和游客量等因素合理配置基本的水电基础设施。

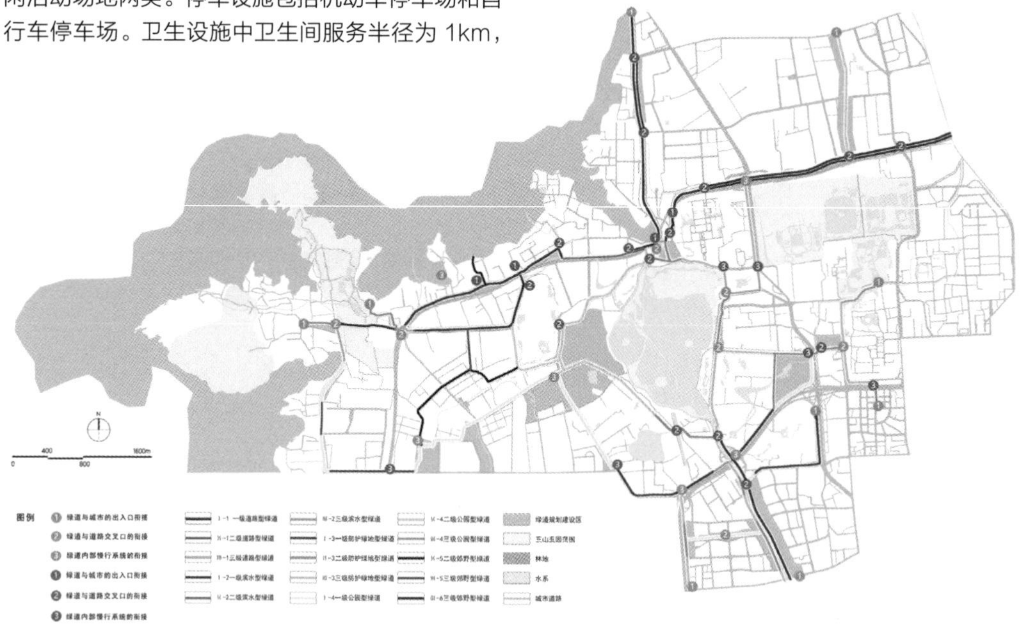
绿道分类分级设计
Tiers and Categories of Green Ways

绿道慢行道路系统保持连贯性和通畅性。慢行道与车行道路交叉时，一般可采取平面交叉和立体交叉两种形式。平面交叉的主要形式包括绿道与城市的出入口衔接，绿道与道路交叉口的衔接以及绿道内部慢行系统的衔接。立体交叉的主要形式包括：绿道与高架轨道交通相交时，宜采用涵洞和高架桥下穿行等通行方式。绿道与河流水道相交时，结合现有的桥梁或新建桥梁跨过河流水面，或利用水上交通的方式通过水面，实现水上交通与绿道的无缝衔接。绿道与高速公路或城市主干道交叉时，宜优先采取安全的立交方式，如高架人行天桥等，以保证绿道的安全性和连续性。

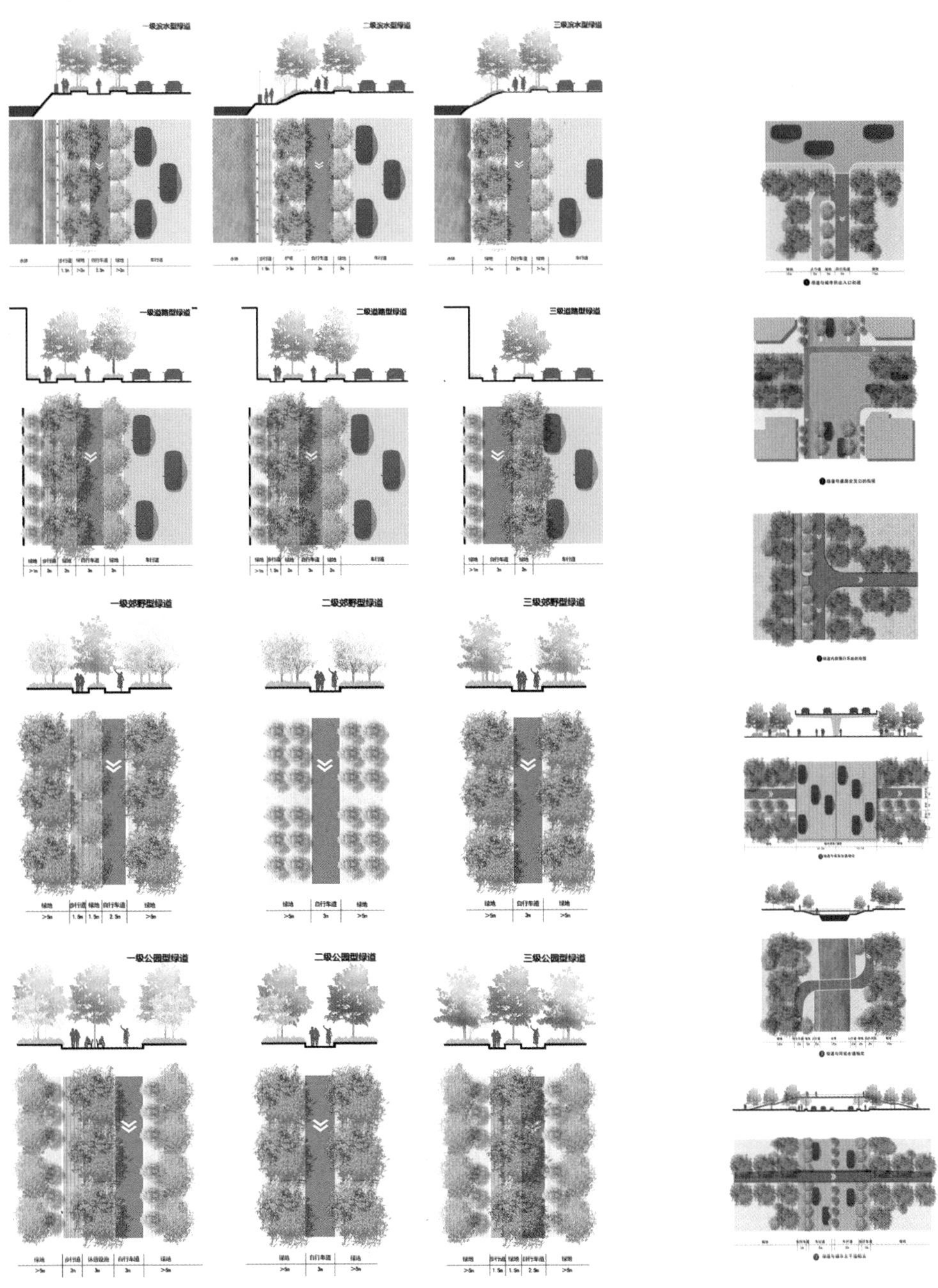

绿道分类及剖面
Green Way Sections by Categories

交叉口组织
Design of Road Intersections

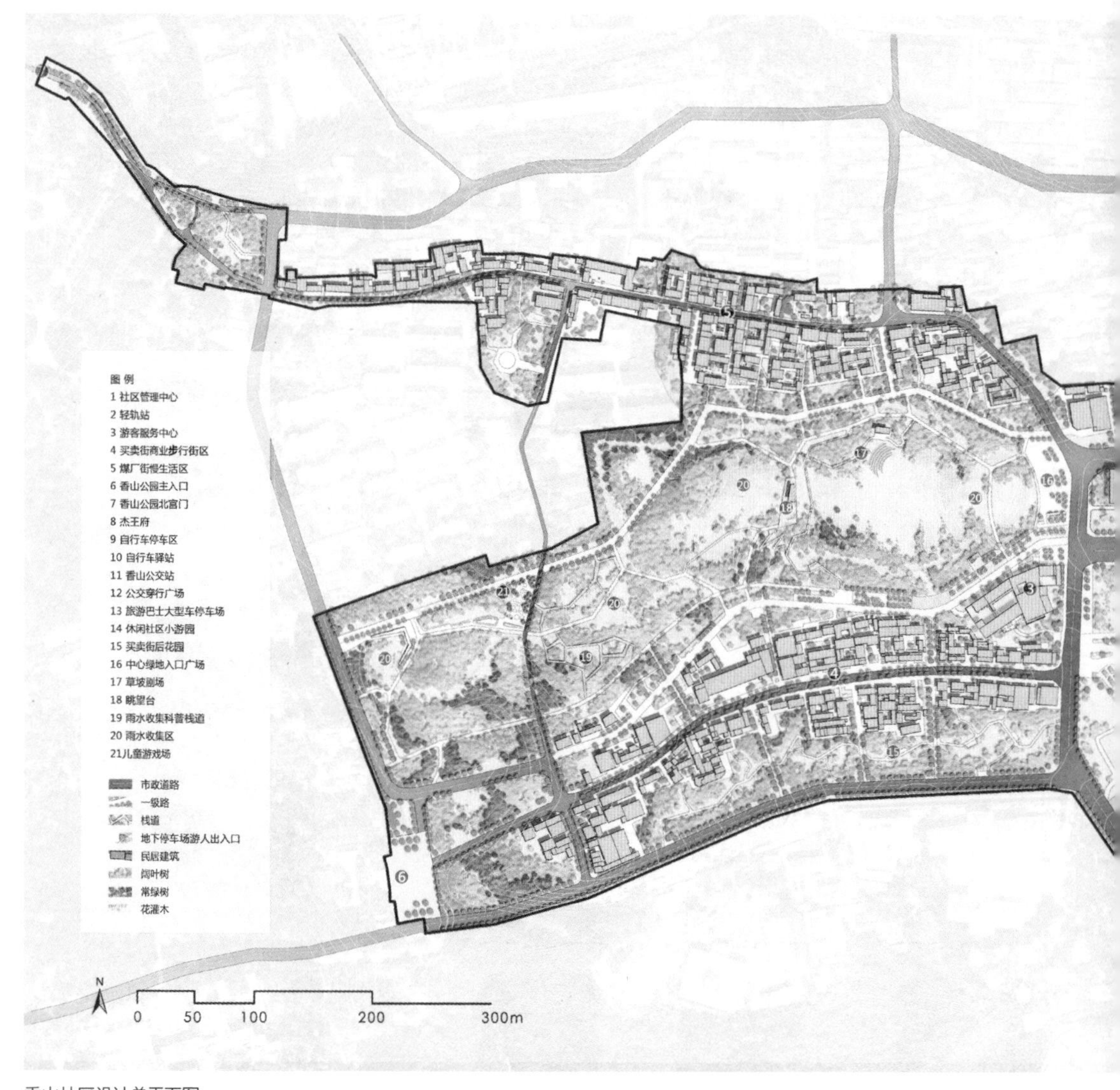

香山片区设计总平面图
Site Plan

设计范围在绿道的位置
Site Location

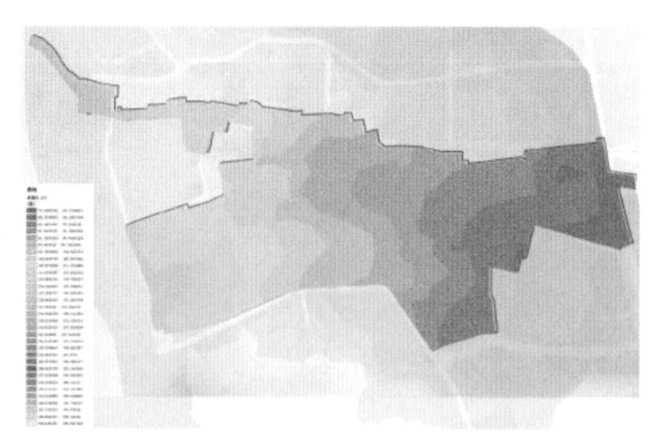

竖向分析
Vertical analysis

设计地块位于北京市海淀区香山片区，西临香山，东临北京植物园和中科院植物园，南临西山，呈现“三面环山一面城”的态势，总面积约 41hm²。设计地段位于绿道总体规划的东侧地段，是绿道从城市延伸到自然的收尾，同时又位于三山五园的轴线上，因此该地段位置优越，是绿道规划的核心地段。设计地段位于香山脚下，是香山山脉和西山山脉的延续，场地内高程由西向东明显降低，山脊线和山谷线明显。现状交通体系稍显紊乱，居民和游人路径相互干扰，道路体系比较复杂。游人对居民的影响主要为游人噪声大，住房与商业活动缺乏过渡和隔离，居民日常休息空间被旅游活动侵占等。机动车、非机动车分布与场地内外，停车场散布其中。现状建筑色彩以灰色系为主，同时还有蓝色系、红色系、黄色系建筑，形成色彩多样的建筑组团。主要以北京民居、现代建筑、中式园林古建等为主。建筑类型多样，主要以北京民居、现代建筑、中式园林古建筑等为主，容积率 0 ∽ 2 不等。游人游览目的以观光、摄影、健身为主，小部分为野营和科学研究。游人流量方面，以 10 月游览旺季为例，月游人量约 97 万人，日均 3.1 万人，超出“旅游环境容量最大值”。

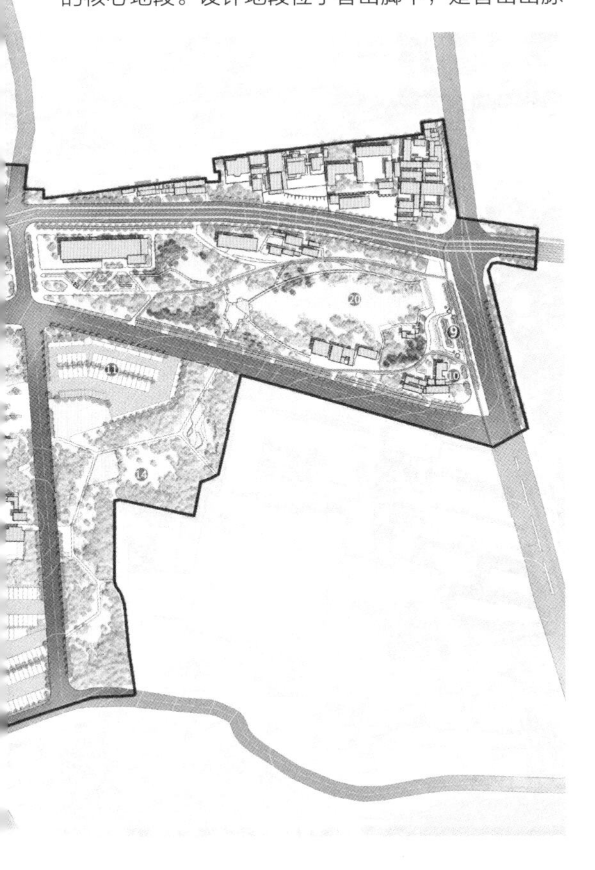

设计结合现状问题进行总结分析，并提出解决策略。总体设计概念以西山山脉为主要依托，东侧有玉泉山和许多采摘园，位于香山公园、北京植物园和中科院植物园之间，是从东部城市向西部自然山体的过渡区域，建立由香山到城市建成区的有机联系。设计后的地形以原有地形为基础，旨在解决香山区域内的暴雨积水问题，在绿地中设置几个汇水片区，对暴雨期的雨水进行拦截和消纳。交通布局结合上位规划进行调整，车行路分布于设计地块外，内部规划多条人行路，地下停车场位于设计范围中南部。通过现状调研与上位规划综合评价，对现状建筑划分为拆除建筑、保留建筑和改造建筑来指导设计，同时新建的地铁站建筑和游客服务中心。设计范围总面积为 42.57hm²，其中绿地占 63.4%，建筑占 15.8%，广场占 10%，停车场和道路占 10.8%。

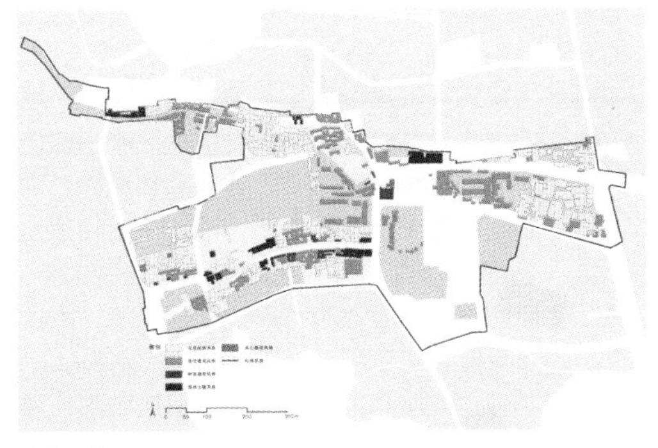

建筑风貌
Architectural Style

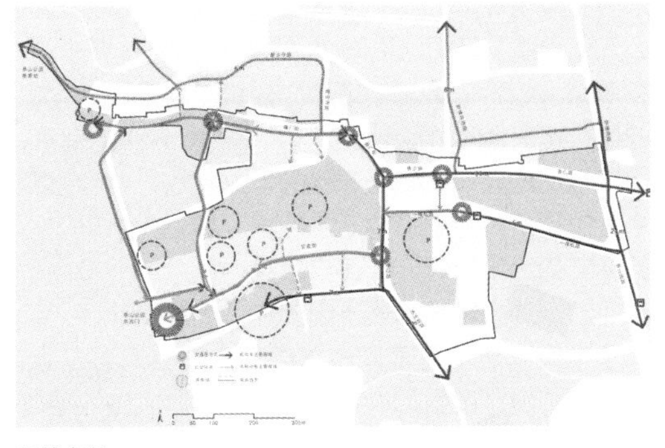

现状交通
Existing Traffic

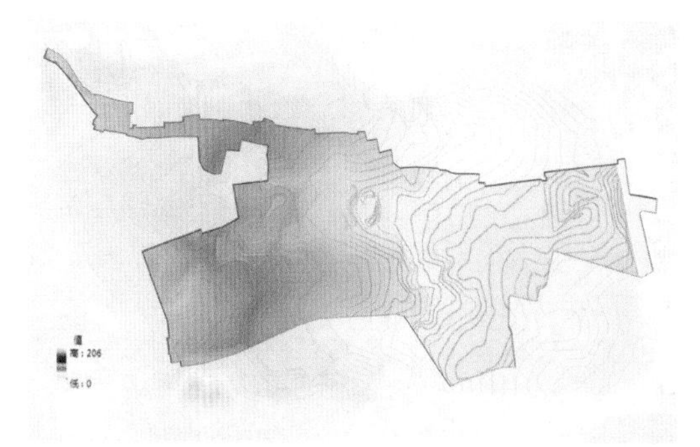

设计地形分析
Existing Landform

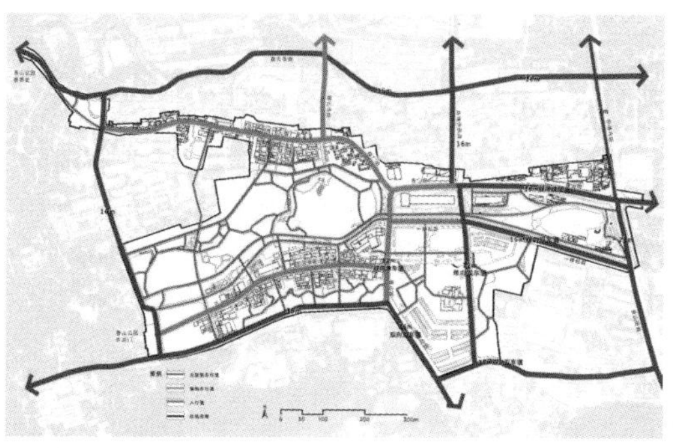

设计交通分析
Traffic System Plan

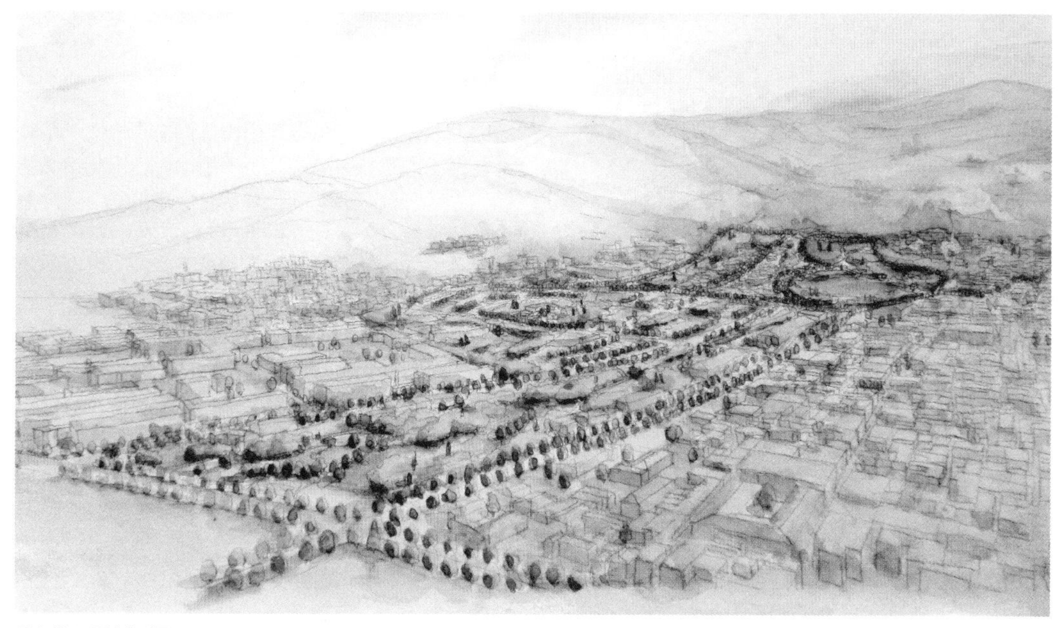

香山片区设计鸟瞰图
Aerial View

节点设计：节点一位于设计范围的东部，是三山五园绿道与香山的接入点。设计的自行车停车场，方便游客的游览需求。场地内保留了部分风貌优良的现状建筑，同时也拆除了大部分风貌较差的建筑，用于绿地的建设。结合现状的地形特点，并考虑整个香山片区内的地形与汇水趋势，场地内设置了两处下沉的雨水花园，并适当抬高中央绿地。节点一西部为规划的轻轨车站，因此在其周边设计了一系列场地来满足停留于休憩需求的广场空间与设施，轻轨站东侧布设自行车租赁点，方便游人进入绿道内游赏。

节点二主要解决交通问题，将公交总站置于地块东侧，为减少东侧车流对地块内的干扰，设置了地上的大巴停车场以及地下车库，东侧有 2 个入口，西侧有 2 个出口，地上设置了 8 个人行出入口。地上主要交通流线为从公交总站到香山片区的快速通行道路与南北向轻轨站和中心广场的连接道路，重点设置了东西向和南北向的广场，可以在高峰期集散人流，还可以提供足够的活动场地。除此之外，地块东南侧的绿地主要服务于周边的医院等，提供了充足的休息与活动空间。由于地下部分是停车场，因此地上部分主要采用以灌木为主的植物群落，减少对地下承重的荷载，地上部分也结合地下的柱网结构设置了临时性休憩的廊架，以便纳凉休息。

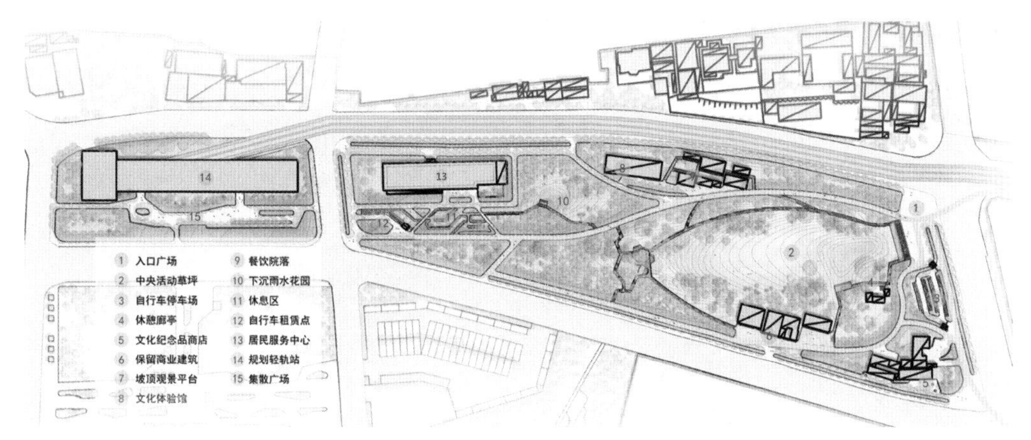

设计节点一平面图
Detailed Site Design 1

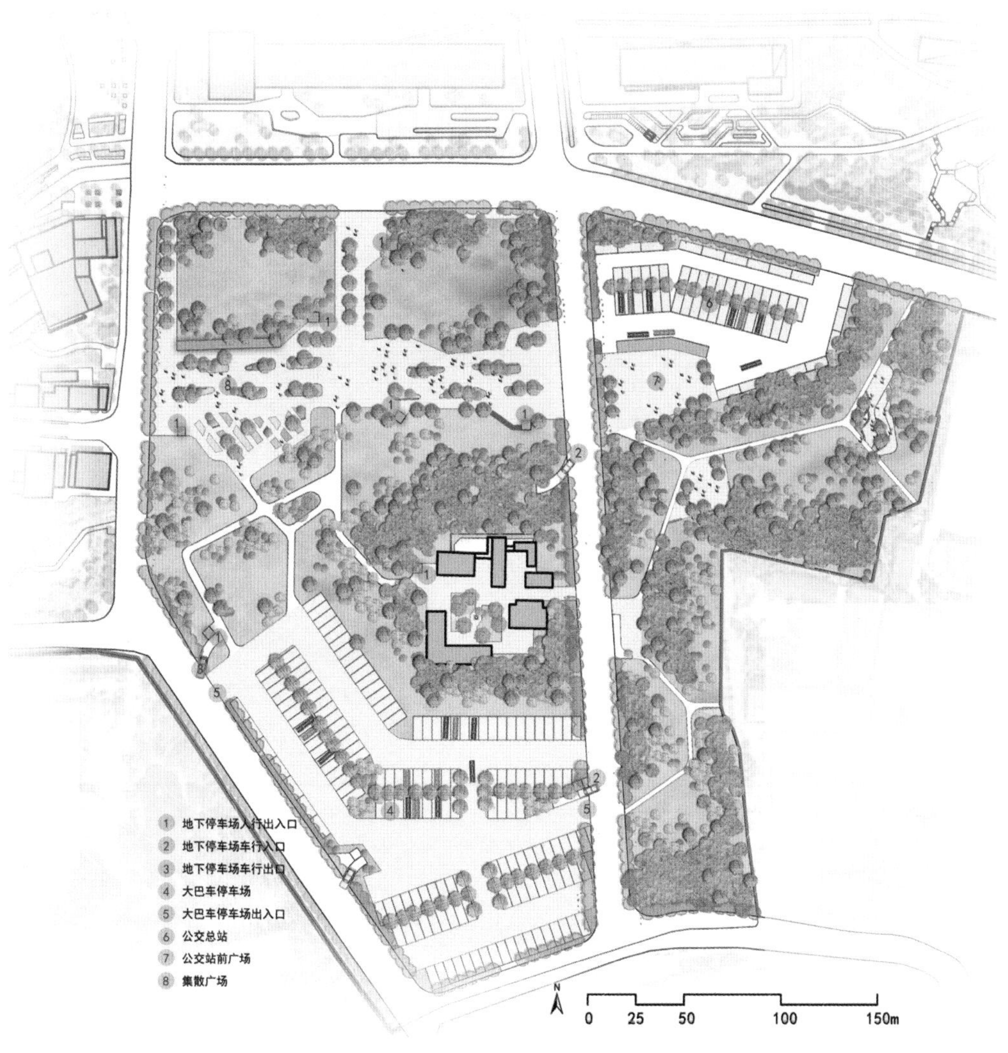

设计节点二平面图
Detailed Site Design 2

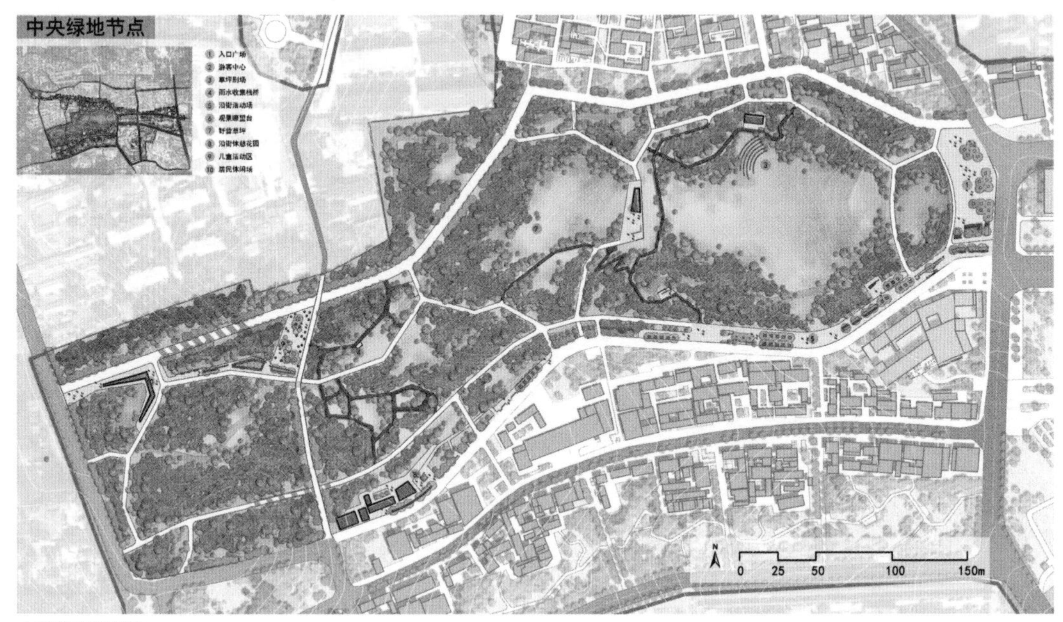

中央节点平面图
Detailed Site Design 3

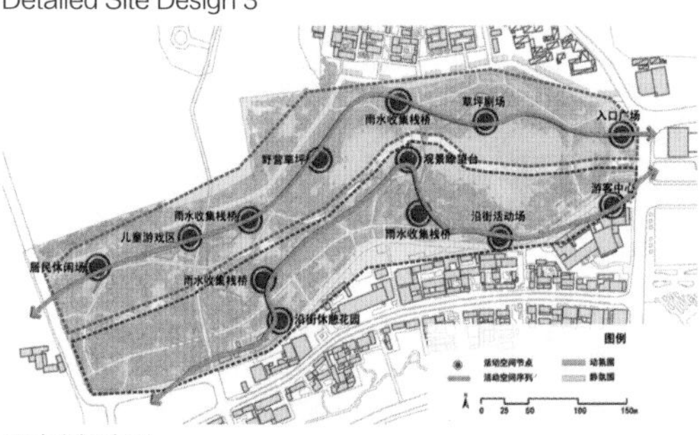

活动空间序列
Activities

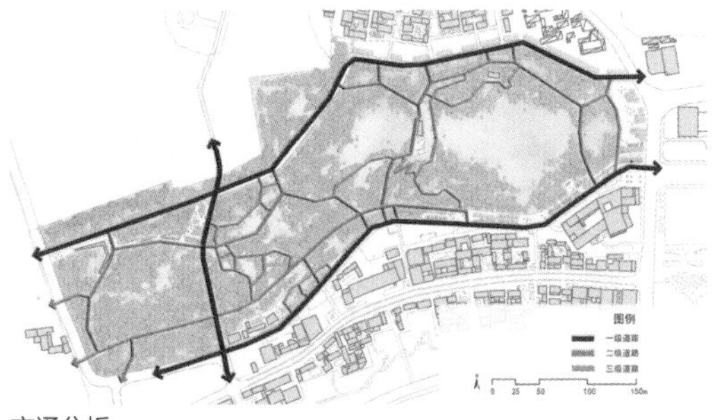

交通分析
Road System

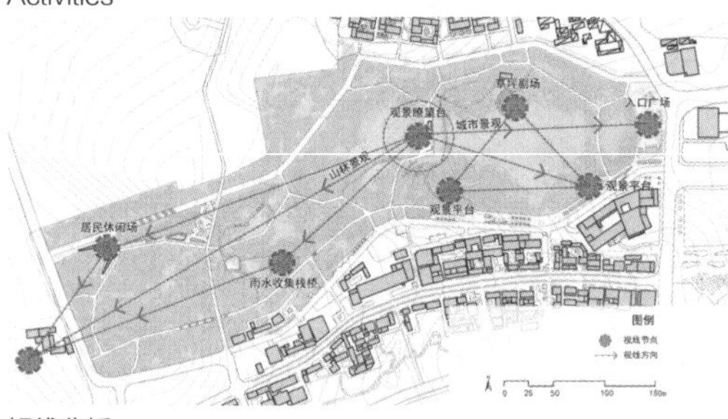

视线分析
Sight Lines

汇水分析
Catchments

节点三是位于买卖街和煤厂街之间的中央绿地节点，现状为少量三类居住用地和荒地，改造之后，中央开拓出整合度较高的绿地空间，设计形成“东城西林，北禅南市”的格局。其中，西侧延续香山自然山林景观，逐渐向东过渡为城市景观。现状居住、商业、交通如同堡垒，阻隔了场地中央绿地与周边绿地的空间联系，因此，中央绿地设计多条连通买卖街和煤厂街的视线绿廊，打破街区的阻隔，建立了中央绿地和周边绿地的联系，将绿地延伸渗透入居民生活中，使香山中央绿地成为一个集历史文化、生活休闲和生态多元为一体的“城市客厅”。根据煤厂街和买卖街“北禅南市”的历史街区氛围，中央绿地也营造出“北静南动”的活动空间。北侧靠近煤厂街设置绿化隔离带，活动空间多且较为私

中央节点效果图
Rendering

中央节点鸟瞰图
Aerial View

密幽静；南侧贴近买卖街，空间丰富，活动类型也富有活力。道路划分为三个层级：一级道路承载东西向和南北向的主要人流，二级道路实现内部环路，三级道路连接重要节点。场地高程西高南低，通过地形和植物引导游人视线，实现步移景异的效果，吸引游人进入景区。场地内通过局部下挖预留阻滞雨水径流的下凹绿地，能够有效缓解自山上冲刷而下的雨洪压力。设计排水方式为道路排水和排水沟排水两种。

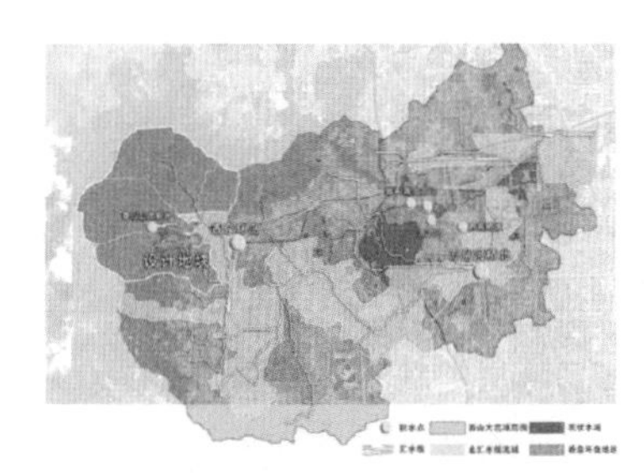

宏观流域面关系分析
Macro Basin Surface Relationship Analysis

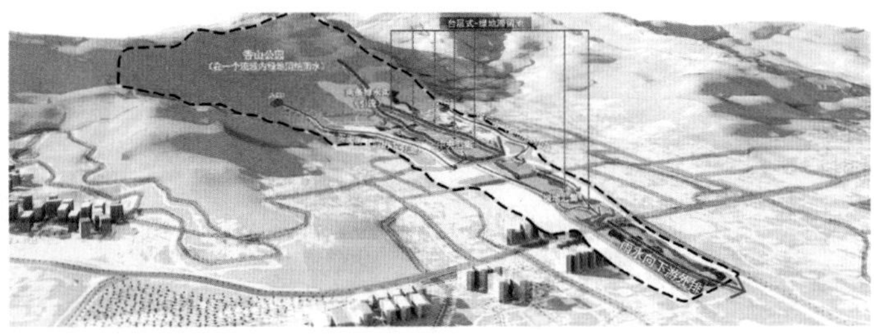

设计地块结合流域面的宏观雨水控制策略
the Macroscopic Rain Control Strategy Combination with the Basin Surface of the Site

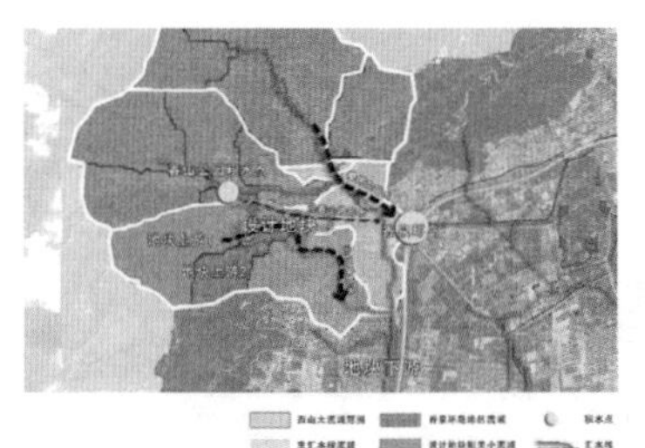

地块周边小流域分析
Surrounding Watershed

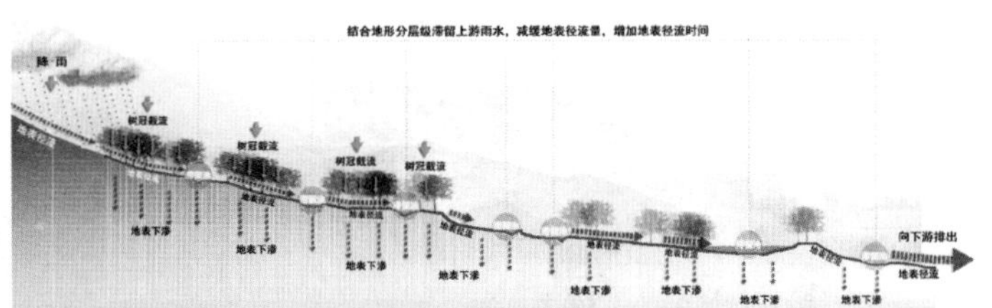

设计地块雨水管理模式剖面示意
Rainwater Management Section

大西山宏观层面流域面的关系为从西山东南侧完整流域面至主汇水流域，再到设计地块相关小流域，它位于大流域的上游，因此设计地块相关区域流域对雨水进行控制会对下游产生有利影响。设计地块地表径流分为两个流向，其一是下游香山南路以南，可以缓解地块下游雨水压力。另一个是沿香山路引流排出至香泉环岛，可缓解香泉环岛桥下积水压力。设计地块有意识地截流和有组织地排水等都是有效的雨水控制管理措施。

设计地块结合流域面考虑宏观雨水控制策略。从子流域角度考虑雨水管理，通过树冠、地表截流、绿地滞留以及绿地和地表下渗，达到减少径流、降低流速和减少排出水量的目的。排水沟纵断面改造包括六项策略：自然状态直排沟、陡坡区沟面改造、缓坡区沟面改造、缓坡区结合滞水改造、排水沟衔接周边滞水池和结合栈道游览的排水沟。实施雨洪管理相关措施后，场地产生的地表雨水流量减少，有效缓解北京暴雨时期雨洪对香山地块的不良影响。设计后场地雨水缓解能力如下：一年一遇暴雨时，地表雨水可全部消纳；五年一遇暴雨时，68% 地表雨水被滞留，向下游排出 9525m³；十年一遇暴雨时，59% 地表雨水被滞留，向下游排出 14229m³；五十年一遇暴雨时，45% 地表雨水被滞留，向下游排出 25150m³；百年一遇暴雨时，40% 地表雨水被滞留，向下游排出 29853m³。

设计地块的种植设计以生态性、景观性和功能性为原则，保留北京海淀区原有生长良好的植被，因地制宜地进行植物配置，延续香山片区的景观特质，提升城市风貌。植被生态格局以西山山脉为骨架，各绿地组团形成生态斑块，由此产生的绿带成为整体格局的沟通纽带，形成整体由山地向城市中心发展的生态格局，融入城市内部。种植分区根据现

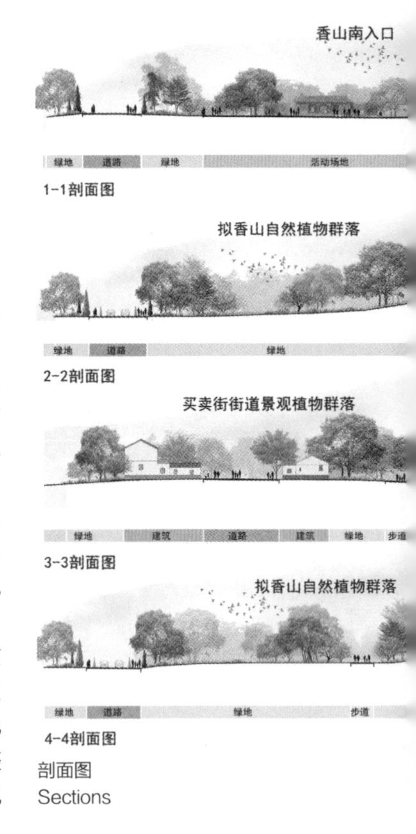

剖面图
Sections

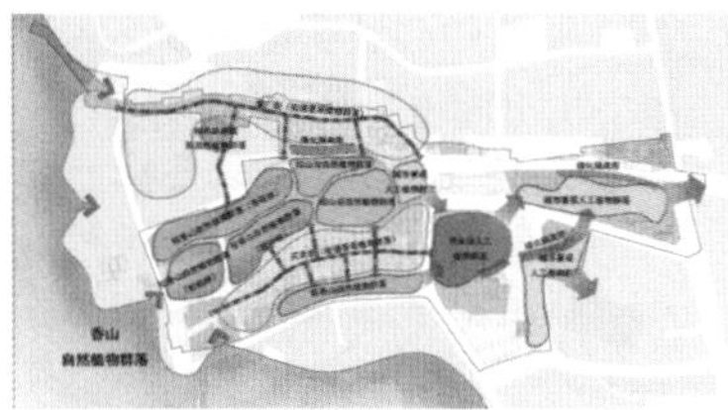

生态格局图
Ecological Pattern

种植分区图
Planting Division

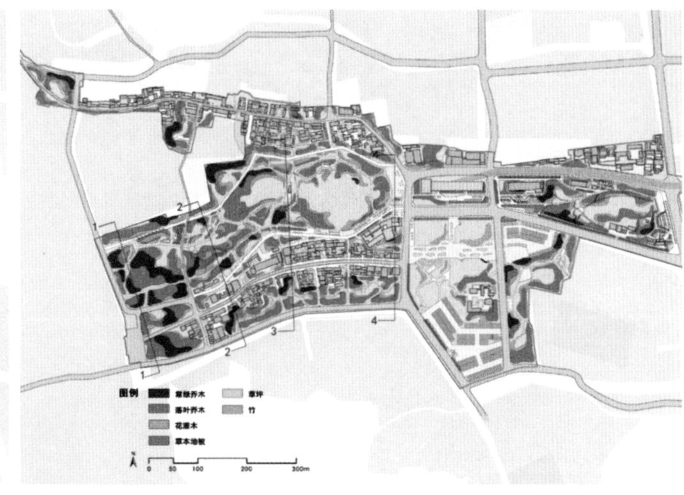

植被规划图
Vegetation Plan

状植被情况和历史文化特色，主要分为槐绿蝉鸣种植区、杏花春雨种植区、松间明月种植区、夏草如荫种植区、层林尽染种植区、溪涧竹韵种植区、绿荫银杏种植区、金栾如锦种植区和桃红柳绿种植区共九大区域。植物空间的营造模拟自然界植物的生长状态，以乡土植物为主，适当引入植物新品种，形成以乔木为主体，灌木、草本植物、藤本植物相结合的复合群落，形成“近自然”植物群落景观，乔木、灌木（藤本植物）与适量的空旷草坪结合，针叶树与阔叶树结合，季相景观与空间结构相协调，充分利用立体空间，构造开敞型、半开敞半封闭型和覆盖型植物空间，增加绿地的绿量，丰富游人体验感。

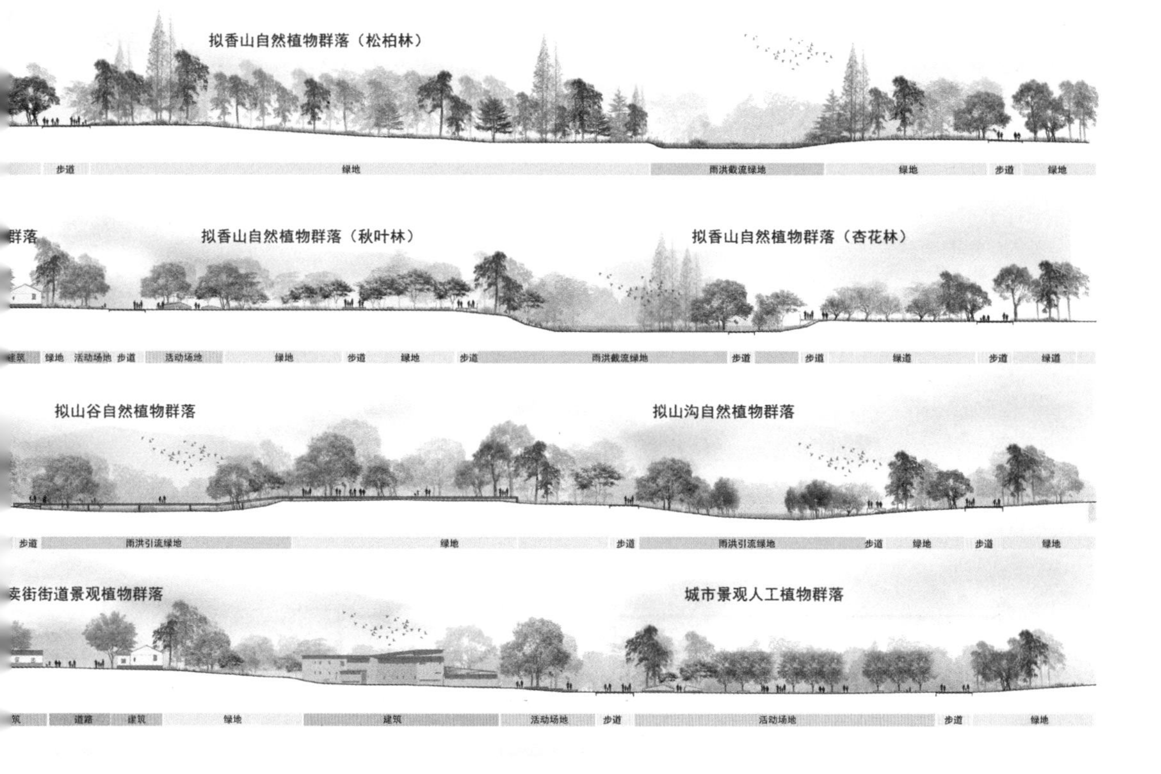

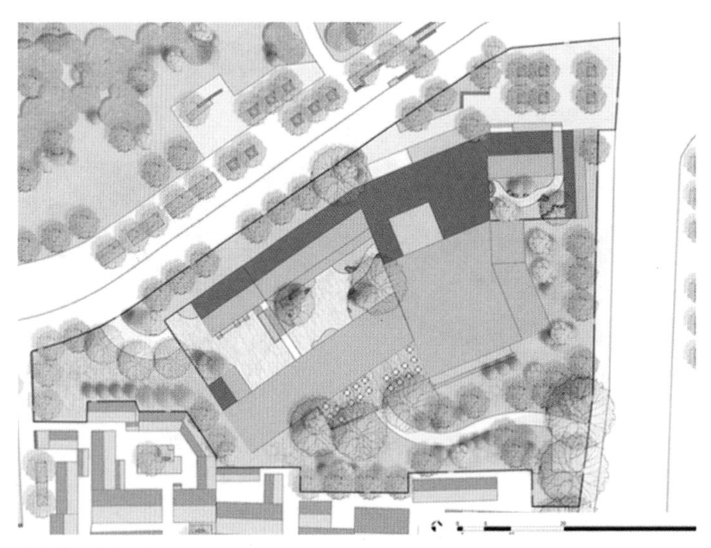

游客中心总平面图
Site Plan of the Visitor Center

游客中心位于香山脚下买卖街街区的东侧，紧邻设计地块的中心绿地及停车区，是三山五园绿道的服务设施，其内设置有自行车驿站，兼具餐饮售卖和书画展览等功能，方便游客使用。建筑设计以院落为概念，通过四合院来布置建筑空间，增加空间的流动性及开放性，与香山买卖街的院落融为一体。

煤厂街位于设计地块的北部，是去往香山碧云寺的必经之路，街道两侧遍布着居住小院、小卖部和佛教主题的商铺等。街道较窄，坡度较大，有一种曲径通幽的感觉。街区设计本着将煤厂街打造为静谧、禅意的“茶禅一味”的生活街巷空间，结合雨水花园，改造居住院落。

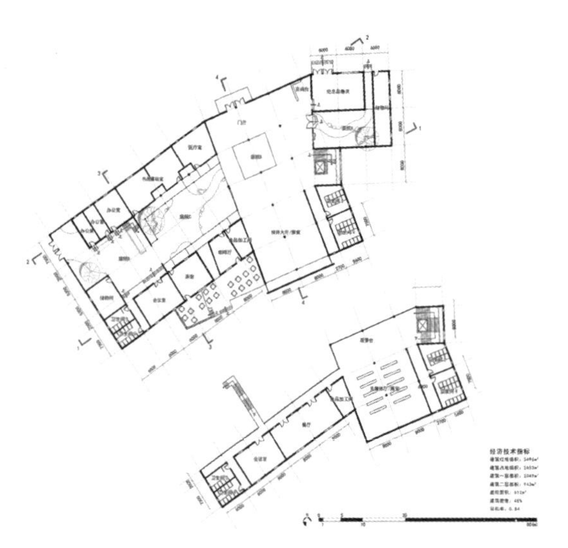

一层、二层平面图
1st and 2nd Floor Plans

鸟瞰图
Aerial View

效果图
Rendering

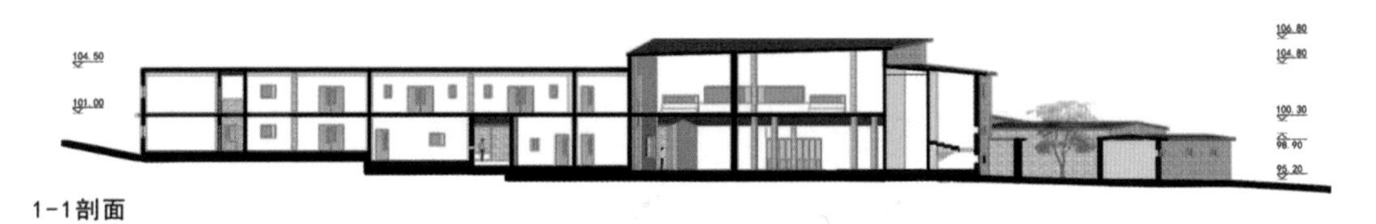

剖面图
Section

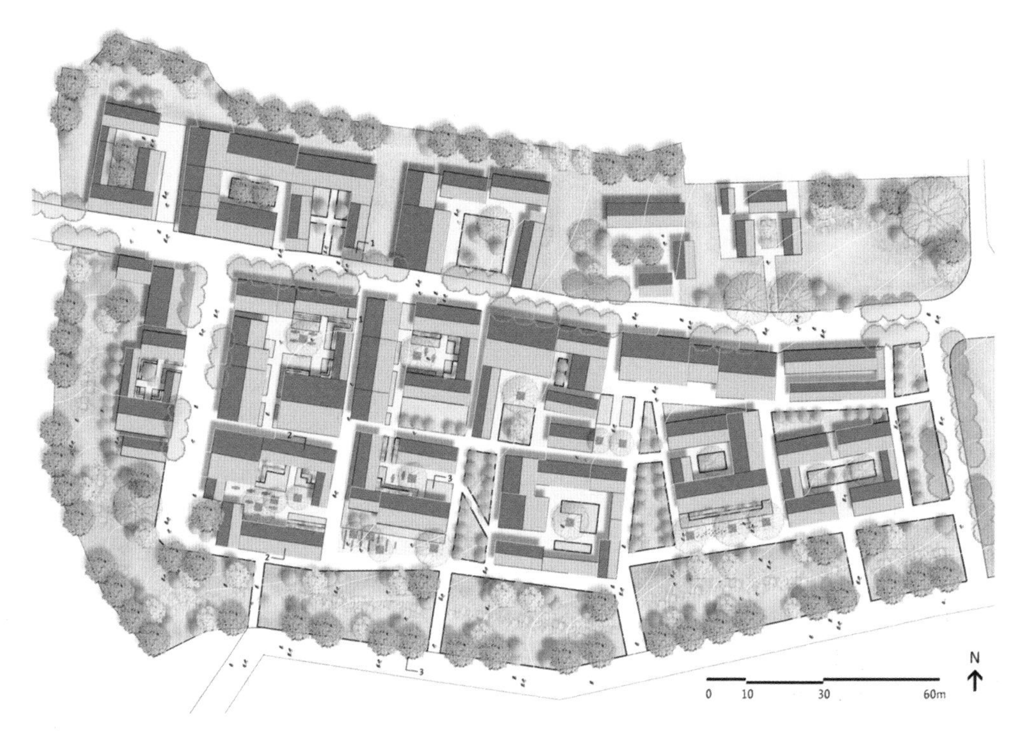

煤厂街街区节点设计平面图
Detailed Site Design 4

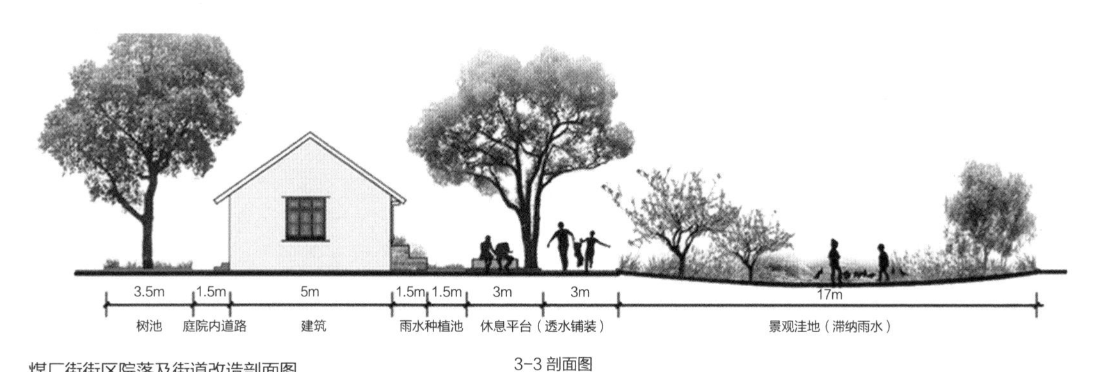

煤厂街街区院落及街道改造剖面图
Section of Yards and Streets in the Coal Street Block

3-3 剖面图

煤厂街及买卖街街道雨水花园改造工程措施剖透视
Perspective of Rainwater Recycle

规划方案三 及 安河桥北片区概念设计

Plan 3 & The Conceptual Design of the Northern Section of Anhe Bridge

李婉仪、葛韵宇、张芬、胡盛劼、李雅祺、赵欢、邓力文
Li Wanyi/Ge Yunyu/Zhang Fen/Hu Shengjie/Li Yaqi/Zhao Huan/Deng Liwen

本规划方案中，将整个绿道体系分为生态型、文化型、都市型三种类别，生态型绿道与山林游步道相结合，文化型绿道与历史遗产廊道相结合，都市型绿道与街道绿化、街头绿地相结合设计。

安河桥北片区位于五环高架桥、清河排污渠、京密引水渠、地铁 4 号线地上联络线和山体之间。设计中加强了本地块与周边地块的步行连接便利度，设置了较多公共活动场地和污水生物净化景观系统，营造一个融合了城市公园和郊野绿地特征的绿色开放空间。

In this plan, we divided the whole greenway system into three categories: the ecotype, the cultural type, and the urban type. The ecological greenway is combined with the mountain forest trails, the cultural greenway is integrated with the historical heritage corridor, and then the urban greenway together with the street greening is united in wedlock.

Our site is located among the fifth ring viaduct, Qing River drainage canal, Jingmi diversion canal, Subway Line 4 and the mountains. We have promoted the convenience of walking connection the local blocks and the surrounding blocks, set up more public activity sites and sewage biological purification landscape system, and built a green open space that integrates the characteristics of urban parks and country green spaces.

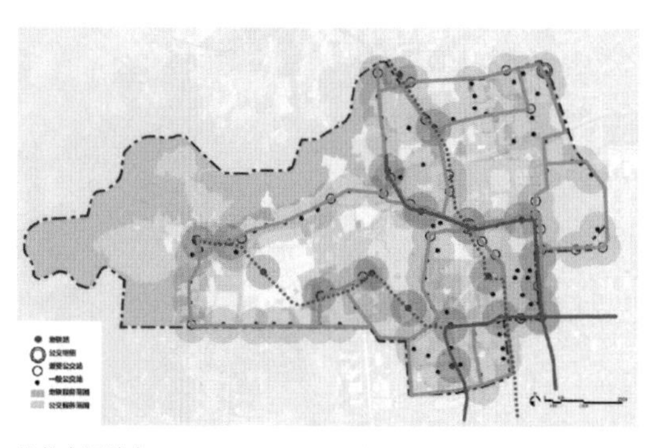

公共交通站点
Hubs of Public Traffic

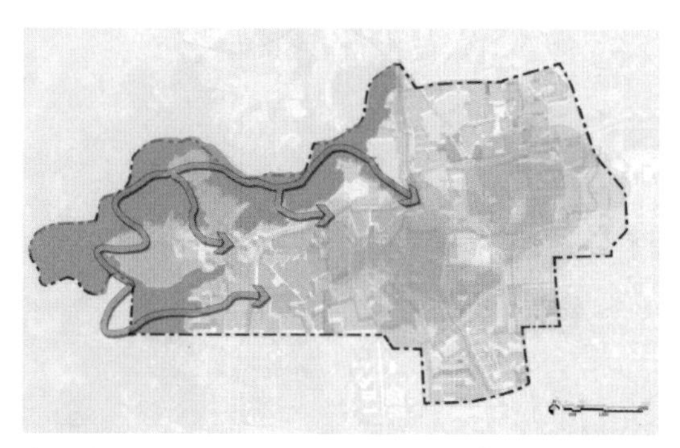

生态型绿道结构
Structure of Ecological-oriented Green Ways

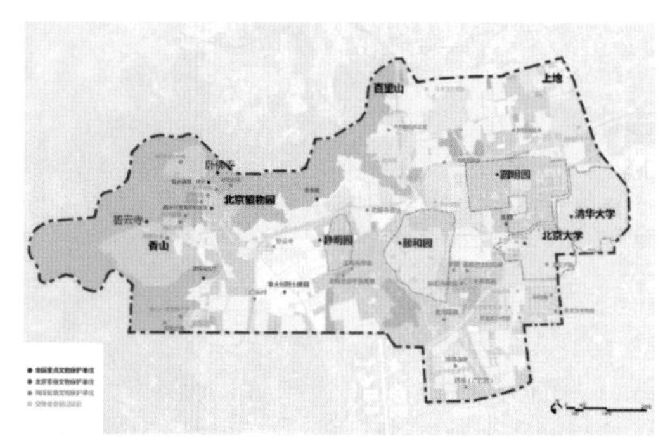

文物保护单位
Heritages

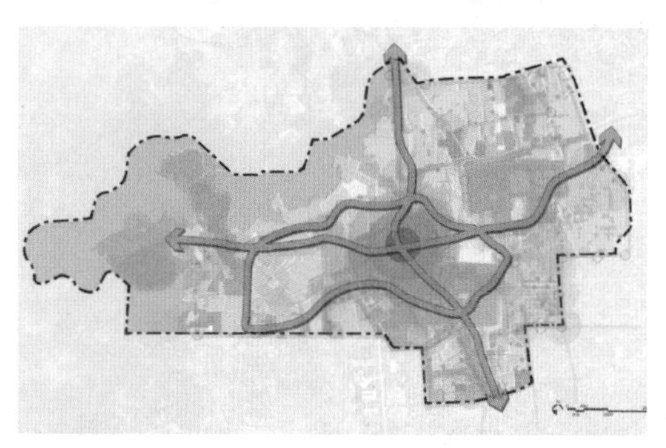

文化型绿道结构
Structure of Cultural-oriented Green Ways

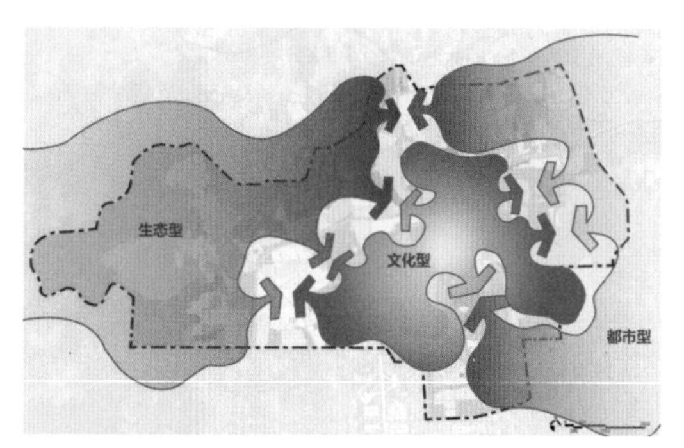

绿道分区分类
Categories of Green Ways

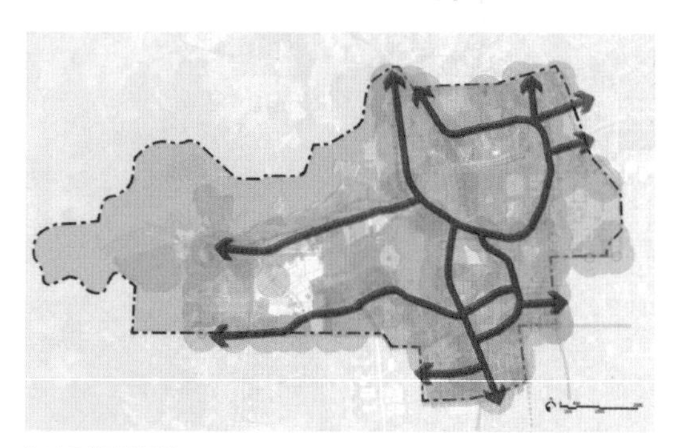

都市型绿道结构
Structure of Urban-oriented Green Ways

根据前期调研，得出三山五园地区有以下特色：绿地、水系数量众多、面积大，生态环境好，然而绿地组团割裂、连通性不足，景观不连续，市民使用不便。文化底蕴深厚，物质文化遗产保护良好，但呈散点式保护，缺乏整体文化氛围。非物质文化资源丰富，但无展示和传承的专属场所。道路交通便利，但快速路的割裂削弱了场地连通性。

根据以上特色，提出了将三山五园绿道建设为文化型、生态型、都市型三种类型，在空间层面整合空间、绿色串联；景观层面以绿色为基础、因地制宜；文化层面整理脉络、继承文化；功能层面以人为本、共创共享。提出了打造以西山为基底、以历史为线索、以人文为核心的绿道体系，突出文化脉络，整合区域绿地，共筑和谐人文空间。

绿道网络规划选线从以下四个方面进行，从而得出绿道因子选线叠加结果。首先是现状要素调查，从生态和人文两个方面对现状及上位规划进行调研；然后进行要素评价指标体系构建，对规划的生态型、都市型和文化型绿道附以量化评价指标，基于地理信息 3S 技术和区域空间发展理论，通过现场调查、

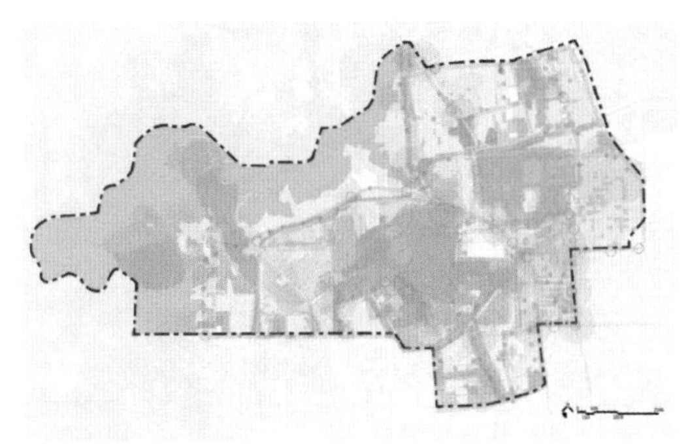

生态型绿道因子叠加
Superposition of Ecology Type

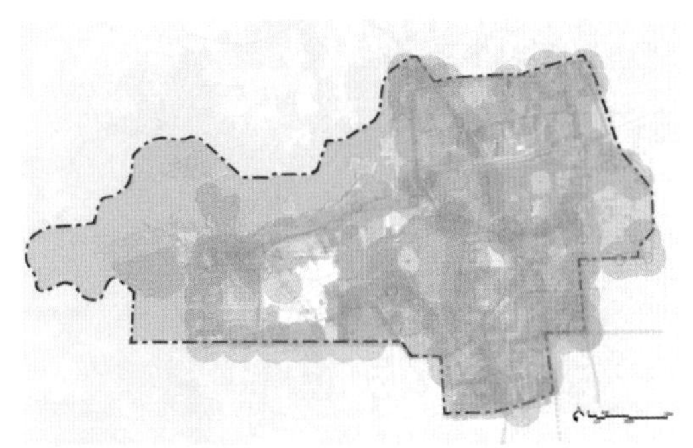

文化型绿道因子叠加
Superposition of Cultural Type

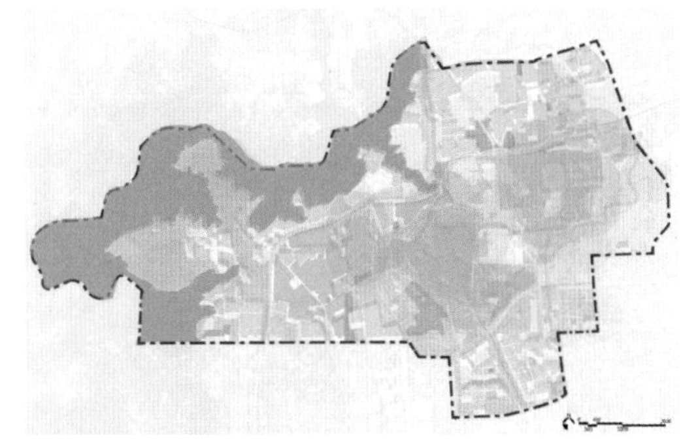

都市型绿道因子叠加
Superposition of Urban Type

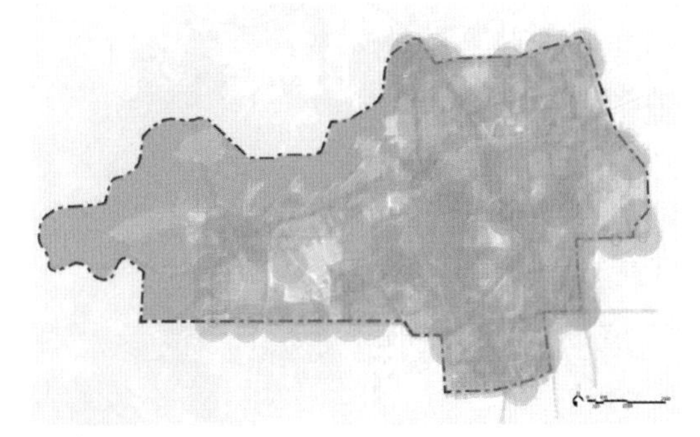

因子叠加结果
Factor Superposition Result

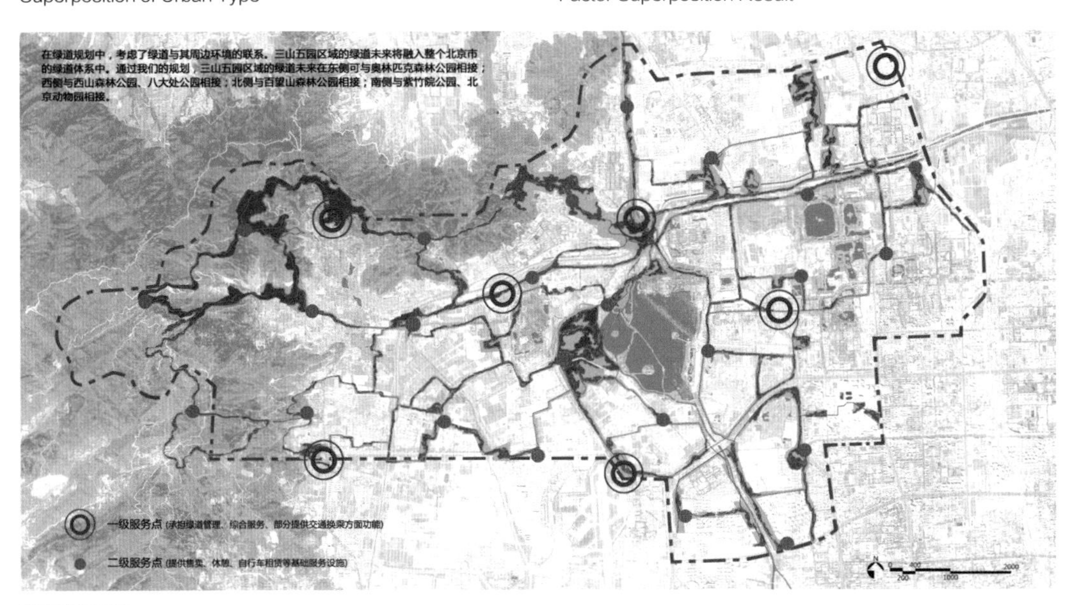

规划总平面图
Master Plan

案例研究、专家打分，建立指标体系；其次是绿道要素评价，采用 AHP 层次分析法进行单要素评价，然后使用 GIS 将单要素加权，进行综合评价，以综合评价结果确定需要连接的景观斑块；最后进行布局选择，对景观斑块进行连接。同时，通过网络结构比选的方法对各个绿道选线进行优化，直至选择

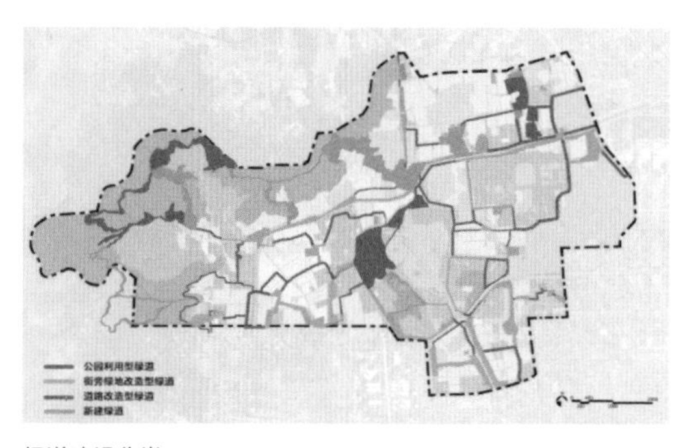

绿道建设分类
Categories of Green Ways

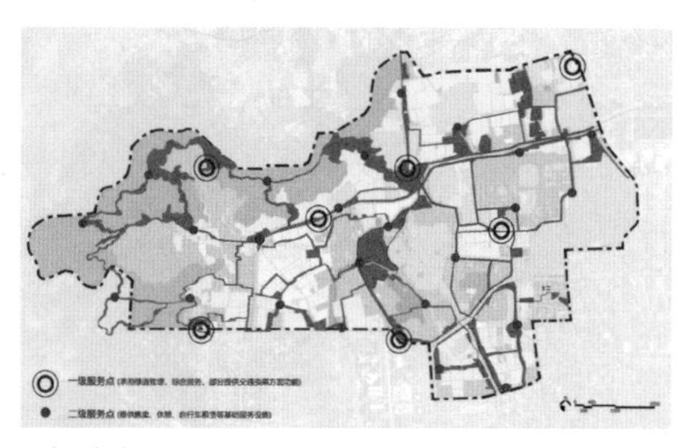

配套服务点规划
Service Facilities Plan

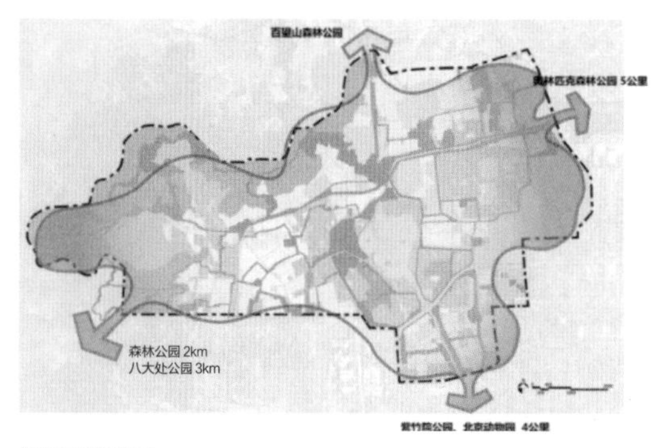

绿道对外联系
Connections with Outsiders

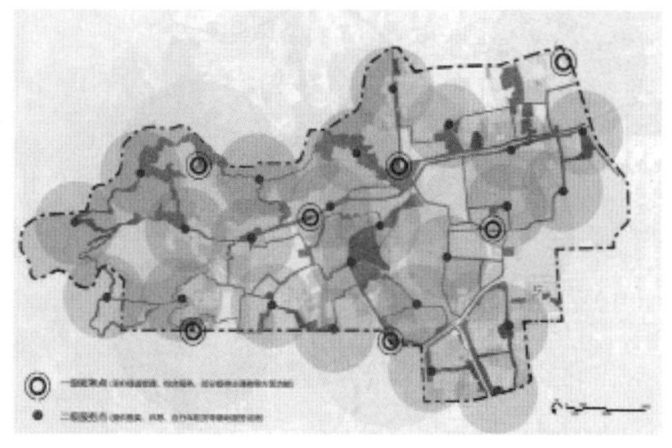

服务范围覆盖
Coverage of Service Facilities

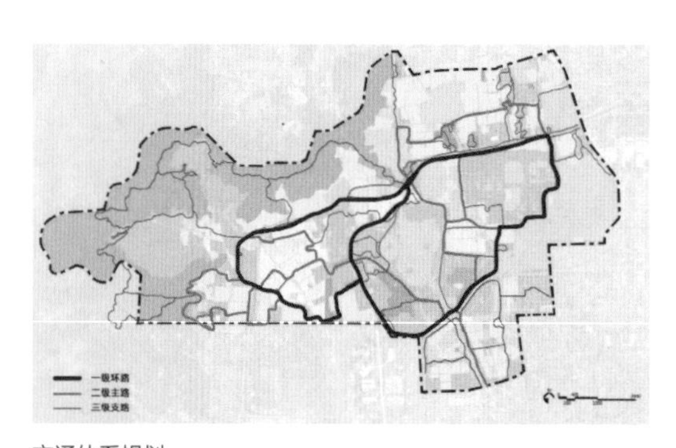

交通体系规划
Traffic System Plan

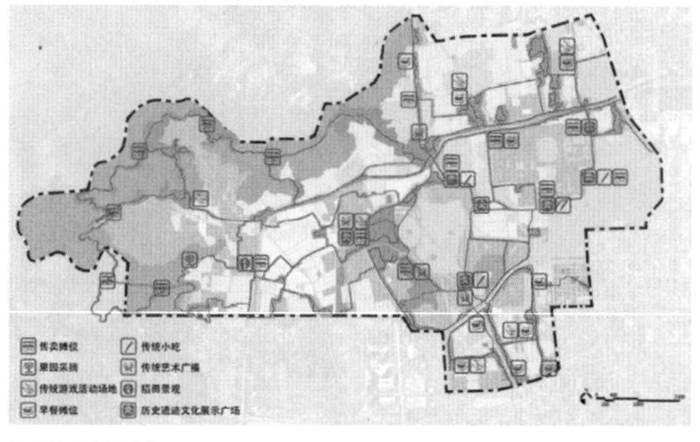

绿道活动规划
Activities Plan

出符合目标的最佳选线方案。

通过对三种绿道选线因子叠加的结果，结合实际情况，将选线落实，得出总平面图。绿道规划中，还具体考虑了绿道与其周边环境的联系。规划从用地类型、现状及规划用地性质、与周边大规模绿地组团的关系等各个方面综合对绿道选线进行综合评估。其中，规划绿道分成三种类型：生态型绿道、文化型绿道、都市型绿道。生态型绿道中自行车道与游步道分离，自行车道分布在地势较为平缓的地区，游步道可以建设在山林中。文化型绿道建议与周边遗产保护相结合,把文化遗迹纳入绿道体系中。都市型绿道建议与居住区绿地、街旁绿地结合灵活设计，寻求可增加的绿地空间。

规划中将绿道交通体系分成三个级别：一级主路成环，贯穿整个区域；二级次路局部成环，串联文化型、生态型、都市型三大区域；三级支

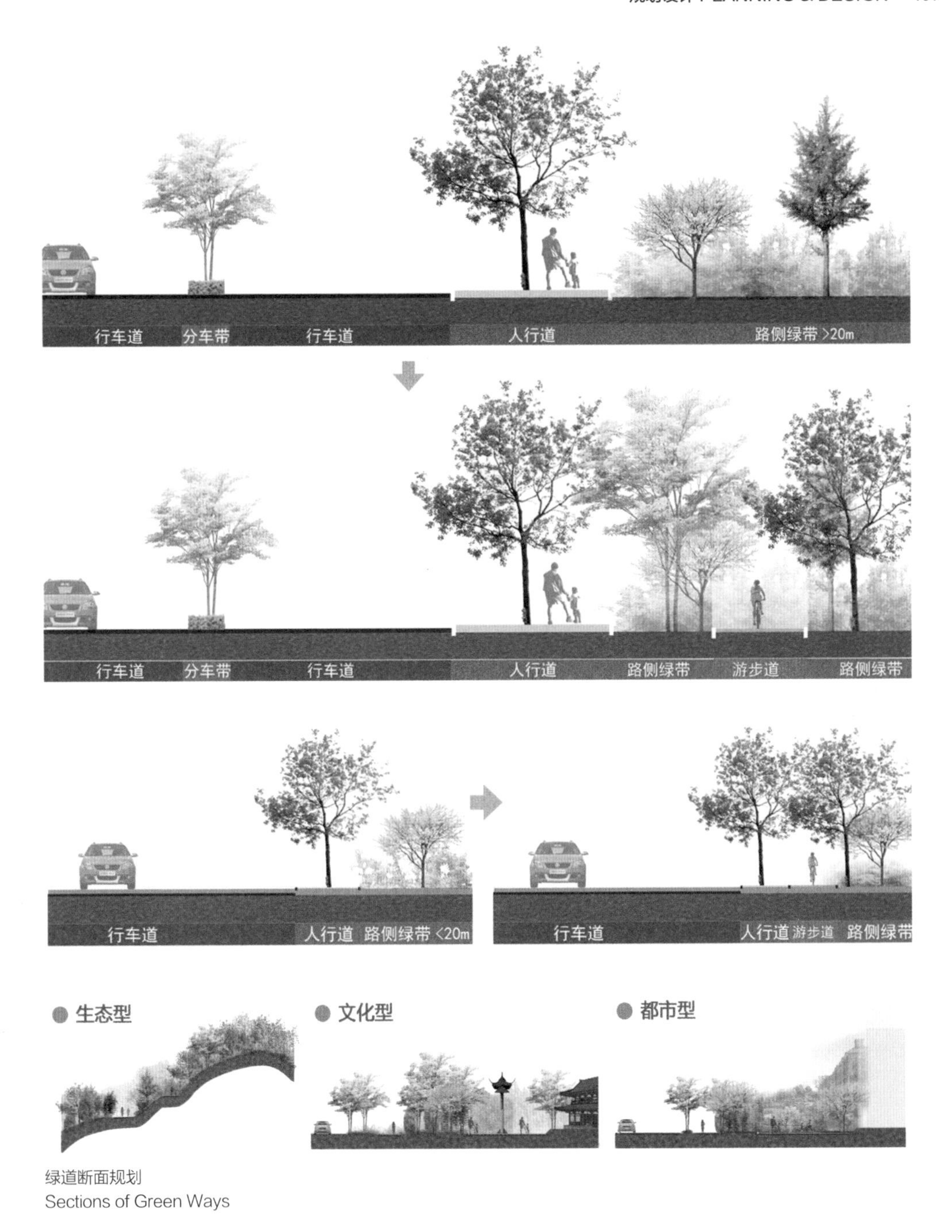

绿道断面规划
Sections of Green Ways

路成网状全面覆盖整个区域。慢行道建设模式根据使用者不同分为步行道、自行车道、综合慢行道。规划将各类活动融入绿道体系中，丰富市民使用绿道的体验。主要沿绿道布置一些历史文化遗迹展示广场、游戏活动场地、果园采摘、早餐摊位等活动内容。

在绿道体系中共设置了 7 个一级服务点，它们主要承担绿道的管理、综合服务、部分提供交通换乘的功能；二级服务点均匀散布在整个区域内，主要承担售卖、休憩、自行车租赁等功能。一级服务点的服务半径为 2000m；二级服务点的服务半径为 1000m。

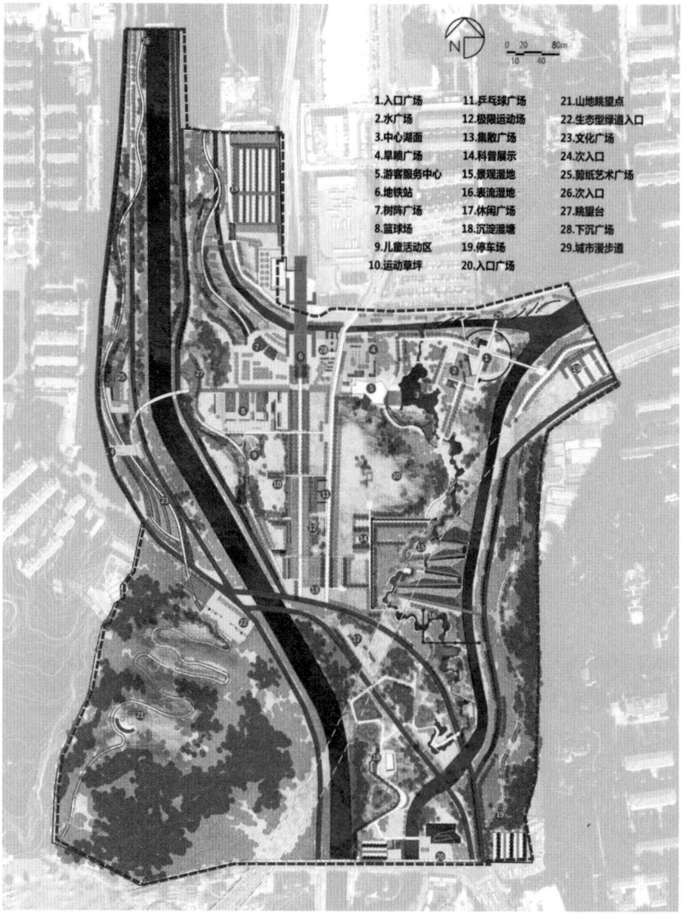

安河桥片区设计总平面图
Site Plan

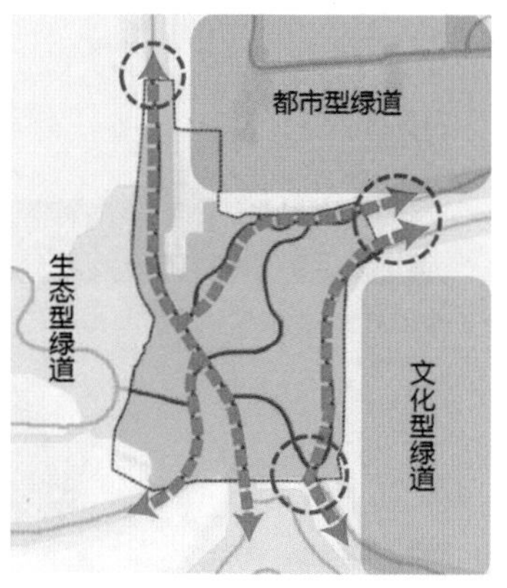

前期规划
Planning

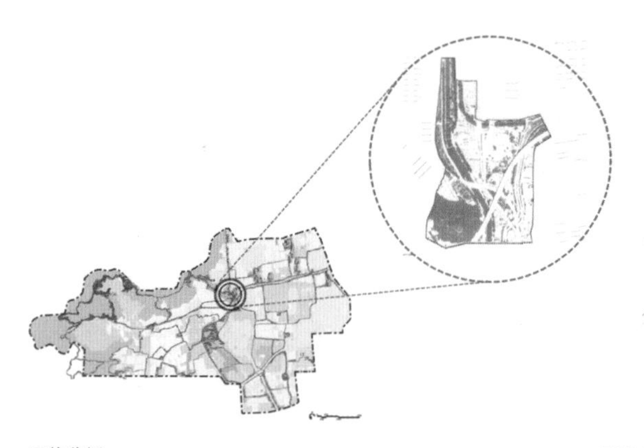

区位分析
Site Plan

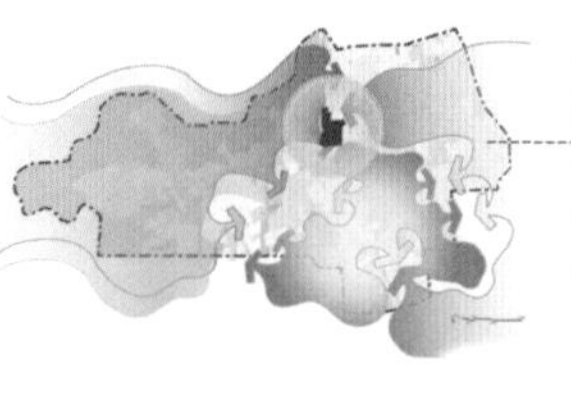

- 在前期规划中本区域是唯一一个**三种类型绿道**交汇的地方
- 生态型绿道的**起始点**，同时衔接都市型、文化型绿道
- 紧邻两条水系，附近较多交通站点，市民可以方便进出绿道
- 现状**割裂问题**突出，亟待解决

区位与构思
Locations and Design Ideas

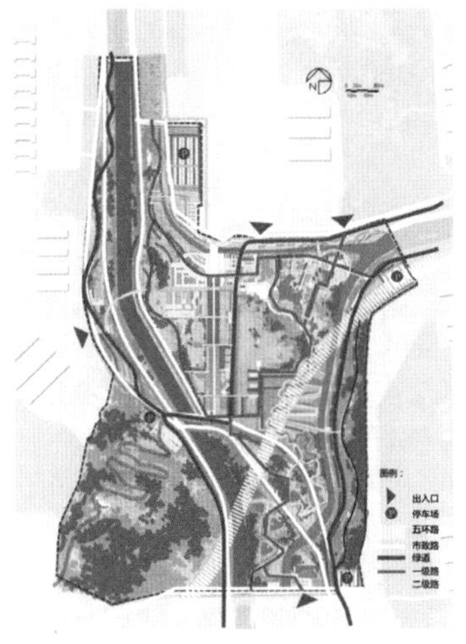

道路分析

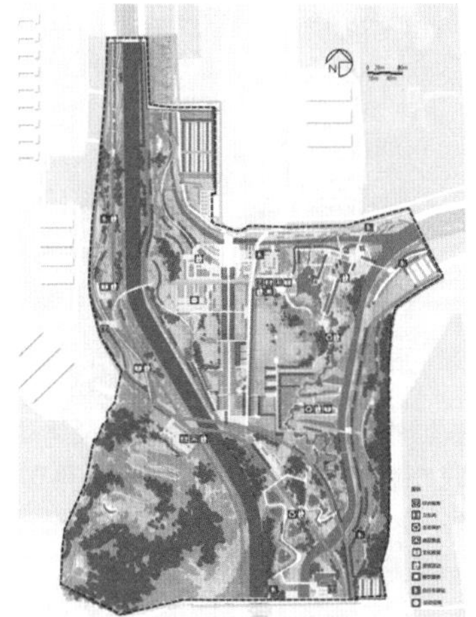

服务设施

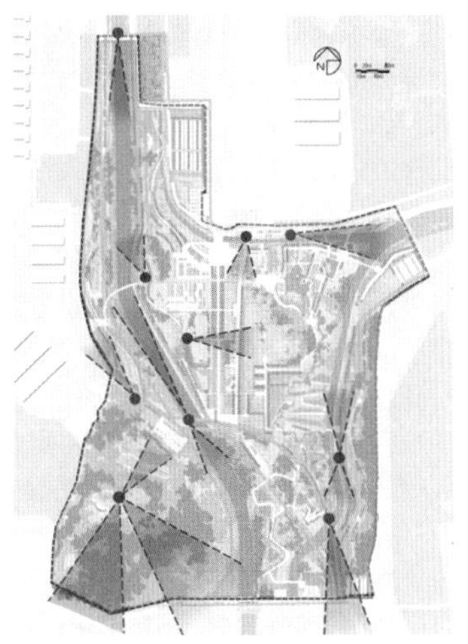

视线分析

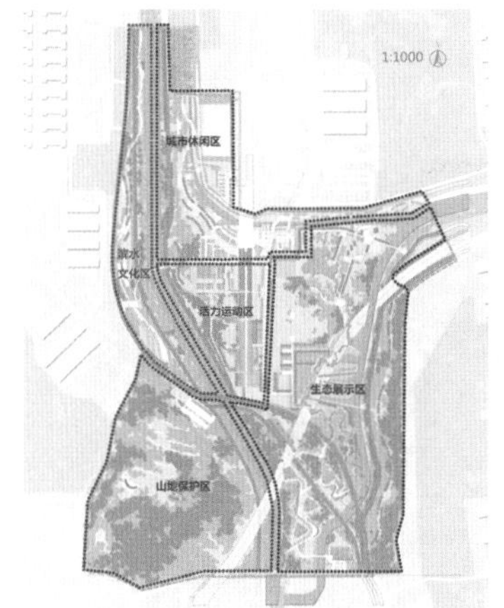

功能分区

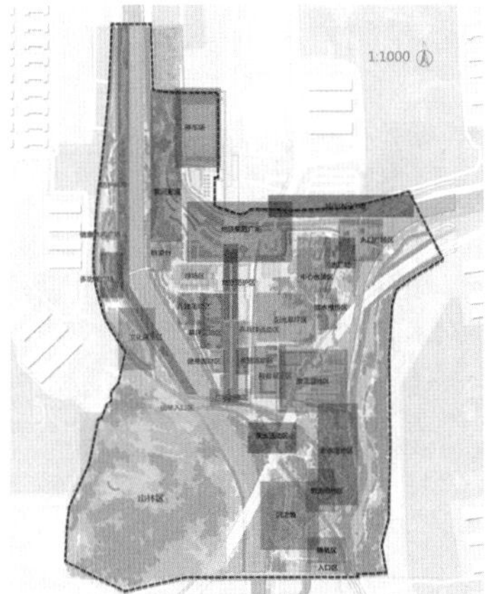

具体分区

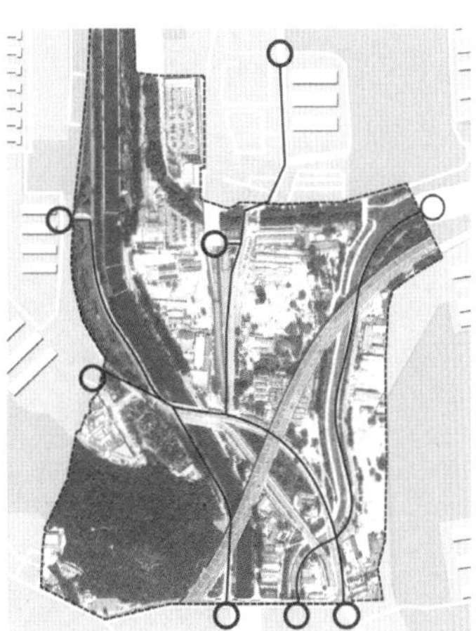

交通现状分析

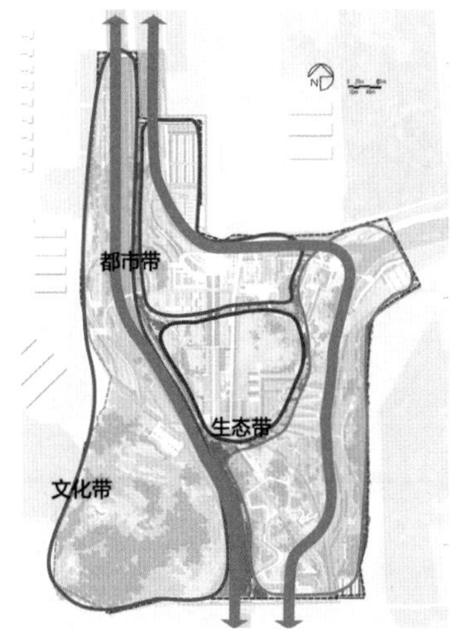

景观结构

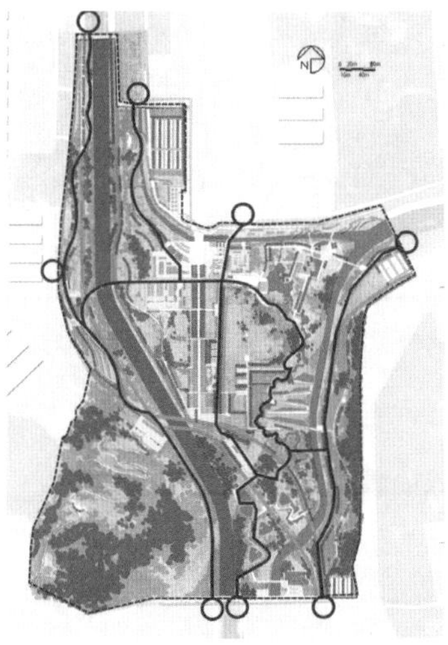

交通分析

各专项分析

Specialized Analysis

节点设计 1——滨水文化区：

场地位于京密引水渠和清河夹合成的三角地段内，西侧与国防大学、平顶山衔接；东侧紧邻中央党校；场地南侧为颐和园，北侧为中高端居住区。

在规划中本地段是唯一一个三种类型绿道交汇的节点地段，是生态型绿道的起始点，同时衔接都市型、文化型绿道；地段紧邻两条水系，附近有较多的公共交通站点，市民可以方便进出绿道；现状空间割裂问题突出，亟待解决。

在规划中，场地作为三种类型绿道的衔接点，包括三条主要的绿道以及三个主要的对外连接入口。

通过综合性绿地来融汇三山五园的三种类型绿道；同时连接现状割裂的地块，组织周边生态斑块。

地段内传统文化元素丰富，可以分为物质文化遗产和非物质文化遗产。其中物质文化遗产并不在设计范围内分布，但是周边的资源特别丰富，可以作为场地内的可借之景。上位规划中规定该区域为非物质文化展示区，因此，该区域内部，我们选择了来自民间文学、传统音乐、传统舞蹈、传统戏曲、曲艺、传统医药、民俗、传统杂技与经济、传统美术作为展示对象。

节点设计 2——活力运动区：

场地现状为建筑已经基本拆除的未利用地，地势平坦，相对场地外围高程较低。现状有少量生长情况良好的乔木，缺乏地被及灌木。设计针对场地周边的地铁站及较高规模的居住区，概念定位为一个集生活及运动为一体的片区。场地内设有足球场、篮球场、乒乓球场、漫步道、康体器械区、阳光草坪、林荫广场等功能多样的区域以满足周边居民的使用需求。

场地功能分区分为：眺望台、篮球场区、儿童活动区、草坪运动区、滨河景观带、健身器材区、健康跑步道、广场休憩区、乒乓球运动区、极限运动区。

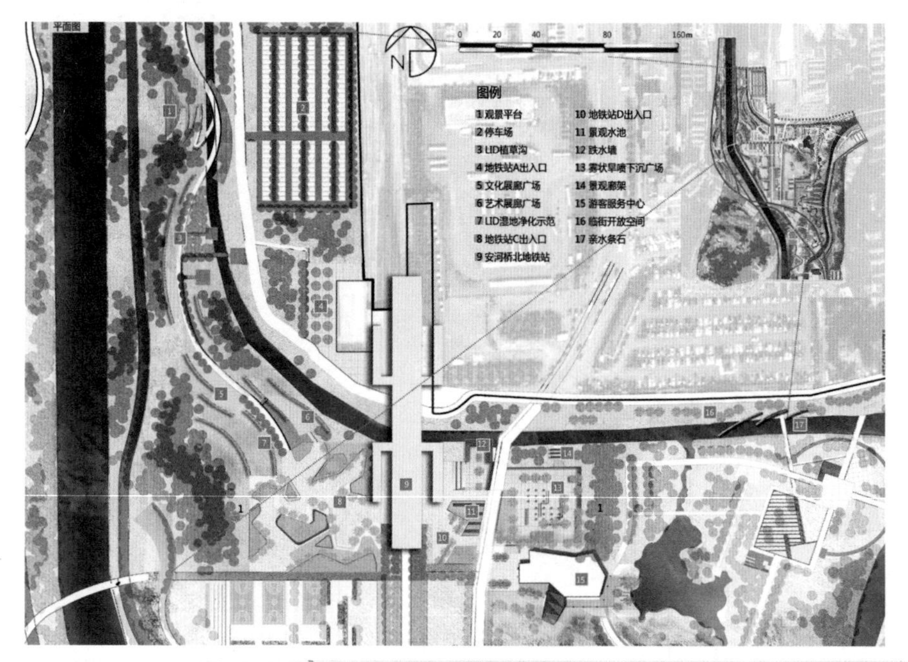

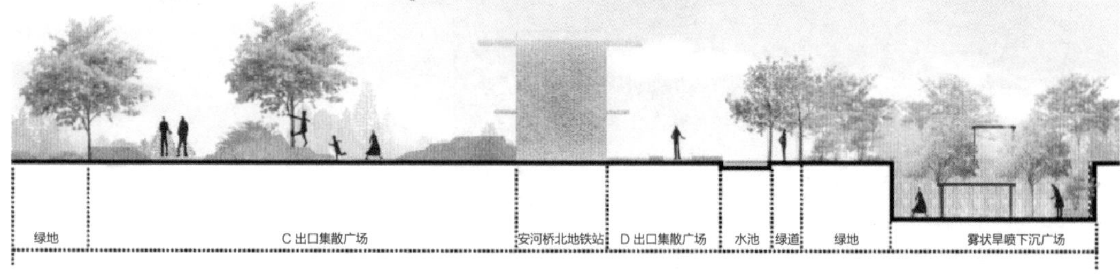

节点设计 3
Detailed Design3

节点设计 3
Detailed Design3

节点设计 3——城市休闲区：

场地周边设置植草沟收集雨水，广场主要采用透水铺装以利雨水渗透，同时周边设置下沉式绿地，收集的雨水通过渗管流入集水罐，集水罐收集的水可以用于驿站的公共卫生间。

路侧还设置湿生植物净化群落展示的区域，一侧场地作为雨水转输设施集中展示的场所，生态滤层由具有净化功能的植物组成，进行第一层进化，然后过滤到吸附层的沸石层，再到吸附层的粉煤灰层，最后到碎石层。碎石层下设有防渗层，可以使储存并净化过的雨水流入雨水管储存利用。

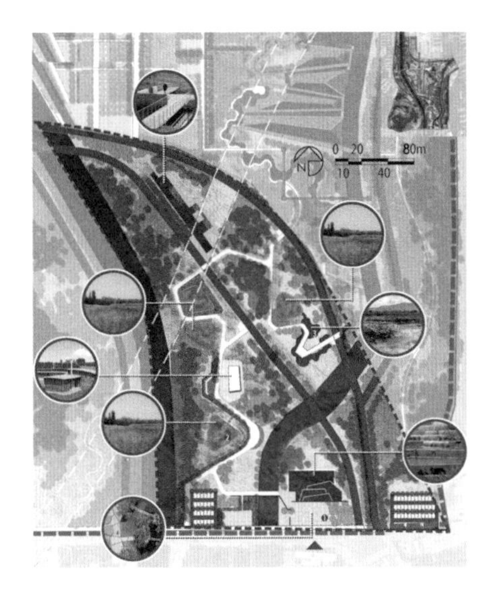

节点设计 4：

该节点位于地块的东南角，是设计地块与城市衔接的过渡段，也是清河水体净化展示的源头，它主要承担的水体净化功能包括：人工曝氧池、潜流净化塘、表流净化区。

当水从场地南侧进入场地后，通过暗管抽取一部分，使用机械泵压到人工曝氧池中，通过层层跌落，增加水与空气的接触面积，与此同时，在每层的曝氧池池底放有生态滤层及碎石层，来进行初步沉淀。之后通过暗管把水引入潜流净化塘，通过水生植物拦截等作用降低流速，更进一步净化水体。最后通过暗管将水导入表流净化区，实现进一步净化。

在此主要进行的是水体前期净化，包括：人工曝氧池、潜流净化塘、表流净化区。人工曝氧池变成多层叠水池，形成声音景观来吸引游人进入场地。潜流净化塘变为水生植物观赏区，并设置了多处观赏平台，可在观花期供游人观赏。而表流净化区则变为蜿蜒的溪流景观，可供游人欣赏水鸟、水生植物。

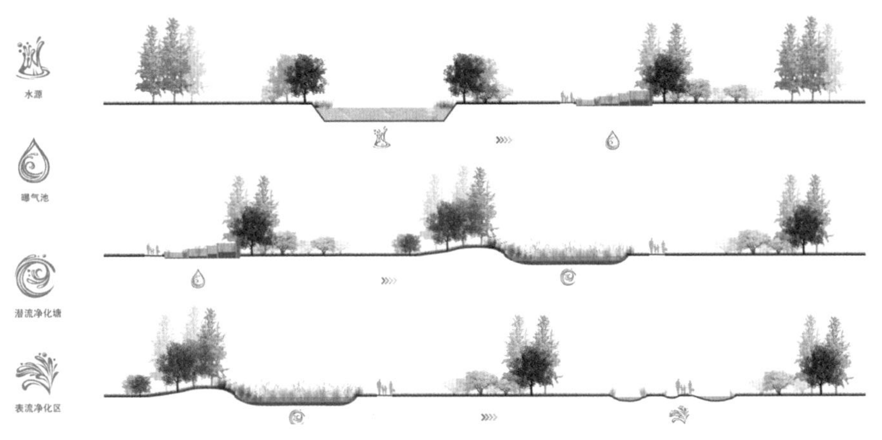

节点设计 4
Detailed Design4

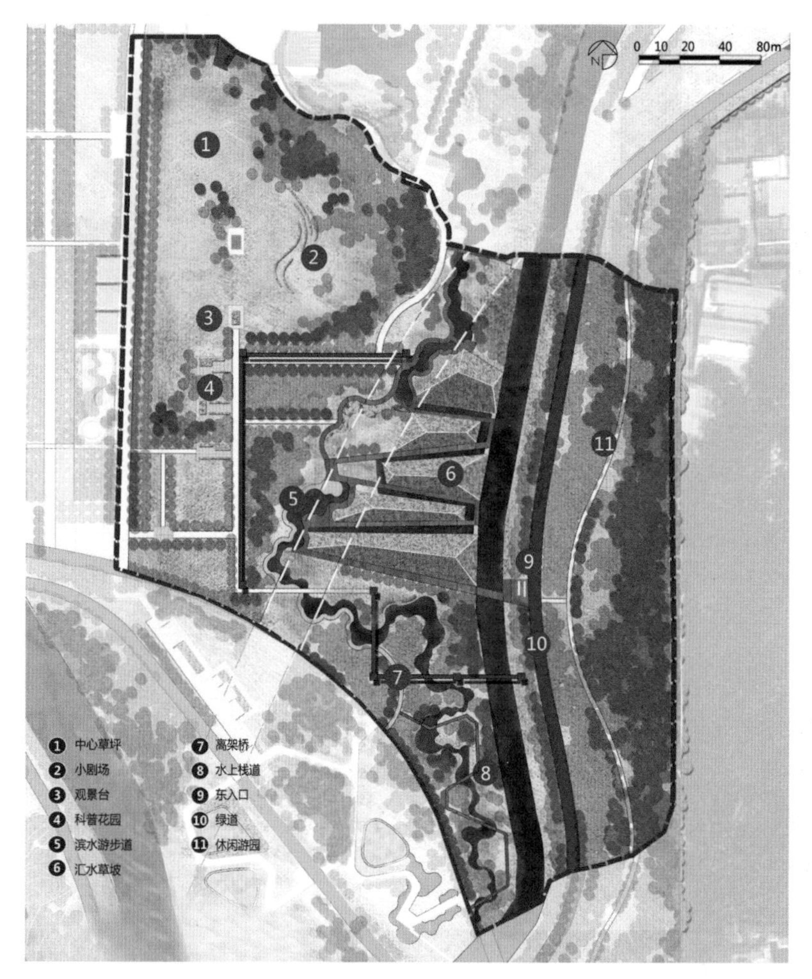

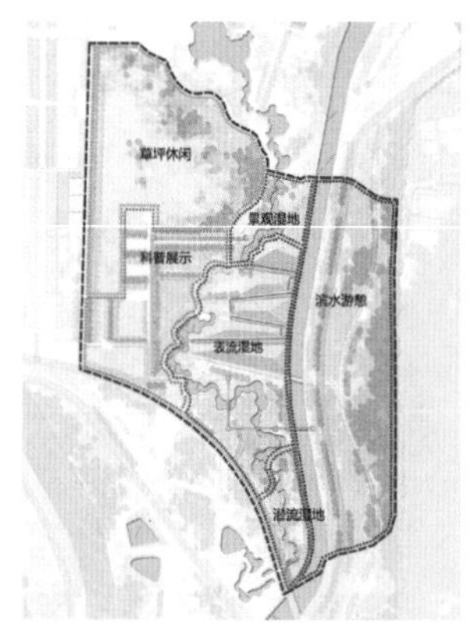

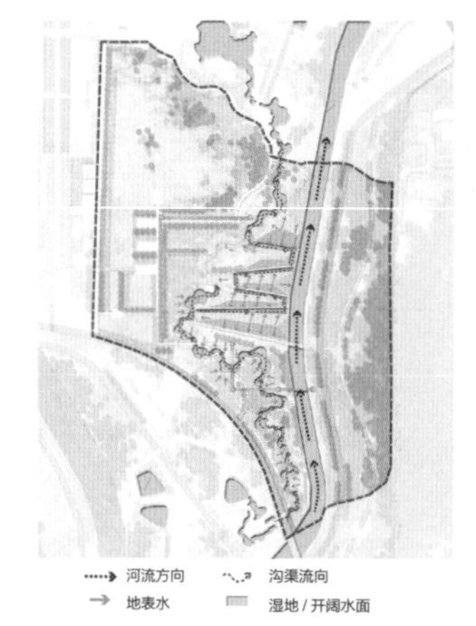

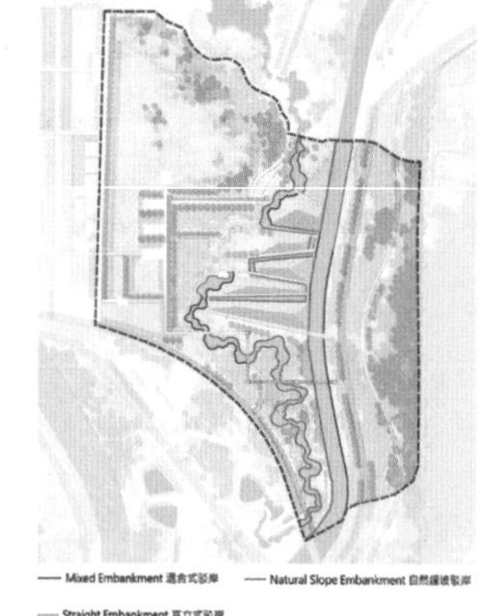

节点设计 5
Detailed Design5

节点设计 5：

该节点位于场地东侧，属于东部生态展示区的表流湿地部分，以植物净化为主体，兼顾科普展示和雨水收集功能。总面积约 10.4hm^2，以一条内河湿地贯穿其中，承担净化功能，净化后作为公园内的景观用水。

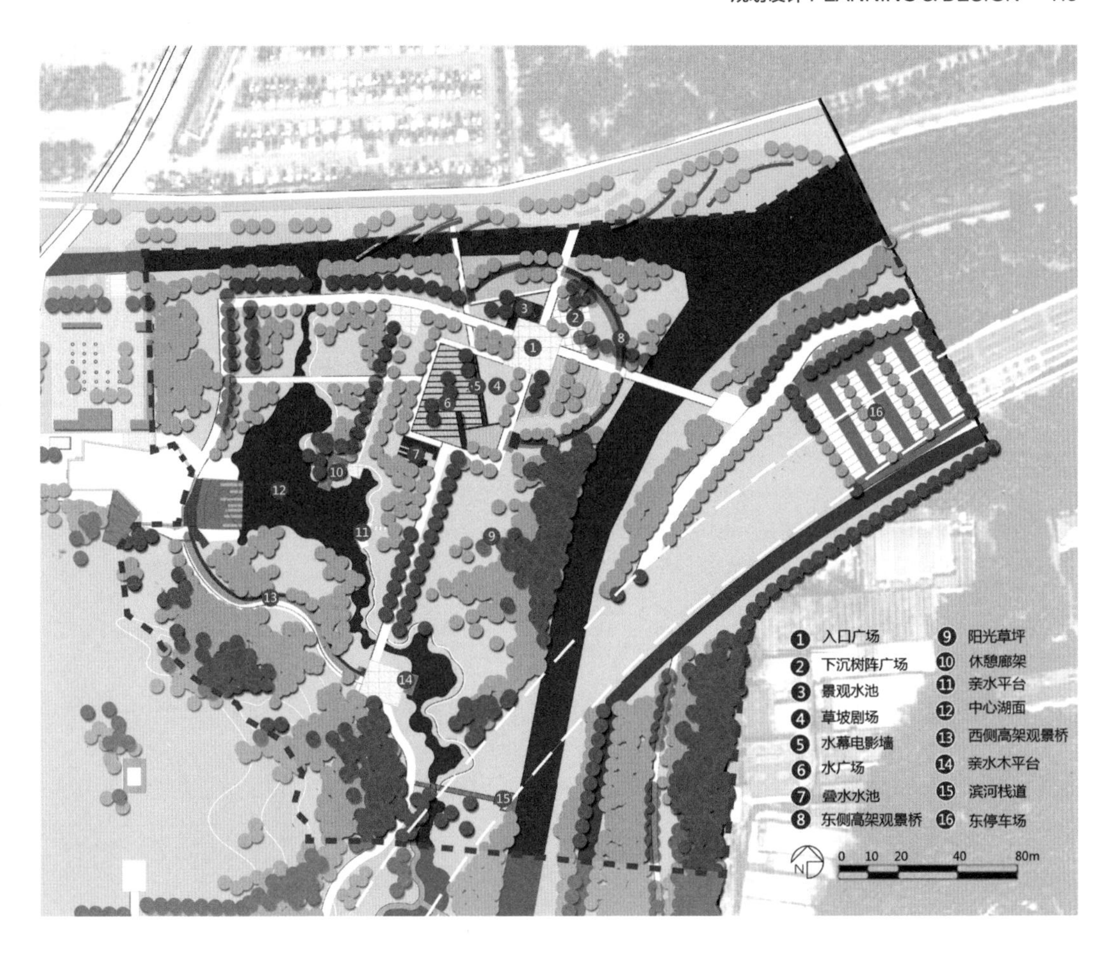

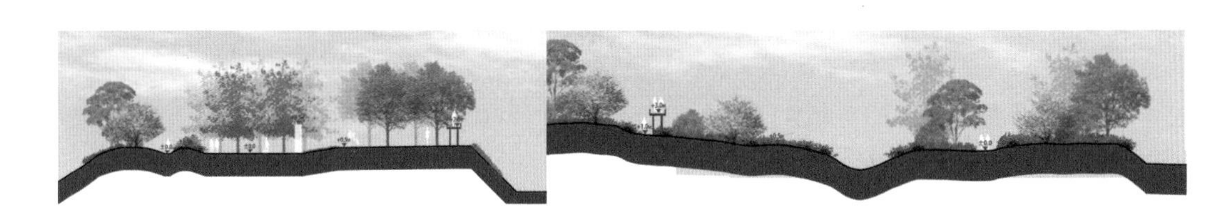

节点设计 6
Detailed Design6`

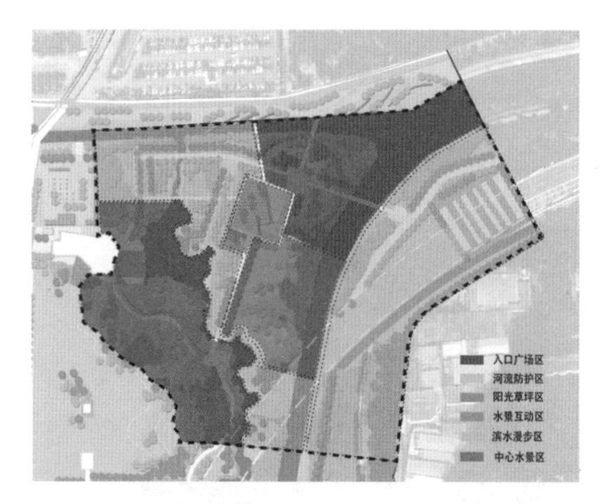

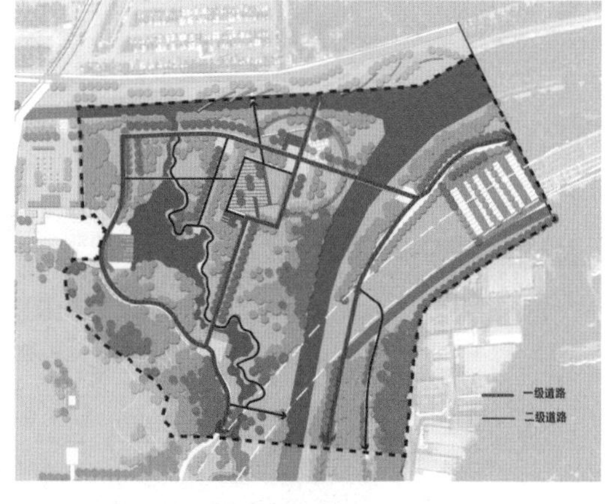

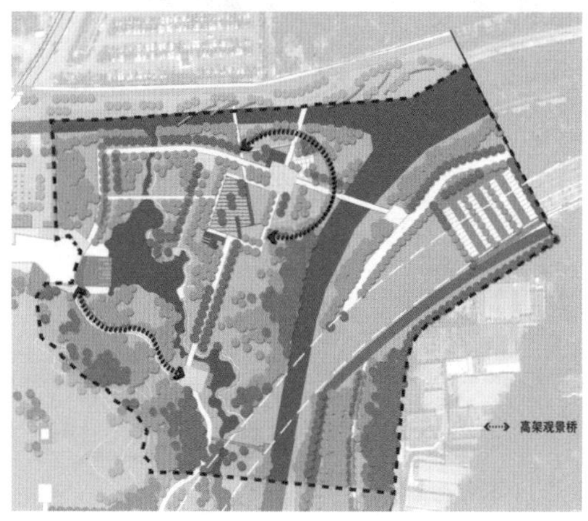

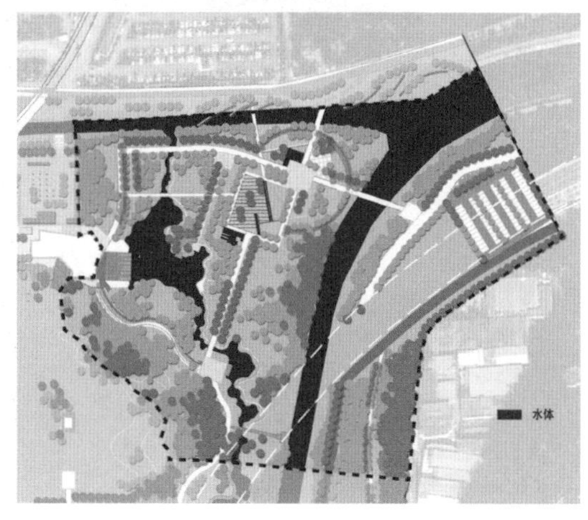

节点设计 6
Detailed Design6`

节点设计 6：

本节点位于整个绿道的东北角，总面积约 5.9hm^2。该地块是整个节点最大的一个出入口，是向游客展示三山五园独特文化的门户展示区，也是上位规划中生态展示区内的“景观湿地区”，此外它还承担着集散大量人流的功能。因此，在此地块的设计中，主要利用生态展示区中“表流湿地区”和“潜流湿地区”净化的水打造水景，其中包括开阔的湖面、景观水池、音乐喷泉广场、叠水幕墙、游戏水池、生态透水材料展示等与水景相关的元素，既为游人提供气氛活跃的水广场活动空间，又为游人创造安静舒适的滨水步行空间。

在上位规划中，绿道节点共有五大分区。其中，本地块位于东侧的“生态展示区”内。东侧“生态展示区”共有 3 个序列，从南向北依次是表流湿地区—潜流湿地区—景观湿地区。本地块属于景观湿地区。

场地东侧与北侧被清河包围。在场地的内部，通过表流湿地区与潜流湿地区净化的水汇集成湖面，形成了场地的中心水景区域。

此外，在场地内的一些硬质广场上设计了水幕墙、游戏水池、生态水槽等，向游人展示净化水的多种利用方式。

在场地内一共布置了两处高架观景桥，一处位于主入口广场区域，另一处位于中心湖面西侧。从东侧观景桥，向东可以眺望清河景观，向西可以俯瞰全园，是整个园区内观景视线最好的区域。在西侧观景桥的西侧以地形结合密林作为背景，主要引导游人观赏东侧的中心水景区域。

高架观景桥可以为游人提供不一样的观景体验。

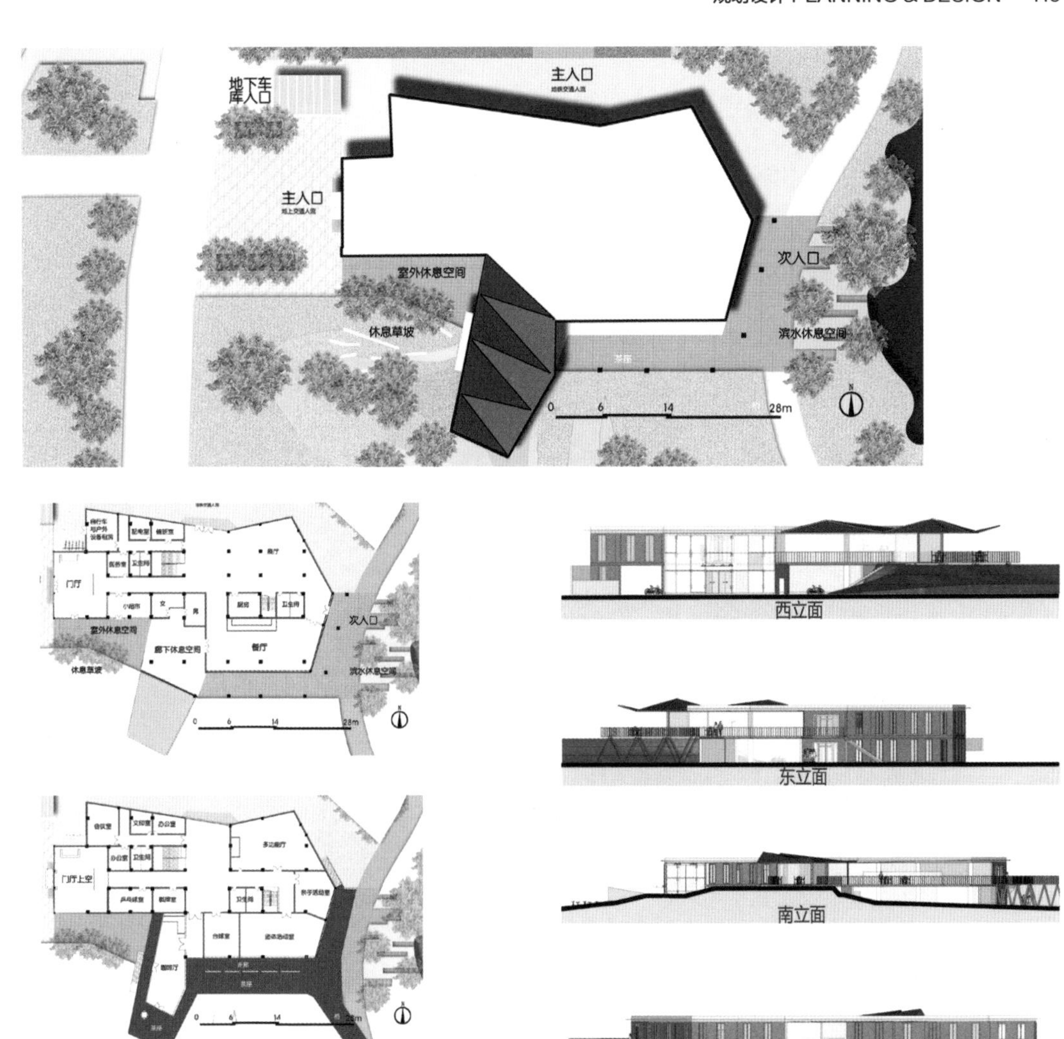

建筑设计
Architectural Design

建筑设计：

该驿站为一级驿站，是三山五园东南部的最大驿站，承载了交通接驳、综合服务与管理功能。与其他 6 个一级驿站一起，覆盖整个绿道的服务范围。该驿站位于西山脚下，临近居民区，附近有圆明园与颐和园等著名的旅游景点，游人数量较多。

该服务性建筑的选址，是整个设计地块范围内的心脏位置。20m 内为地铁枢纽，且为终点站，客流量较大。30m 内有大型的安河桥北公交枢纽。四周为居民区与科教单位，该建筑服务的人群比较复杂，也比较广泛。首先要满足公交与地铁的接驳，其次要满足周边居民的休闲活动。还需要满足绿道管理的要求。

建筑内部的功能布置，西侧为门厅，主要承担游客服务咨询、物品寄存的服务，门厅南侧是自行车租赁与户外运动设备租赁，同时配有库房，满足绿道健身游人的基本需求。南侧结合地形做开敞的休息空间。

二层南侧架空的咖啡厅，为游人提供较为安静的休闲场所，并与园中的桥相结合，做了整个的二层架空平台，休闲的同时，可以有良好的视线，与一层滨水空间一起，上下错落，既满足游人需求，更能享受与自然融合的美景。

游人流线与服务流线基本分离，主流线主要连接建筑出入口与上下楼梯，并与建筑外部相接，串联起建筑内部的所有功能；支流线通向各个功能分区室内。

规划方案四 及 青龙桥－玉东片区概念设计

Plan 4 & The Conceptual Design of the Qing Long Bridge-Yudong Section

刘加维、吴明豪、王茜、刘童、张琦雅、夏甜、韩冰
Liu Jiawei/Wu Minghao/Wang Xi/Liu Tong/Zhang Qiya/Xia Tian/Han Bing

本规划方案中，除考虑到绿道的连贯性和景观效果外，还着力打造一个雨洪管理手段完善的绿色基础设施系统，同时也考虑营造了一系列生物栖息地和迁徙廊道。

青龙桥－玉东片区既包括了颐和园的北侧游客出入口，也包括了青龙桥商业古镇遗址和“京西稻”的重要原产地。设计中不仅打造了古镇风貌的相关旅游和商业服务设施，也恢复了京西稻田风貌并增设了游憩设施，丰富了万寿山和玉泉山之间的景观格局。

In this case, in addition to considering the continuity and the landscape effects of greenway, we also put forth effort to build a green infrastructure system with perfect rain and flood management measures, and constructed a series of biological habitats and migration corridors at the same time.

The Qing Long Bridge-Yudong section covers not only the north exit of the Summer Palace, but also the site of Qing Long Bridge commercial ancient town and the important source area of Jingxi Rice. In the design, we not only built related tourism and business service facilities of ancient town style, but also restored the style and features of the rice field in the western district of Beijing and set up more recreational facilities, which helped to enrich the landscape pattern between Wanshou Hill and Yuquan Mountain.

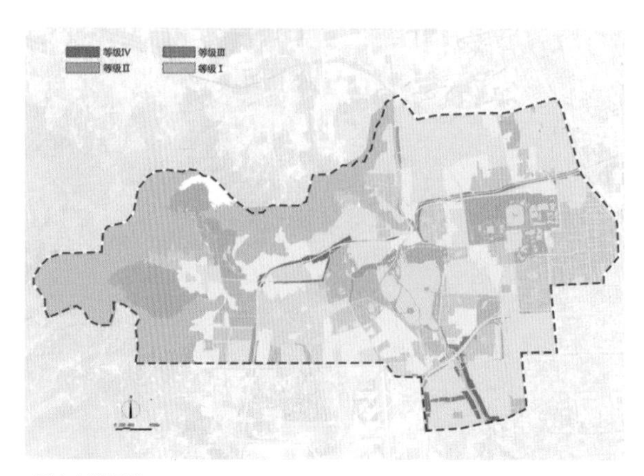

绿地廊道评价
Evaluation of Existing Green Corridors

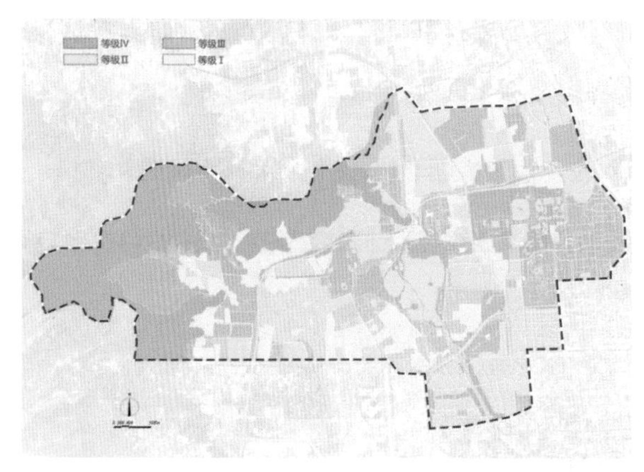

绿地基质评价
Evaluation of Green Matrix

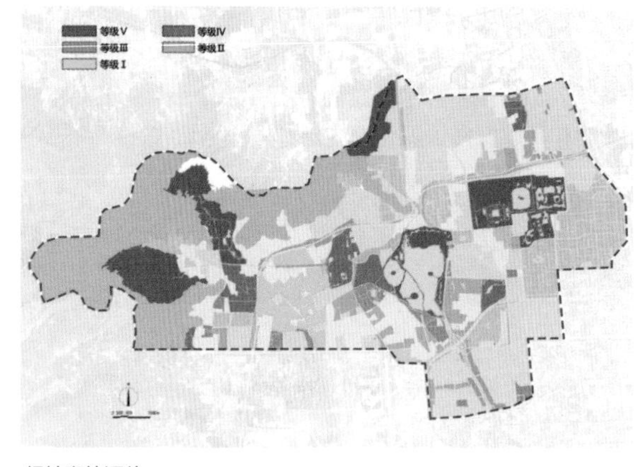

绿地斑块评价
Evaluation of Existing Green Patches

欧美城市绿道的建设经历了百余年的探索，19 世纪后期，公园道和绿化带融合建设的绿道仅单纯提供休闲功能，发展至今，绿道越来越要求多目标，多功能价值的实现。绿道功能的演变是伴随着人们对于其认识的加深以及赋予各种使用诉求而不断变化发展的，从最初的注重景观功能的林荫大道发展到注重绿地生态网络功能的综合绿地系统，不仅强调绿地空间的连接性，而且具备一定规模的线性绿道才能发挥其应有的生态、游憩、社会和景观价值。根据上位规划，三山五园地区连接西山与中心城区，整体可看作城市与山林的过渡地带，也连接了北京市绿地系统的西山片区，尤其是北京城市西北方向的两条楔形绿地。绿道连接海淀区东西向皇家园林廊道及南北向水系景观廊道，形成东西向历史文化绿道与南北向生态水景绿道，成为三山五园片区绿道骨架；同时串联各历史名胜，田园山林城市河道等共同组成绿道网络。本次规划采用模式化与针对性相结合、文化遗产的保护与功能辐射相结合、生态恢复与拓展相结合的策略，重构水系网络，构建兼具文化教育、生态涵养、休闲游憩等多种功能于一体的综合型绿道。

运用 AHP 法，对场地的文化、生态、游憩三个方面分别进行评价，通过将生态、游憩、文化三个评价结果进行加权叠加，将评价等级高以及较高的地块纳入三山五园绿道选线范围，形成了初步的绿道选线。选线包括了三山五园以及大部分公园、滨河绿地、街旁绿地、田园、林地等；结合现有的基础设施，分级设点、适度补充，串联散布的文化遗产，拓展游憩空间，使得绿道的交通与城市公共交通及静态交通空间进行对接，进一步与城市空间相融合。

水系统规划通过对现状水系与西山片区汇水分析叠加，结合规划绿道调整梳理，连通该片区水系脉络，追寻历史皇家水乡田园记忆；交通衔接系统则通过开辟绿道出入口、完善停车设施配套、设置交通换乘点等措施，提高绿道的可达性，同时根据节点功能的侧重点不同，分为文化、生态、游憩三类，进行分级规划设计；绿道特别对该地段的游线进行了整理，一条为历史文化游览路线，串联香山、北京植物园、静明园（可远眺）、颐和园、畅春园遗址，最终到达圆明园。依照历史上的慈禧水道，增设一条水路游览线路，从颐和园码头上船，经由昆明湖到达昆玉河，一路向南。通过绿道的

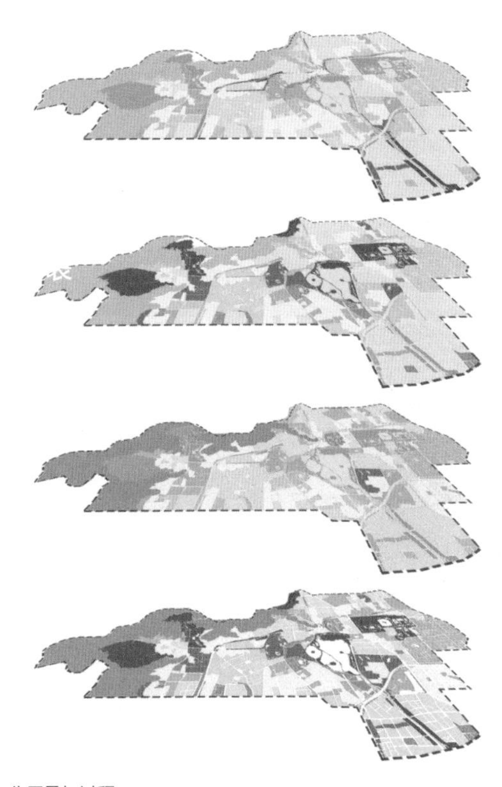

分层叠加过程
Process of Overlay

建设，以三山五园地区文化为主要脉络，串联和梳理该地区原有的重要面状文化遗产区，同时形成绿色基础设施网络，系统体现该地区的生态功能，为城市和郊区环境作串联，同时为外来游客和本地居民提供休闲游憩的良好环境，优化三山五园地区的交通出行环境。

绿地廊道评价：通过层次分析法对绿地廊道的类型、生态稳定性和自然条件进行加权分析，对三山五园地区的绿地廊道进行评价。

绿地斑块评价：评价的内容细分为斑块面积、斑块与基质结合度、服务半径、开放程度、绿化覆盖率、水域面积、群落层次结构、物种多样性、适生树种比例、林地空气污染减少率、负氧离子浓缩增加率、降温效应指数、环境容量、绿地空间与郁闭度、景观多样性、绿地多样性、公园绿地景观破碎度、对自然条件的利用状况等。

绿地基质评价：通过层次分析法对绿地基质的自然（第一自然）、母体（城市建成）两方面进行加权分析，对三山五园地区的绿地廊道进行评价。

分层叠加过程：将廊道、斑块与基质评价结果 1∶1∶1进行叠加，得到区域内地块的生态性评价。这一评价结果将与后面的历史文化评价和休闲游憩评价进行加权叠加分析，最终得到初步的绿道选线。

绿地生态评价
Ecological Evaluation of Existing Green Spaces

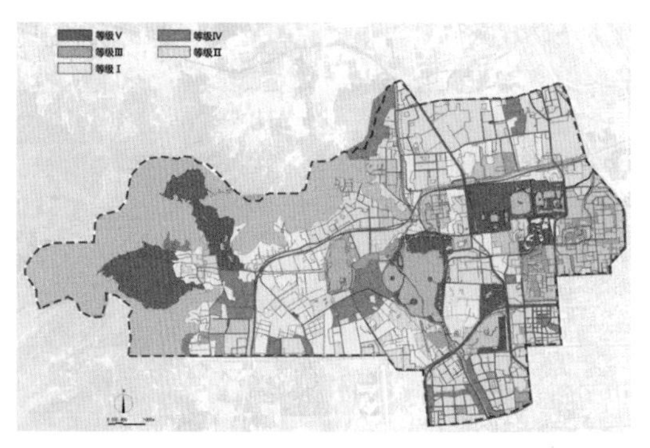

休闲游憩评价
Recreation Evaluation

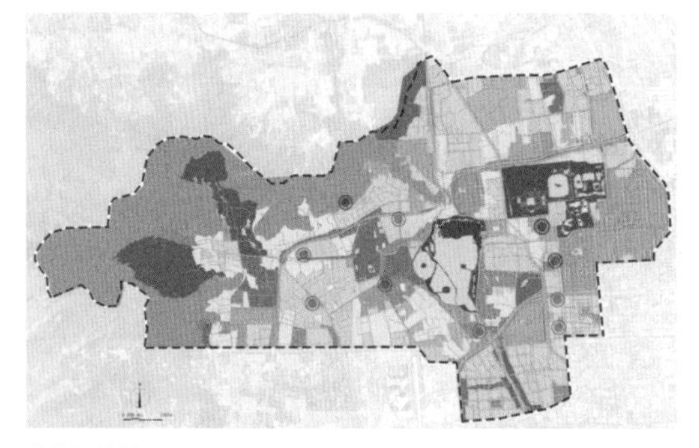

整体评价结果
General Evaluation Results

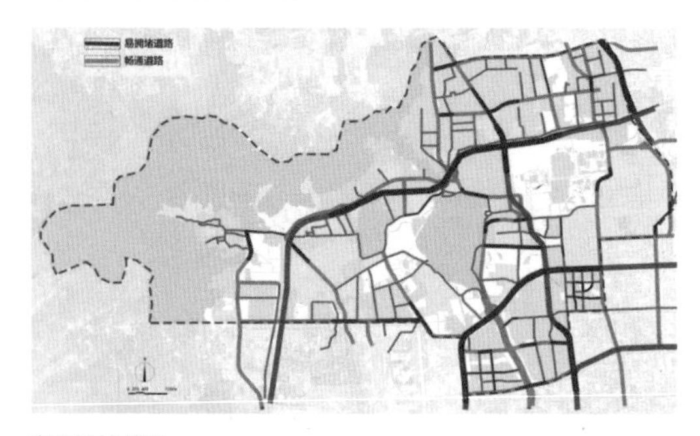

交通拥堵状况
Traffic Congestion Situation

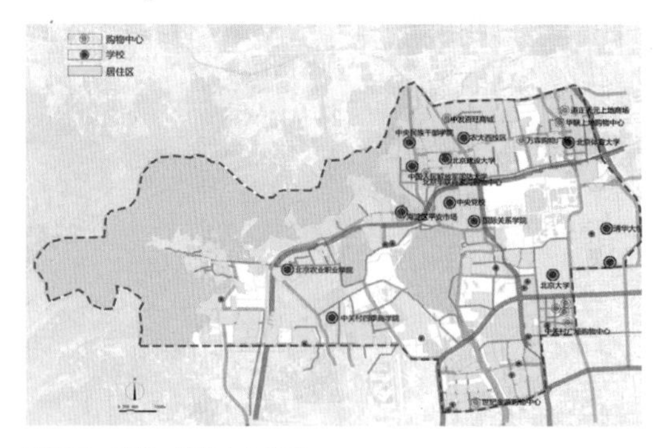

居住区、学校、购物中心分布
Distribution of Residential Areas, Schools, Shopping Centers

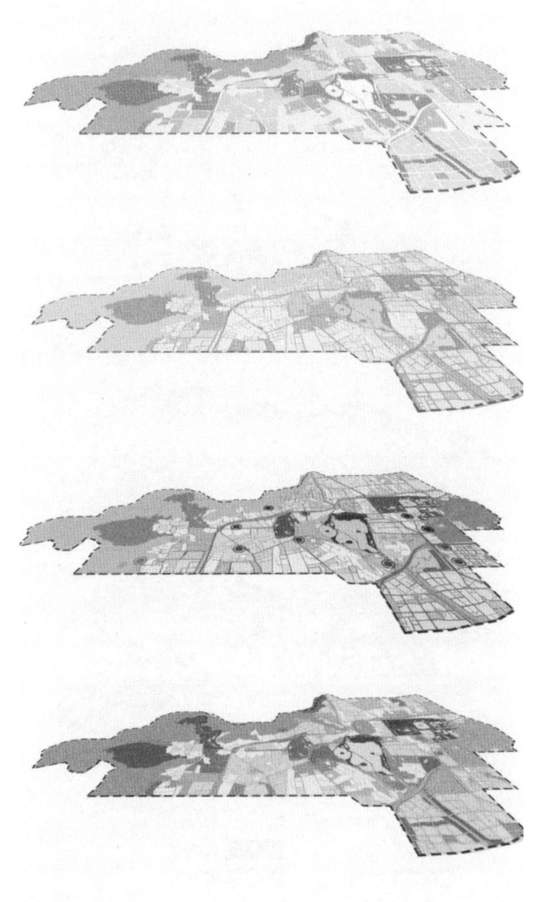

分层叠加过程
Process of Overlay

历史文化评价：通过层次分析法对绿地的历史文化进行评价。

休闲游憩评价：通过层次分析法对绿地中游憩地、游憩廊道进行评价。评价指标有区位条件、空间适游度、设施完备度、环境舒适度、景观优美度、服务质量、连接度、观赏价值、交通条件等。

分层叠加过程：将生态、休闲游憩与历史文化评价结果以 1 ∶ 1 ∶ 2 进行加权叠加，得到对区域内绿地的整体评价。

整体评价结果：图中颜色越深的区域表示该区域作为绿道的适宜度越高，颜色越浅则表示该区域作为绿道的适宜度越低。通过此图初步得出三山五园绿道选线。

绿道初步选线：通过将生态、游憩、文化三个评价结果进行加权叠加，将其评价等级高以及较高的地块纳入三山五园绿道选线范围，形成了初步的

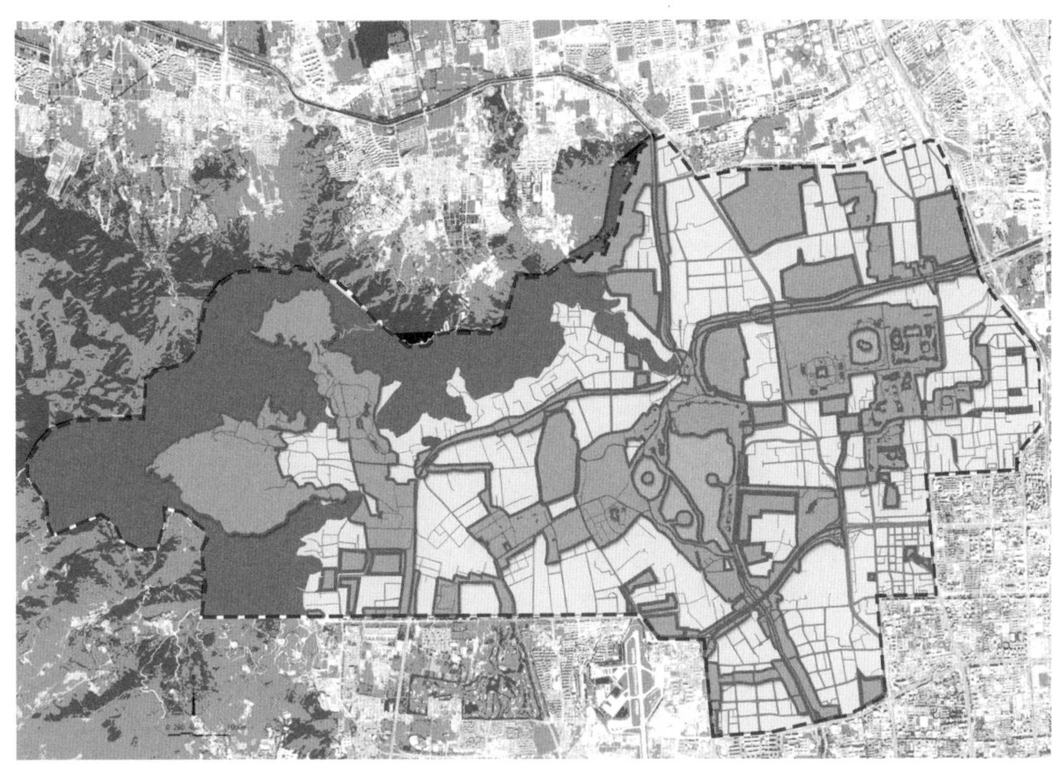

绿道初步选线
Preliminary Greenway Selection

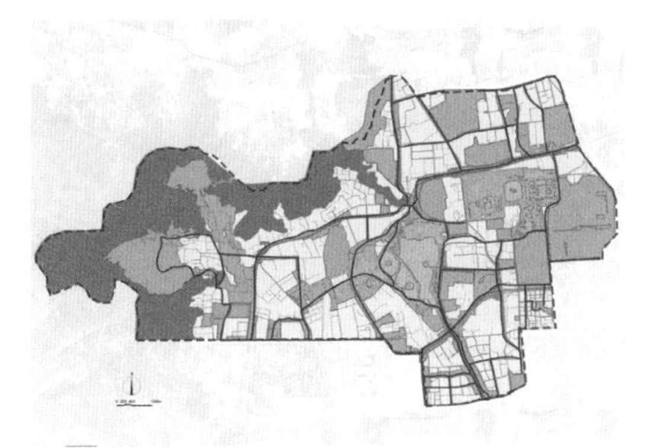

选线调整结果
Adjustment Result of the Greenway

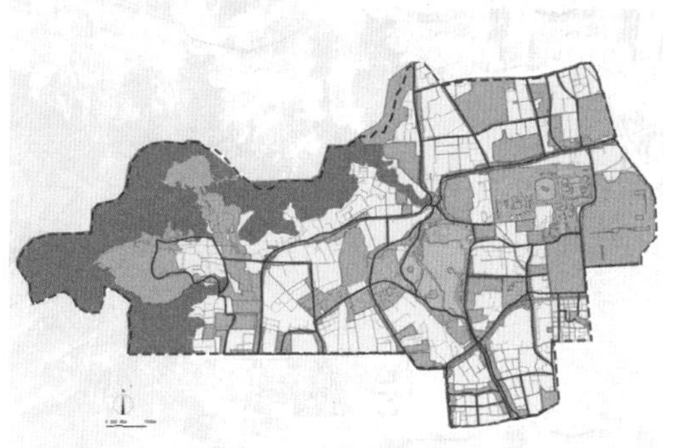

可行性调整
Feasibility Adjustment

绿道选线。选线包括了三山五园以及大部分公园、滨河绿地、街旁绿地、田园、林地等。

交通拥堵状况：通过对三山五园地区拥堵状况的监测和分析得出区域内易拥堵道路和畅通道路路段的情况，以此作为调整绿道选线的依据之一。

居住区、学校、购物中心分布：分析了三山五园地区居住区、学校、购物中心的分布情况，以此作为调整绿道选线的依据之一。

选线调整结果：通过对该地区交通拥堵状况，居住区、学校、购物中心分布情况以及该地区汇水情况和清代御道路线的考虑，调整了初始的绿道选线。

可行性调整：在上一步调整的绿道选线的基础之上，对绿道沿线进行可行性分析，找出了封闭的大院、学校、公园等不可穿行的地段，作为最终的绿道选线的参考因素。

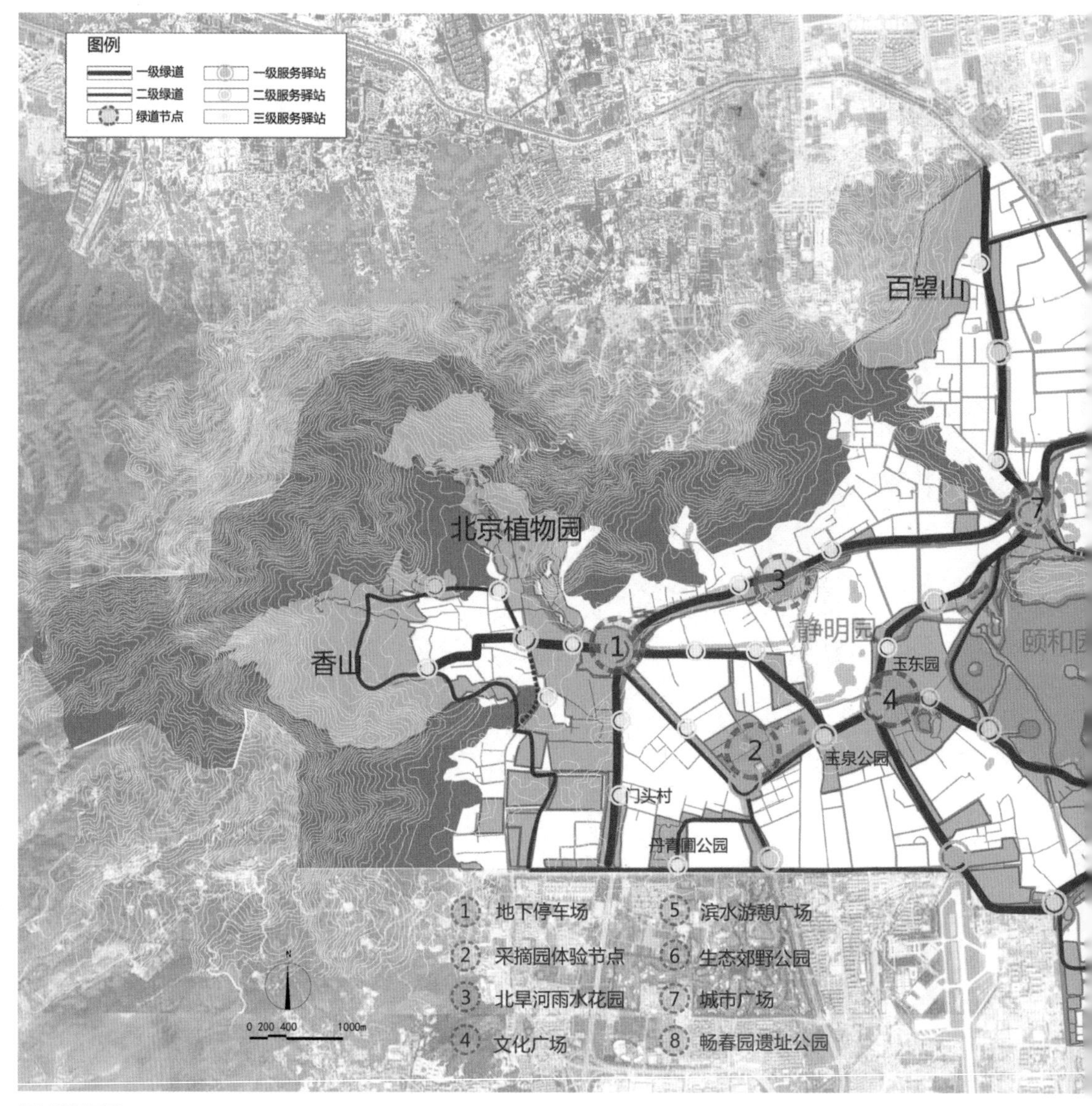

规划总平面图
Master Plan

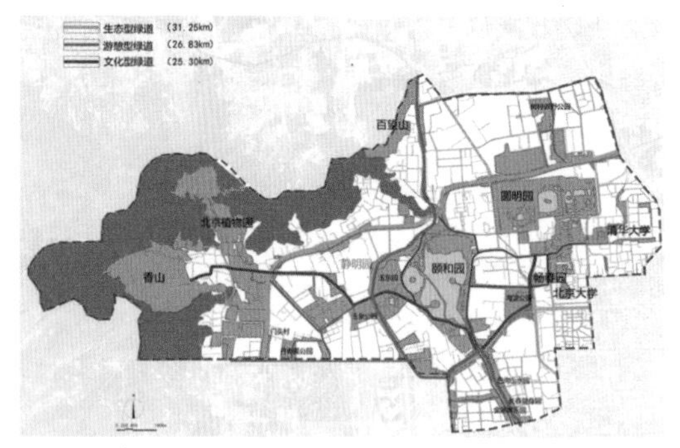

绿道分类
Categories of Green Ways

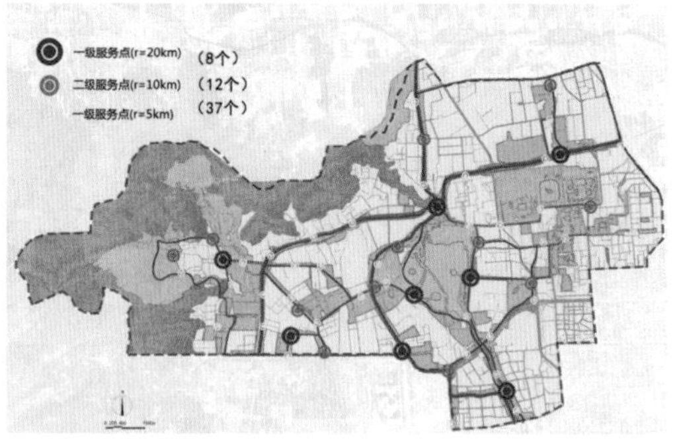

绿道服务站点规划
Service Points Plan

专项规划 Specialized Plans

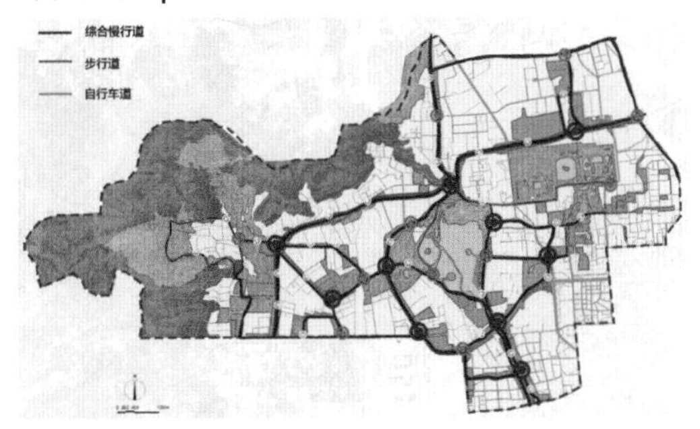

慢行系统规划
Slow Traffic Plan

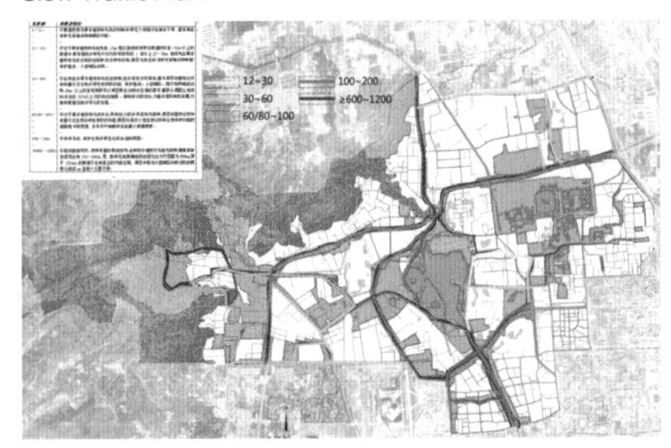

生物迁徙廊道规划
Biological Migration Corridors Plan

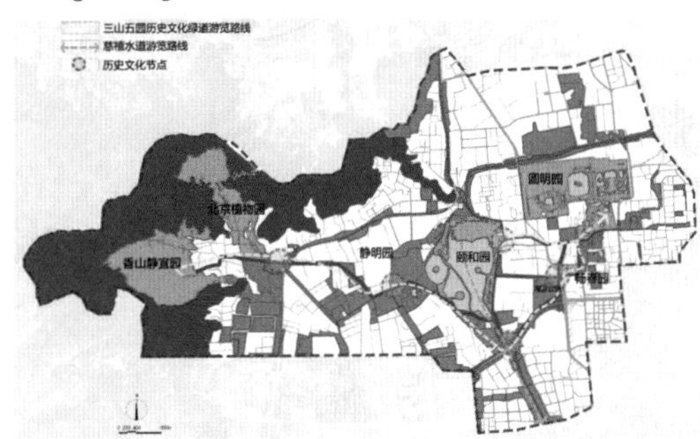

游览路线规划
Tourist Routes Plan

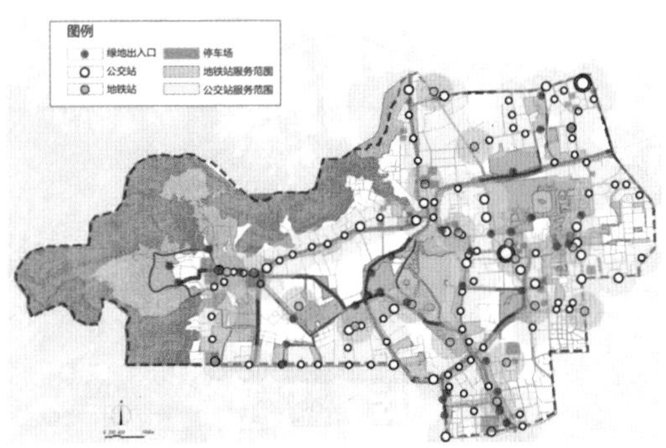

公交衔接站点规划
Public-transit Hub Plan

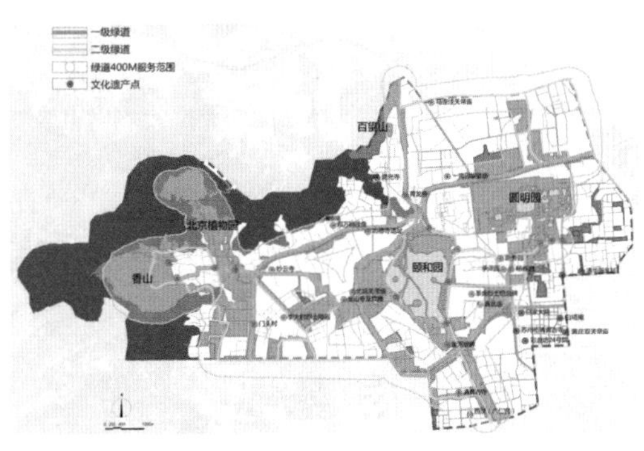

绿道服务半径
Covering Radius of Green Ways

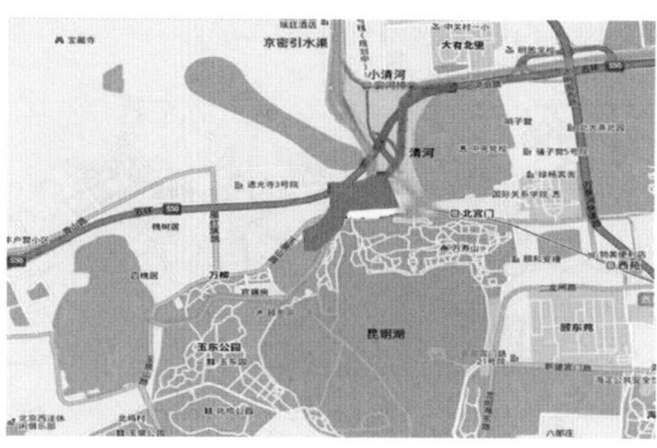

区位分析
Site Location

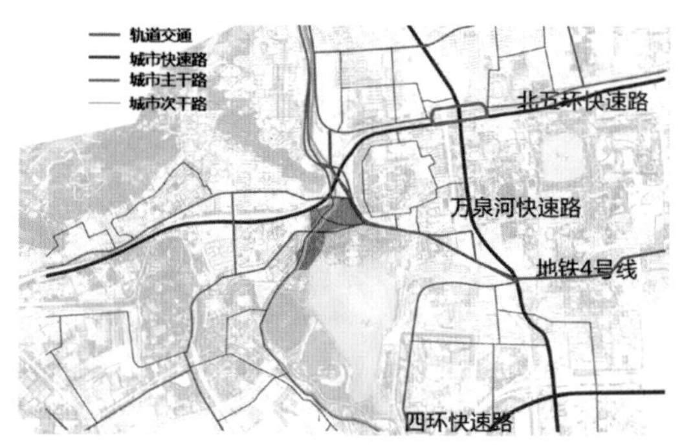

道路交通
Surrounding Road Traffic

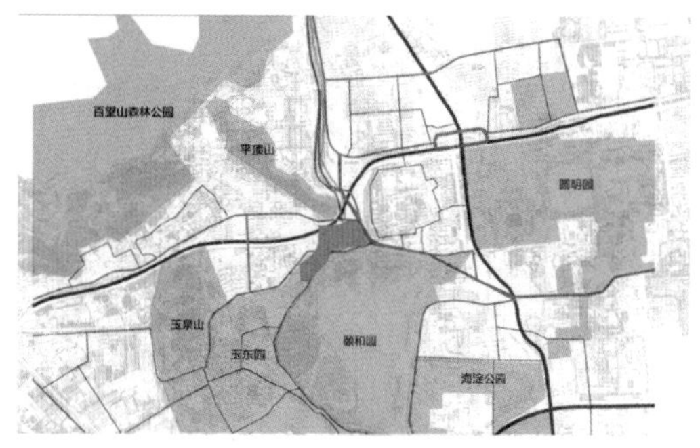

绿地分布
Surrounding Green Land

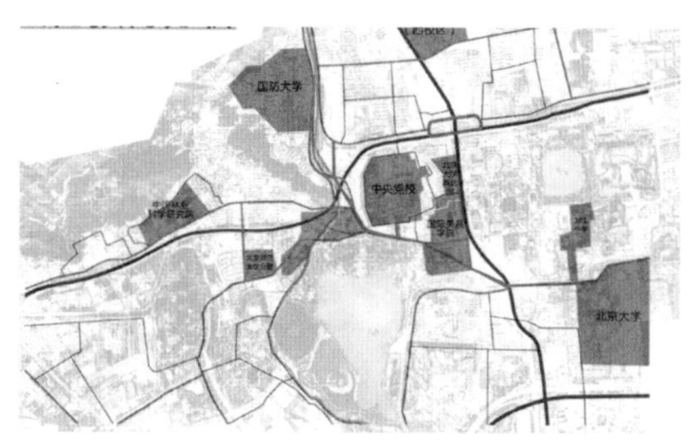

高校分布
Surrounding Universities

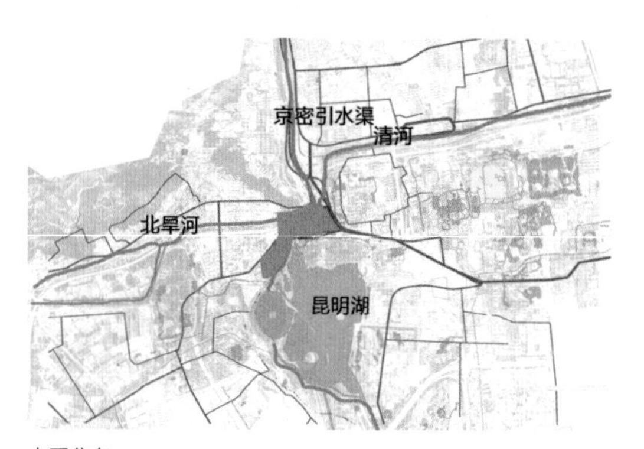

水系分布
Surrounding Water System

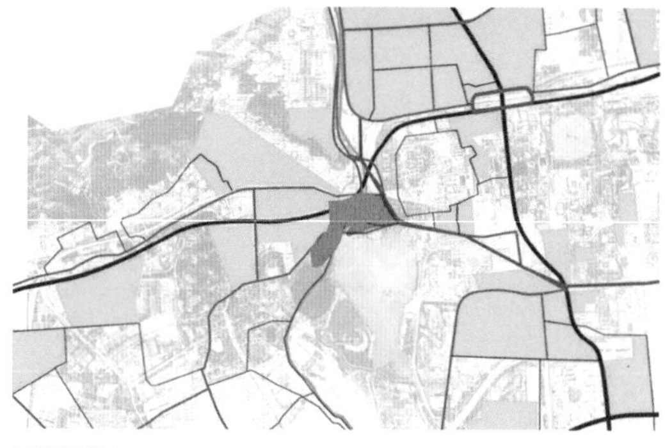

居住区分布
Surrounding Residential Areas

在已完成的“三山五园”绿道总体规划的研究成果的基础上，选择颐和园北宫门外地块作为重点设计的地块。

地块内部交通比较混乱，且人车混行，安全性不够。由于北五环和京密引水渠的切割，车辆只能从香山路穿越五环，常常形成拥堵。地块内机动车和自行车停车场地不足，车辆随意停放缺乏管理。

设计地段内，北五环、京密引水渠和城中村等对生态起到了严重的隔离与割裂作用，切断了场地、颐和园与北部山区的连接。同时，场地内部植物种

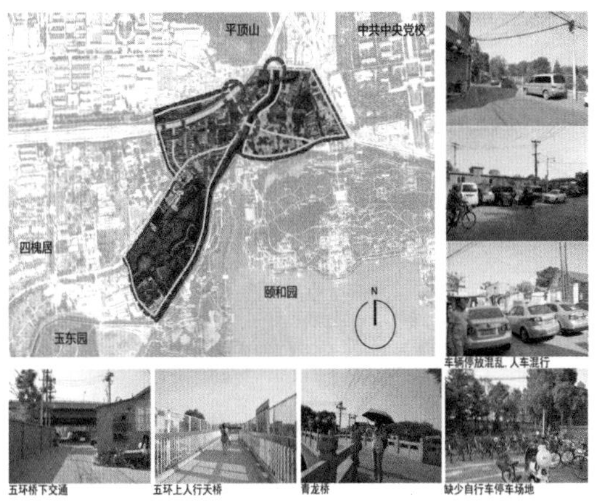

场地内交通现状
Traffic Disconnection on Site

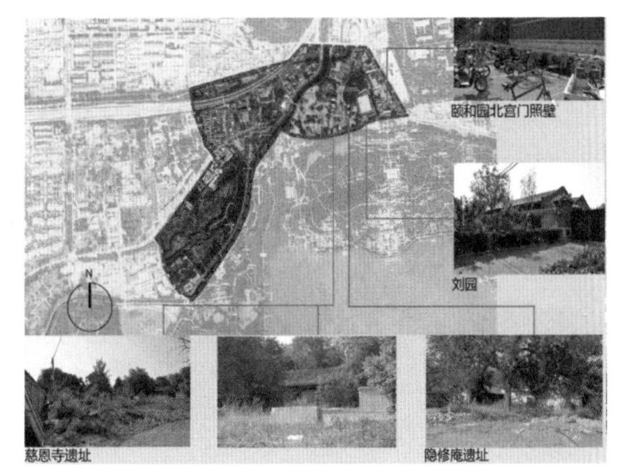

场地内历史文脉
Historic Context on Site

植较疏散，缺少空间和层次，植物种类单一。地段位于颐和园北宫门外，北至香山路，中央党校西侧、地铁四号线安河桥北站南侧，京密引水渠西侧，西至五环路、玉泉山路一线，总占地面积为 36hm^2。地段位于西山山脉、皇家园林及学校片区、科技园区的环绕中，也处于北京中心城区连接西北山区的交通要道。

场地内部和周边拥有悠久的历史文化：

历史园林。场地东南与颐和园相接，西南远望玉泉山，是连接颐和园、玉泉山和北部山区的重要节点。

自然村落。青龙桥村从元代开始便有记载，是这一区域最早出现的村庄。如今该地区大部分村落都已消失，只留下一小片城中村。

商业文化。元代建立大都之前，中都城通往居庸关的一条大道从海淀经过，这条大道向西北经过青龙桥地区，给当地村落带来了商机，买卖街逐渐形成。1860 年，西郊园林被英法联军焚毁后，商户纷纷外迁，村落愈加衰败，至今，买卖街已不复存在。

设计中结合规划的生态型、文化型、游憩型三条绿道，将场地分为三大片区。其中，沿京密引水渠和北旱河一线为生态景观，建立了万寿山和北部山区的联系；北宫门外，青龙桥村一带为历史文化区，延续这一区域的历史；沿玉泉路一线为休闲游憩区，与南部玉东园、玉泉公园一道形成京西城市公园群。

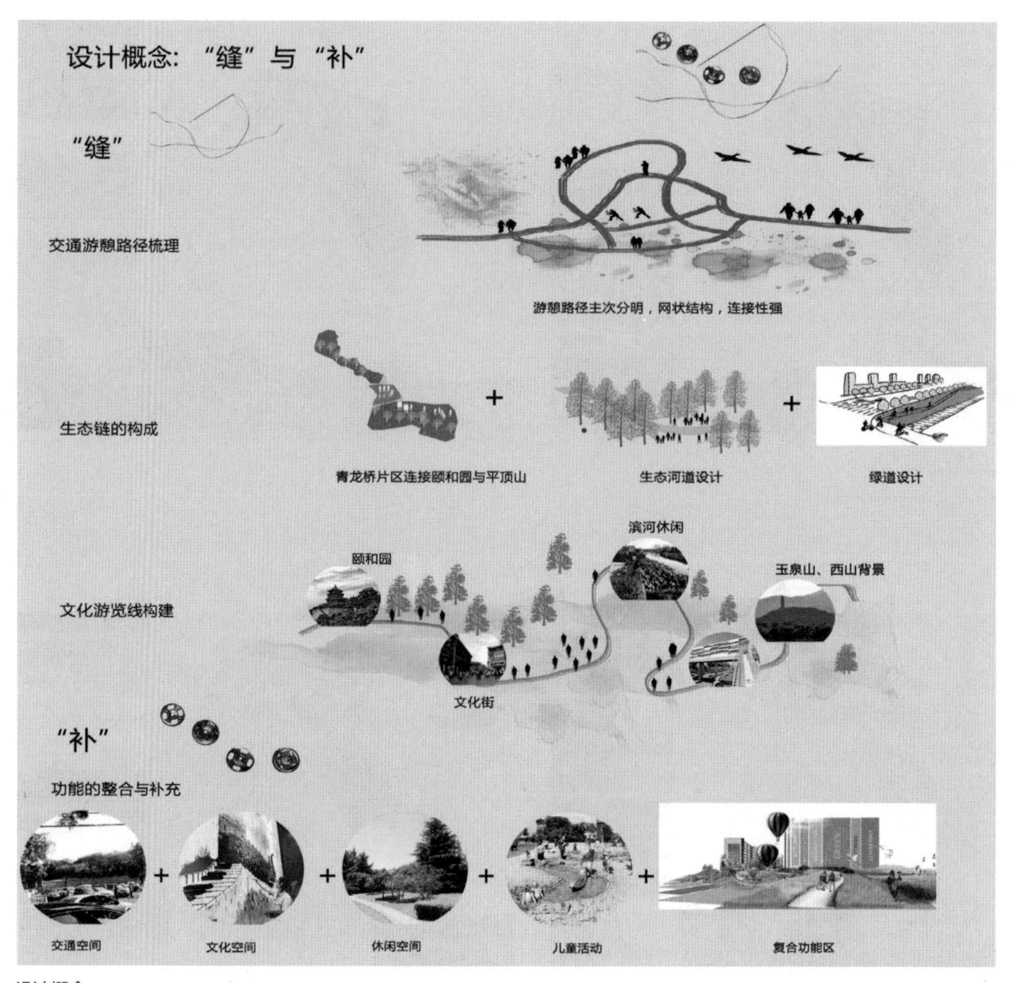

设计概念
Design concept

设计策略
Design Strategies

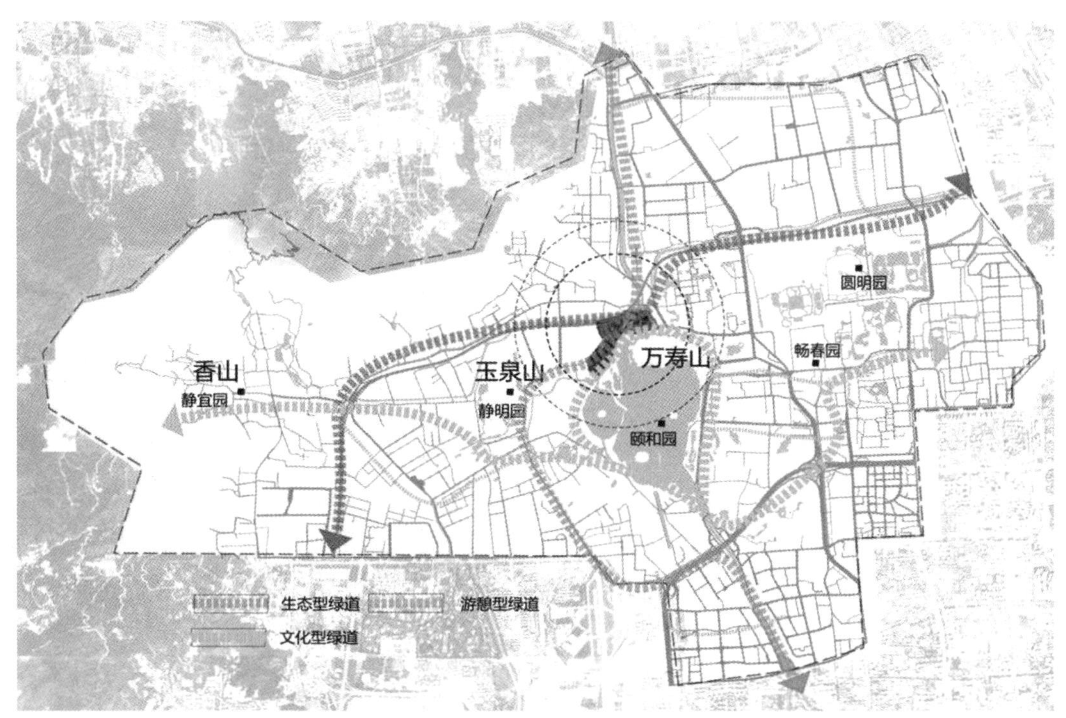

绿道分类
Categories of Green Ways

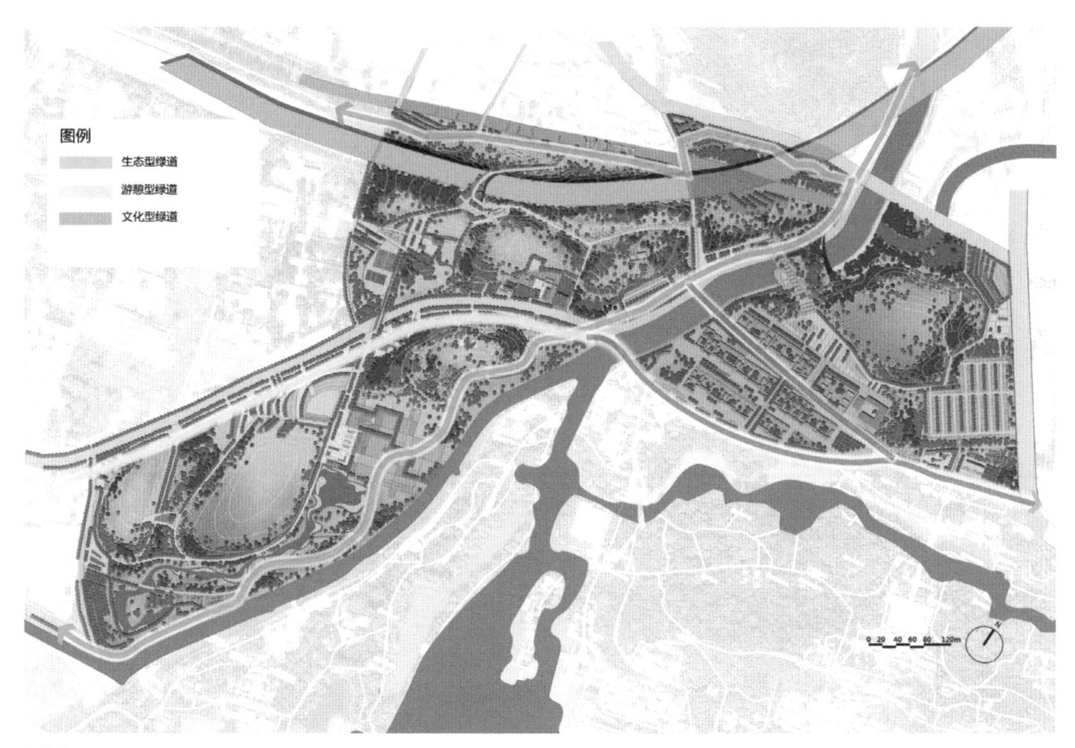

功能分区
Functional Areas

安河桥－玉东片区总平面图
Site Plan

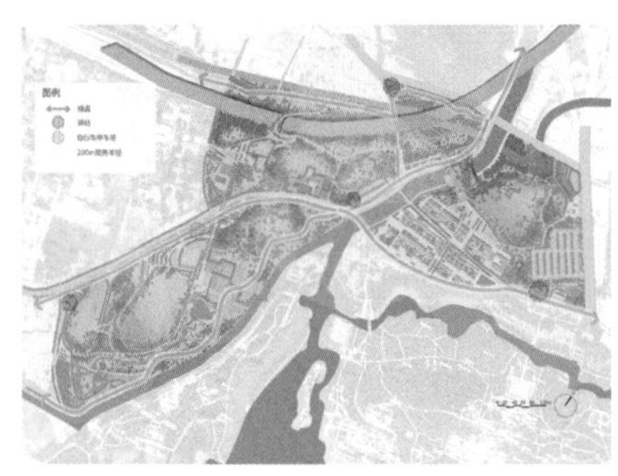

绿道服务半径
Service radius of Greenway

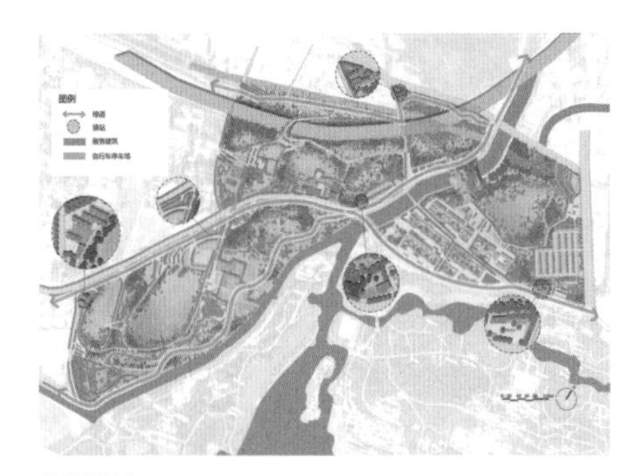

接驳设计
Connection design

交通设计
Traffic design

设计地块位于三山五园的交通枢纽与文化中心节点片区，是规划的生态型、文化型、游憩型三条绿道的交汇处。

该地区特色突出、公园景区遍布、连片分布整体性强、南侧与万寿山上的宝塔形成对景、西北部的百望山与平顶山也处于视野之中，设计中也有重要影响。然而场地也存在不少问题，包括北五环、京密引水渠、机动车道、城中村等对周边环境质量的负面影响；京西稻田不复存在、买卖街消失、周边遗存没有妥善保护的文脉延续问题；周边居民和外地游客的游憩需求得不到满足等问题。

因此，本次设计采用“缝补”的概念构想，对交通游憩路径重新梳理，缝合生态链，构建文化核心，整合与补充场地的功能空间。

根据本地居民及外地游客需求，合理组织旅游路线，串联多种功能活动空间，丰富游憩体验；保留颐和园和玉泉山之间的视线通廊，通过设置适宜的观赏点来充分引借周边优美风景；历史文脉延续方面则充分尊重原场地的历史记忆和空间肌理，重建北宫门历史商业街区，以部分保留和改造的方式保存青龙桥村落的院落肌理，形成一片稻田种植景区，以表达对历史上御稻的隐喻和怀想。

1.入口广场
2.综合停车场
3.照壁广场
4.登高望远亭
5.缀花大草坪
6.休闲广场
7.综合服务建筑
8.历史商业街区
9.如意门休闲绿地
10.生态廊桥
11.林荫广场
12.生态景观池
13.停车场
14.绿道服务建筑
15.森林小憩台
16.山顶观景台
17.休息平台
18.村落拾忆院
19.景观桥
20.健身球场
21.西入口广场
22.下沉儿童乐园
23.休憩木平台
24.活动广场
25.休憩木平台
26.自行车停车场
27.入口广场
28.玉泉万寿亭
29.趣味广场
30.御稻流芳园
31.下沉剧场
32.林荫广场
33.树阵广场
34.入口广场
35.休息平台
36.宠物乐园
37.阳光大草坪
38.水院
39.湿地观赏区
40.露天茶座
41.水边观景台
42.景观廊架
43.旱喷广场

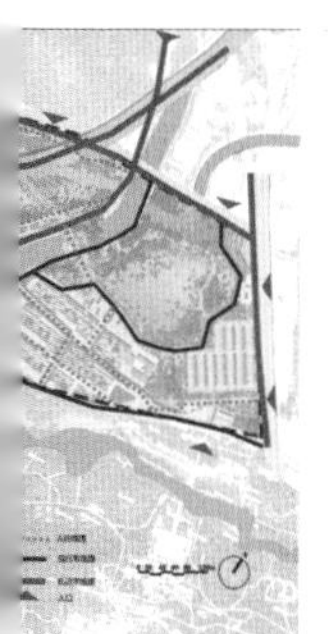

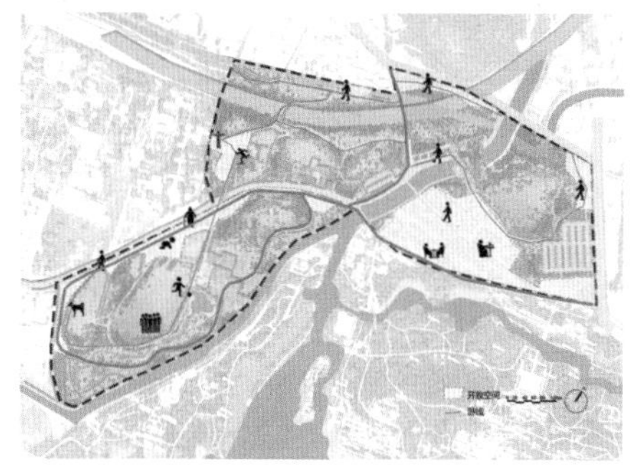

功能设计
Functional design

景观视线设计
Sight design

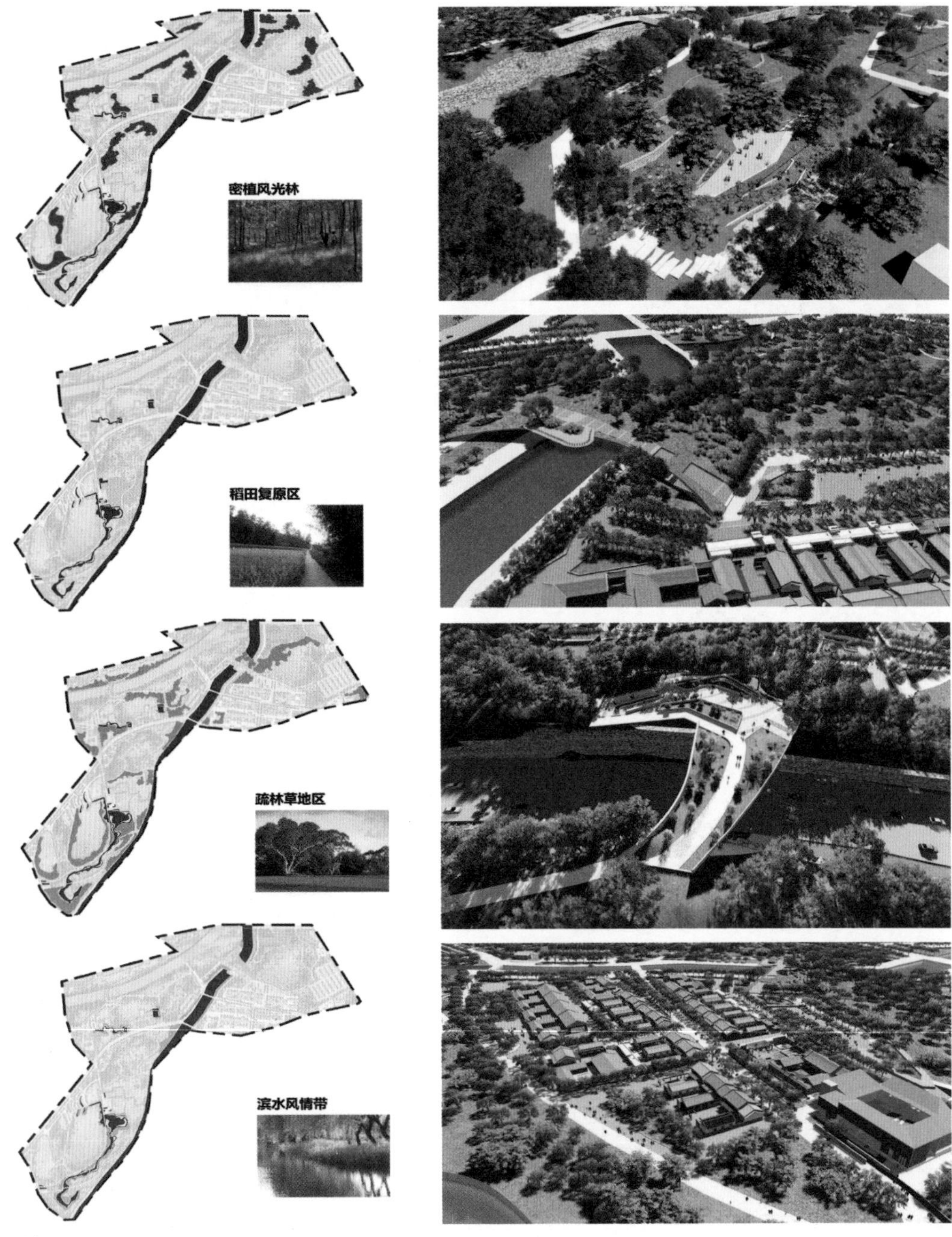

详细设计
Detailed Design

主要植被类型依据生态习性及景观场景营建主要分为三种，即稻田复原区、疏林草地区及滨水风情带。稻田复原区目的是为了追忆古时"稻花香里说丰年，听取蛙声一片"的自然田园景观意趣，设计时主要以网格分割场地，形成形态不一的种植圃，营造生动活泼的休闲教育体验空间；疏林草地在保留原生植被的基础上，以草坪为基底，草坪边界处种植花灌木及草本花卉，形成丰富多样的景观空间；滨水风情带以京密引水渠及园中自然式水景边界为主要依托，创造丰富多样的滨水生态空间。

生态方面，改造现状河道，改建驳岸形式为台地，通过生态廊桥创造野生动物迁徙廊道等；交通

鸟瞰图
Bird eye's View

剖面图
Sections

方面，分析现状道路及区域之间的可达性，使得机动车行驶道路和慢行交通互相分离，在公园出入口附近设置合适的停车场，方便游人使用；历史和文化主要表现在北宫门片区商业街的复兴、村落肌理的保留、村落印象景观的展示及稻田的肌理与情景再现等方面。

丰富的地形变化不仅创造出多样的空间形态及景观效果，而且对雨洪管理有着重要作用。其中高速路下方有人行通道，两侧地形坡度较大，但利用挡土墙等形式使得植被种植丰富，形成错落的植物景观层次；商业街区建筑立面形式统一，屋顶形式以硬山式坡屋顶为主，与周边环境融为一体。

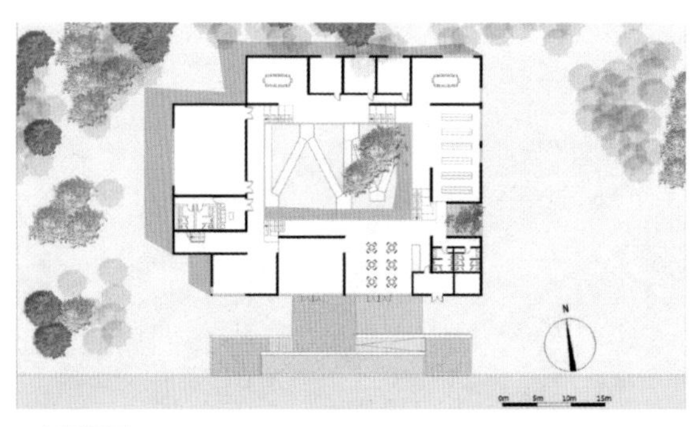

一层平面图
1st Floor Plan

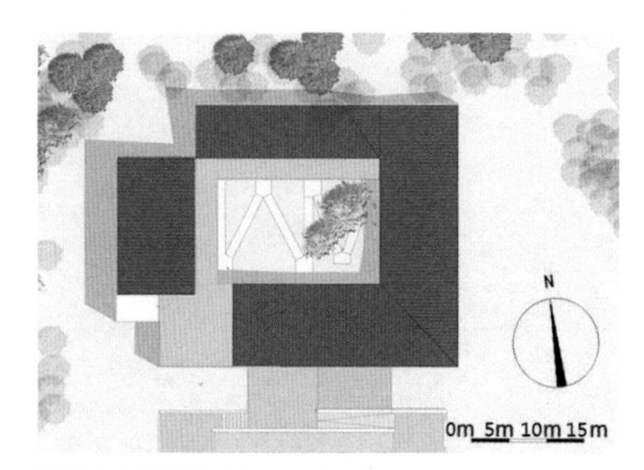

游客中心总平面图
Site Plan of the Visitor Center

北立面图
North Elevation

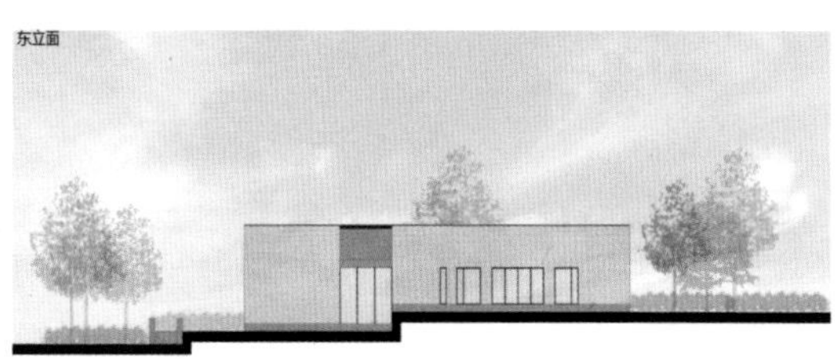

东立面图
East Elevation

南立面图
South Elevation

西立面图
West Elevation

游客中心为方便使用及管理，选址位于商业街入口附近，主要用于为公园管理人员提供办公空间、为游客提供咨询及餐饮娱乐等服务项目，同时作为绿道慢行体系的一个中转站。

建筑以单层为主，局部二层，总面积为1500m^2，大部分面积用于门厅、咖啡厅、展厅等公共服务空间。建筑房间布局南侧主要为咖啡厅、阅览室、门厅等服务于大众的开放空间，会议室、管理用房等内部使用空间主要分布于东西两侧或北侧。整体功能分区明确，使得游客与管理人员交通流线相分离。从形式及结构上看，建筑内呈凹形的“砖盒子”造型，屋面凹形空间向中心汇聚，与中心的庭院连成一个整体。屋顶雨水也将汇聚到庭院中部绿地以灌溉植被。

另外，凹形的设计让建筑整体上呈现安宁、内敛的氛围，同时凹形屋面空间向中心汇聚，也暗示了中国传统的“合”这一个概念。

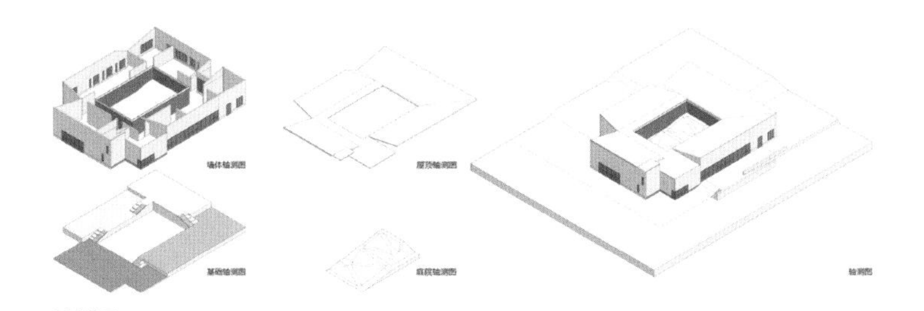

轴测图
Axonometric

功能分区
Function Division

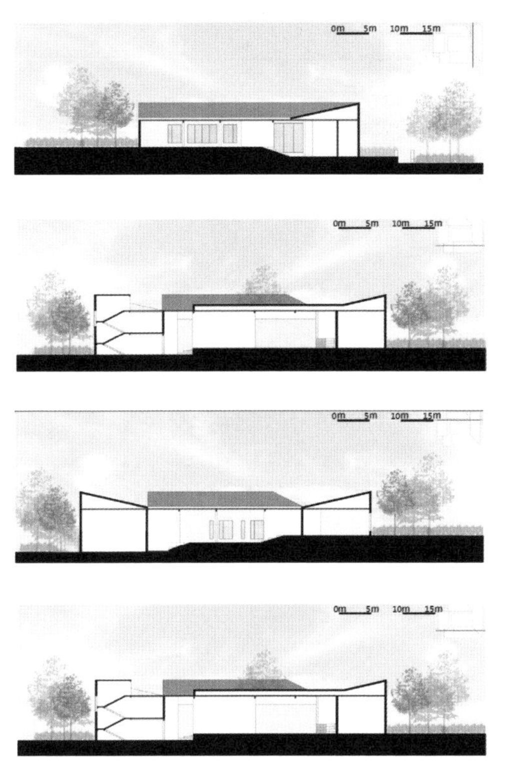

剖面图
Sections

效果图
Rendering

建筑面朝主要商业街，与道路间有绿带及矮景墙相隔，既方便游客使用，也削弱了道路的繁杂与喧嚣。

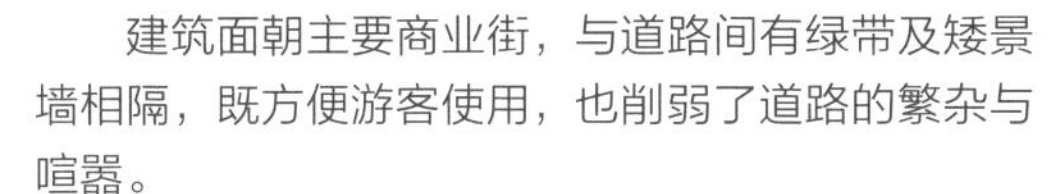

建筑墙体多用灰砖，与周边商业街区风貌协调，其体量比周边合院大一些，原因是其功能复杂，空间面积要求较大，同时大体量也使其成为商业街巷的标志建筑。建筑四周林木葱郁，北面开阔的大草坪与小山丘上繁茂的林木成为建筑的自然背景，使建筑融入林木之中，给人舒适自然的休闲活动体验。

建筑立面大量运用通透轻盈的玻璃材质，结合敦厚朴素的灰砖墙体，形成鲜明的对比，展现出现代材料与传统材料的异样美。大量玻璃材质的使用也使得建筑通透灵活，从室内到室外有良好的景观视线。

设计时充分利用现状地形创造不同高程的建筑空间，通过坡道或台阶连接不同高差的空间，使得内外部空间形式多样，内容丰富，建筑整体错落有致，层次分明。

规划方案五 及 香泉环岛－颐和园西门概念设计

Plan 5 & The Conceptual Design of the Xiangquan Rotary Island-the Summer Palace Western Gate Section

王博娅、丁小夏、郑慧、张司晗、郭沁、刘健鹏

Wang Boya/Ding Xiaoxia/Zheng Hui/Zhang Sihan /Guo Qin/Liu Jianpeng

规划方案中，绿道的覆盖范围更为广泛，并在绿色开放场地设计中包含了高空栈道、景观高架桥、嬉水空间、露天剧场、观景平台等一系列趣味空间，丰富了景观形式和游憩体验。

通过水系整理，从香泉环岛到颐和园西门形成了完整的滨水绿色走廊。在设计中通过明确的景观功能分区，结合精细的雨洪管理策略以及颐和园西门新的游客集散空间的打造，构建了三山五园地区一条新的游憩体验路线。

In this plan, we found the coverage of greenway is more expensive, so we set up a series of interesting space in the design of green open space, including upper plank roads, landscape viaducts, paddle space, amphitheater and viewing platform, so as to enrich the landscape forms and recreational experience.

Through water system consolidation, we established a completed waterfront green corridor stretching from Xiangquan Island to the west gate of the Summer Palace. In the design, we successfully built a brand new recreational experience route in the 'Three Hills and Five Gardens' area through explicit landscape function zoning, combined with elaborate rain and flood management strategy, as well as new tourist distribution space of the west gate of the Summer Palace.

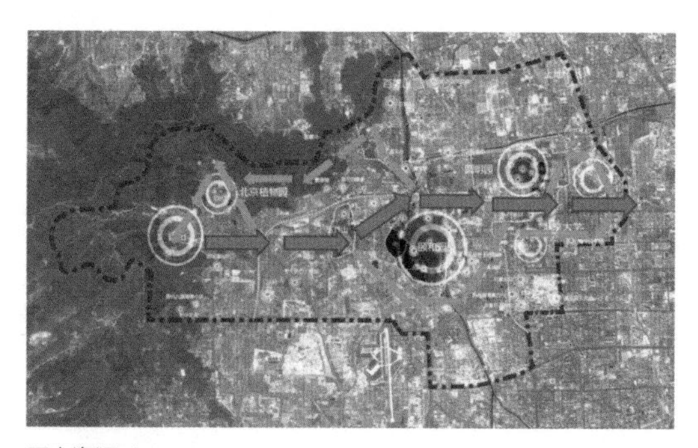

历史资源 1
Historic Resources 1

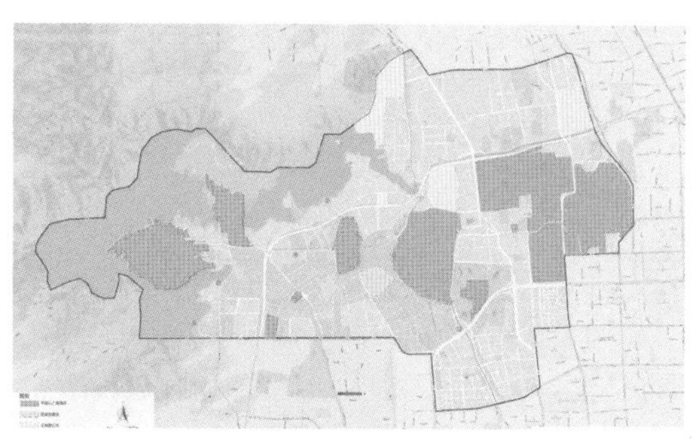

历史资源 2
Historic Resources 2

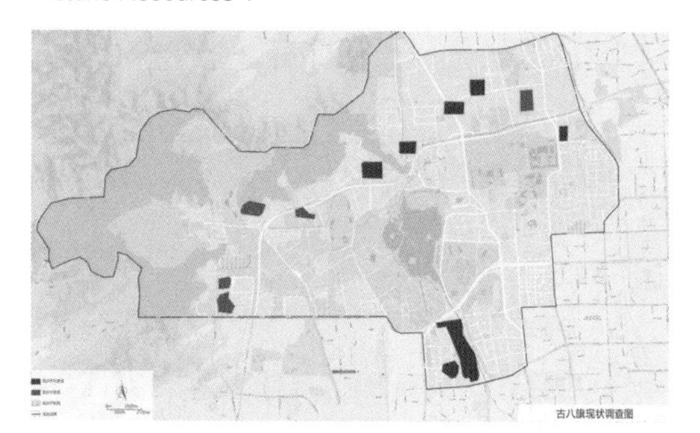

历史资源 3
Historic Resources 3

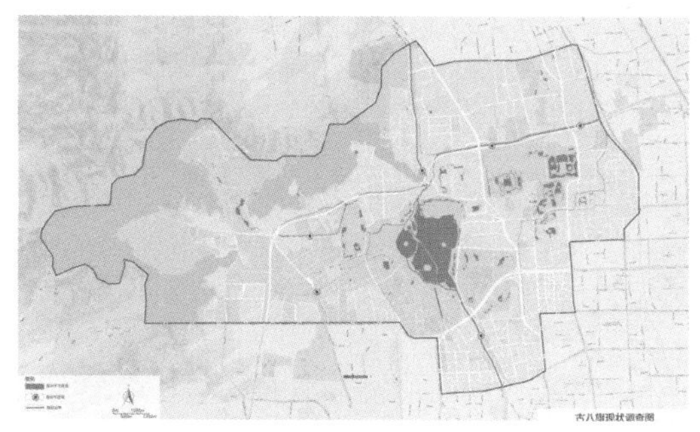

历史资源 4
Historic Resources 4

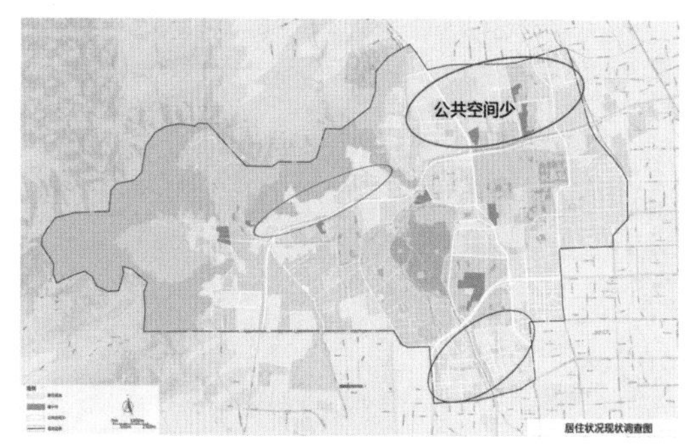

休闲游憩场所
Existing Recreational Venues

现状生态结构
Existing Ecological Structure

1 现状分析

根据调研，三山五园地区的现状问题主要有以下三个方面：历史文化资源缺少整体尺度的保护和利用、风景园林建设未能很好满足居民休闲游憩需要、生态体系破碎化。

历史资源现状主要问题有："三山五园"之间的联系较弱，没有形成整体；点状分布的历史文化资源，保护不力；具有很高历史价值的资源点，利用不够充分；八旗村落遗址不复存在，部分地区场所记忆缺失；历史河道段体系遭到破坏；河流水质差，水量少；驳岸形式单一，景观效果较差；缺少对河流历史文化价值的挖掘；京西稻田被不断侵占，面积不断缩减，独特的稻田景观消失。

地区内的休闲游憩问题有居住区的分布密集，公共空间少；慢行系统不完善、管理不力，机动车和非机动车混行等。从大西山到奥林匹克森林公园的生态体系破碎化，生态基础设施不完善。

规划依据《北京市城市总体规划（2004-2020）》、《北京市绿地系统规划》、《海淀区空间发展战略》、《城市道路交通规划设计规范》、《城市绿地分类标准》、《海淀区文物列表》等，制定如下规划原则：

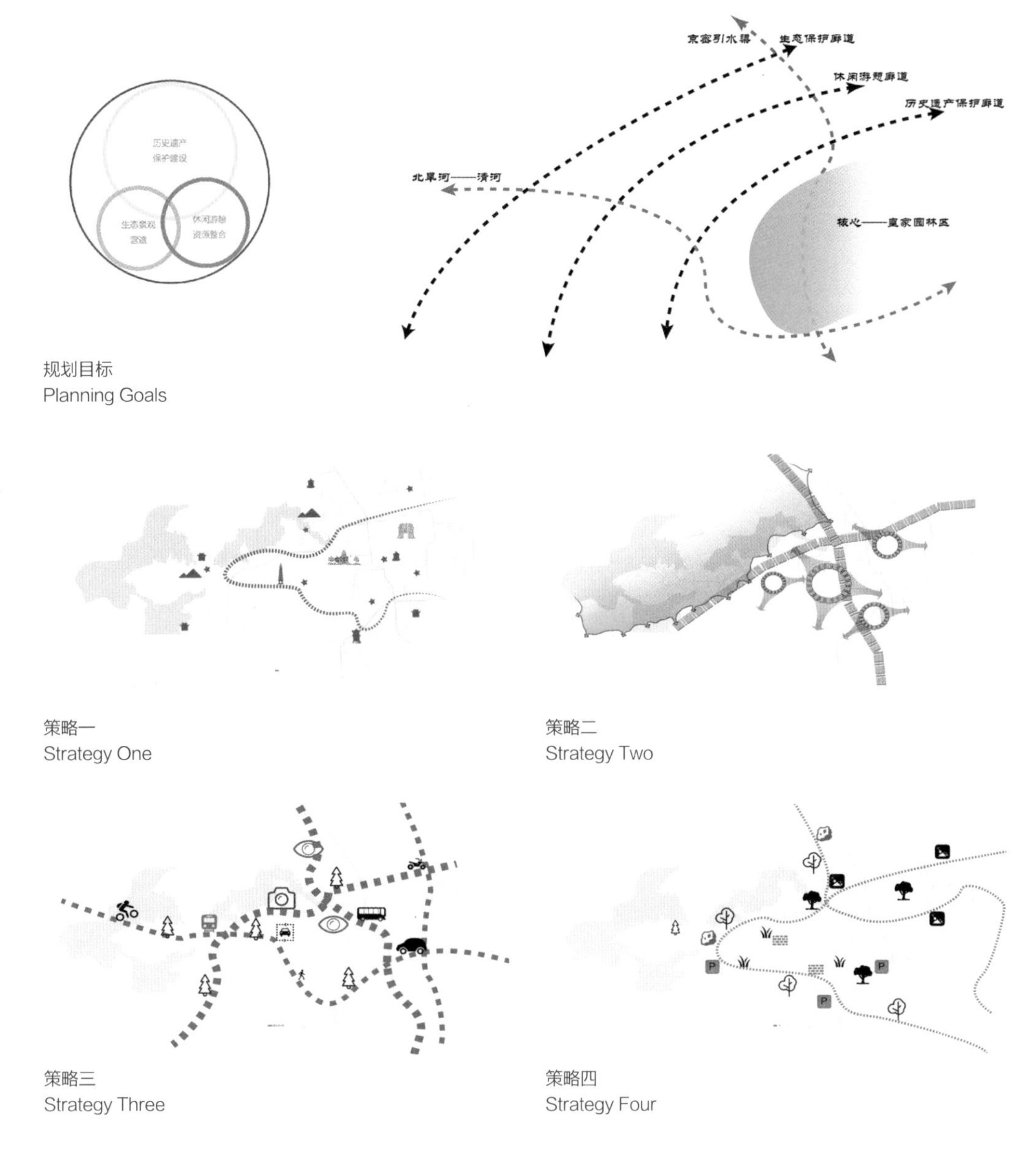

规划目标
Planning Goals

策略一
Strategy One

策略二
Strategy Two

策略三
Strategy Three

策略四
Strategy Four

（1）特色化原则：充分发挥绿道对各类发展节点的组织串联作用，以历史文化资源为依托，尽可能多地发掘并展示本地具有代表性的特色资源。

（2）生态性原则：尊重山水格局，结合现状，使分散的生态斑块得以有机连接，从而发挥绿道作为生物廊道的作用。

（3）人性化原则：指城市绿道在适应于人类的感官和意愿的尺度之上，鼓励绿色出行并建设具有活力的可持续城市绿道，全面运用人性化的、朴素适中的尺度。

（4）共享性原则：指城市绿道系统不能游离于城市结构之外，应该与城市公共空间、景观绿地、滨水空间、主要场所、城市结构性景观廊道相结合，建设共享绿色出行空间。

（5）便捷性原则：指为方便游客进出，应提供与绿道相适应的机动交通支撑体系，可结合城市公交系统设置出入口，方便城市人流进出绿道网络，并考虑配套设施的方便适用。

2 规划目标

建立以整合和保护历史文化资源为核心、以休闲游憩和生态保护为辅助的绿色开放体系，恢复三

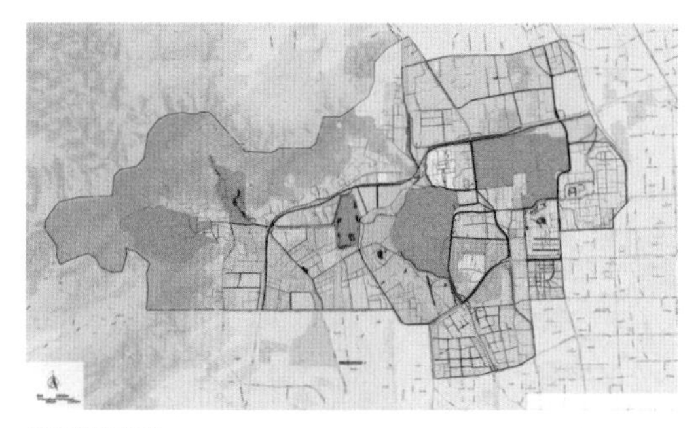

遗产保护用地
Historical Land

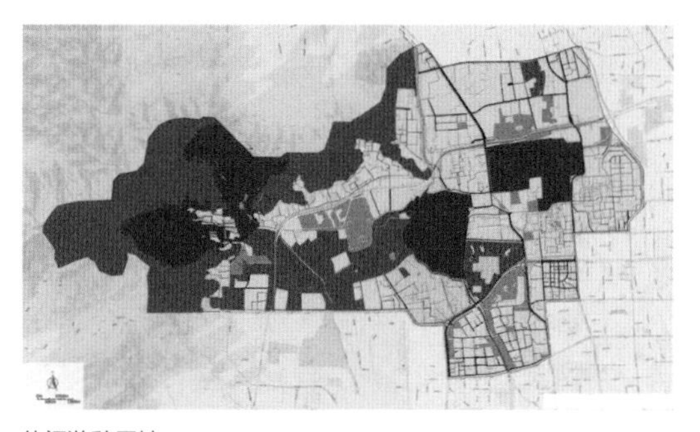

休闲游憩用地
Recreational Land

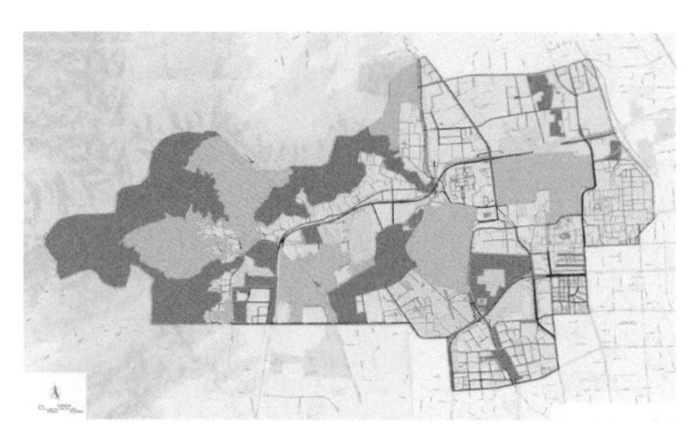

自然生态用地
Ecological Land

绿道选线
Location of Green Ways

山五园的历史空间格局和原有的水乡田园风貌，唤醒场所记忆、恢复场所精神；构建完善的休闲游憩基础设施和开放空间体系，以满足居民的需求，完善生态基础设施，形成完整的生态环境体系，将自然引入城市。整个绿色开放空间体系包括历史文化廊道、休闲游憩廊道和生态廊道三部分。

3 规划策略

（1）串联历史资源——以三山五园为主线，串联资源点，解决资源点散落分布的问题，从而对三山五园进行整体保护，形成整体性的历史文化景观。

（2）整合城市开放空间——将线性要素如城市河流、文化线路、道路系统等融入整个绿道系统当中，将功能分区与市民兴趣点分布结合，活化绿道功能内涵，从而满足居民的休闲游憩需要，改善居民生活品质。

（3）复合道路功能——使道路体系由单一交通功能向交通、生态、游玩休憩和保护等复合功能转变。

（4）增强基础设施的可观可游性——沿道路和河道种植多种类观赏性植物，因地制宜地选择铺装的图案和材料等。

根据规划目标分别制定历史资源适宜性、休闲游憩适宜性、生态保护适宜性的评价标准，并进行叠加，得出结果。具体方法是：由点到线，先甄选有影响力必选的资源点，再遴选有绿道潜力的线形通道。除了按照点线的方式以外，还根据目标的三条廊道进行选择，筛选出适宜构建不同廊道的地块。

基于适宜性评价分析，得到了最终的绿道选线结果，并绘制绿道概念性规划总平面图。整个绿道体系包含历史遗产廊道、生态廊道、休闲游憩廊道。

规划中主要结合绿地和水系为主，并充分考虑道路、慢行系统、建筑、游客服务中心、驿站、广场、停车场等设施，形成完整的廊道格局。规划绿道系统总面积为 596.7 万 m^2。

历史遗产廊道：用遗产廊道将分散的历史资源点连接起来，形成整体保护效应。选线：连接区域内主要历史文化资源点。节点：圆明园、颐和园、玉泉山静明园、香山静宜园等。要素调整：为保证历史文化遗产核心保护区的风貌，限制周围建筑物的高度和新建建筑的风貌。

生态廊道：用生态绿道连接破碎的生境，保护和修复城市中的动植物栖息地，引导人类社会的开发远离重要的动植物栖息地，并为后者预留空间，同时保证物种以及物质能量的流动。选线：连接西山大型生态斑块及区域内其他主要生态斑块。主要形式有：依托西山的山地型生态绿道；依托清河、

规划总平面图
Master Plan

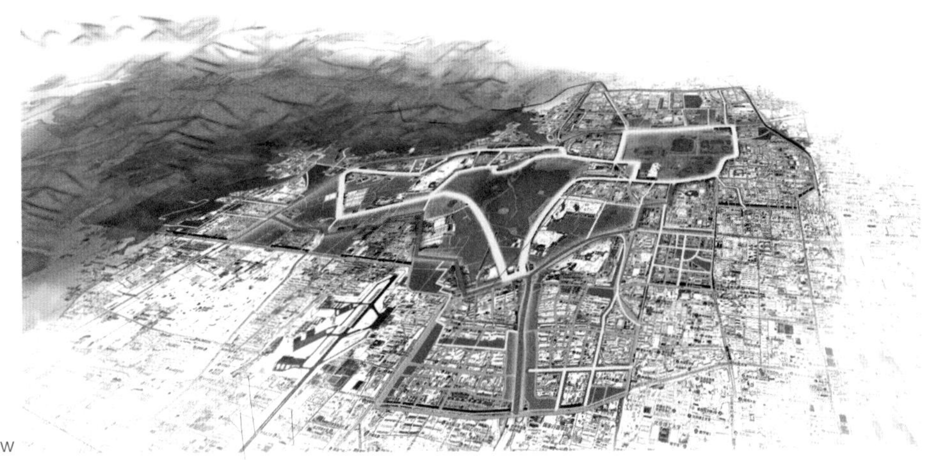

绿道鸟瞰图
Bird Eye's View

京密引水渠等的河流型生态绿道；依托旱河路、香山路等的道路型生态绿道。

休闲游憩廊道：整合城市公共空间并构建功能复合型休闲游憩体系，因此选线的标准是连接游憩设施基础良好的区域，如颐和园、圆明园、植物园等地。休闲游憩廊道的主要形式有交通型游憩绿道、防护型游憩绿道、商业型游憩绿道，颐和园、圆明园、植物园等是绿道上的重要节点。

改造后的河流型生态廊道的植物种类更加丰富，并增加亲水设施；改造后的道路型生态廊道设置天桥连接，以构建完整的廊道体系；改造后的旱河型生态廊道的植物种类更加丰富，并设置浮桥，在旱季作为休闲娱乐设施，雨季作为亲水设施。

此外，还设置了交通型休闲游憩廊道，便于游人通行；防护型游憩绿道（步行街型），起防护作用；商业型休闲游憩廊道，兼顾商业发展。

规划的绿道体系的结构是“一心、两轴、三环”，“一心”是皇家园林核心区，“两轴”指由京密引水渠和旱河、金河组成的轴，“三环”是指由内向外的历史遗产保护环、休闲游憩环和生态保护环。

规划主要节点如下：

双园堆秀——串联圆明园和颐和园。园外园——位于颐和园与玉泉山之间，利用现有林地，形成郊野公园体系。御道夕峰—沿北旱河，借景玉峰塔，形成良好的视觉通廊。玉河浮金——恢复金河水系，以油菜花、采摘园和农家乐为景观特色。长河明波——沿北旱河展开。八旗烟柳——依托八旗古村落和清河流域，恢复田园风光。野旷桑麻——位于北部，以桑蚕农事为主题，打造多层次的现代观光农业体验。云树参差——主要位于中关村，连接商业用地，改善城市居住环境。香泉探幽——连接西部各山体，以生态廊道为主。

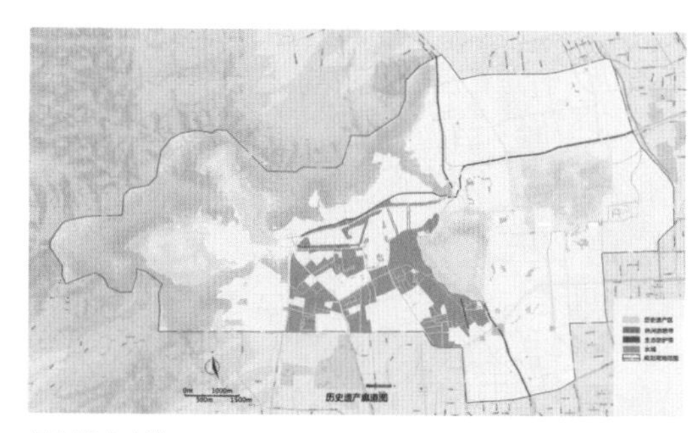

历史遗产廊道
Heritage Corridors

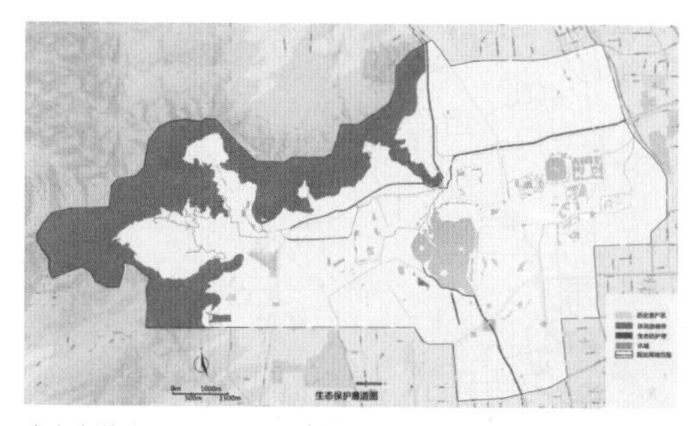

生态廊道
Ecological Corridors

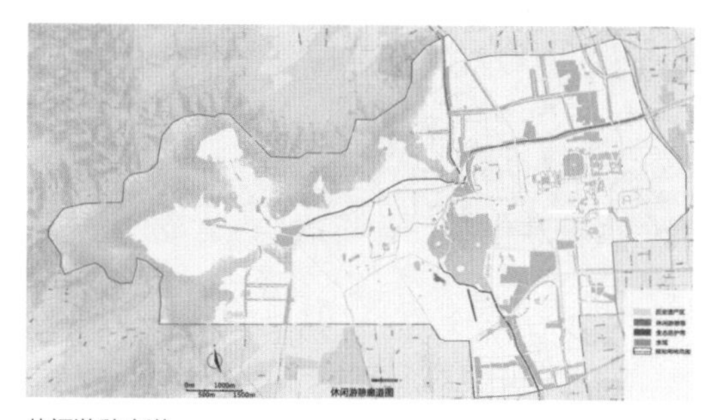

休闲游憩廊道
Recreational Corridors

4 分类分析

规划绿道根据重要程度分为三级。一级绿道是指连接三山五园核心景观的绿道。二级绿道是重视游憩功能与慢行系统功能。三级绿道是连接社区公园、小游园和街头绿地，主要为附近社区居民服务的绿道。

绿道经城市地铁站点、公交站点时，均设置换乘点，实现绿道与城市公共交通系统的有效衔接；绿道与铁路、快速路、河流等连接时应采用立交连接；绿道与城市道路连接时，杜绝人车混行现象。

规划创造完整的慢行系统，为居民提供良好的休闲游憩体验。

对不同的区段水系实施相应的改造策略。

对服务设施进行了分级设置，一级驿站每 5 ∽ 7km 设置，占地 600m^2，游客中心、医疗点、餐饮、停车场等；二级驿站每 1.5km 设置，占地 50m^2，餐饮、停车场、休息设施等；三级驿站每 500m 设置，提供自行车存放空间，有公厕、坐凳等。

植被规划体现了生态优先的原则，规划了环保型、观赏型、生产型、生态型植物群落，应用于相应区域。

将散布的文化遗产用绿道体系连接并保护起来。

设计范围自东向西主要可分为三个片区，分别为香泉环岛区、廊道区与颐和园西门区。

香泉环岛区设计定位：以解决城市洪涝问题为出发点，疏导上下游旱河道的流向与河岸形式。围绕着中心环岛的三块场地，以大面积绿地为主，设计了一系列地形以解决不同排洪量时的雨水问题。通过收集上游、道路与居住区的汇水，在场地内形成雨水景观，同时进行雨水收集与再利用。

生态廊道剖面图
Section of ecological corridor

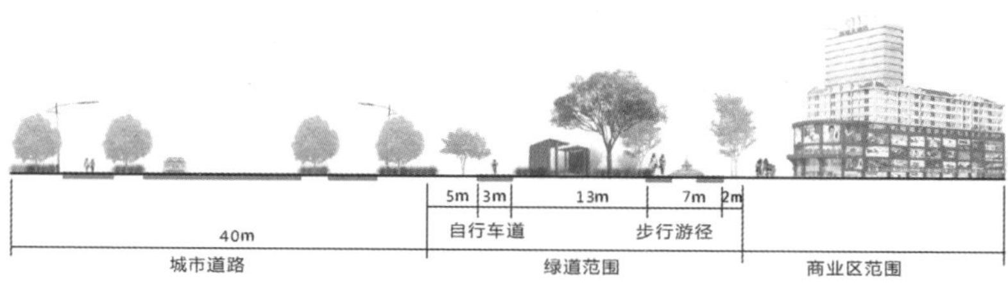

商业型休闲游憩廊道剖面图
Section of commercial recreation corridor

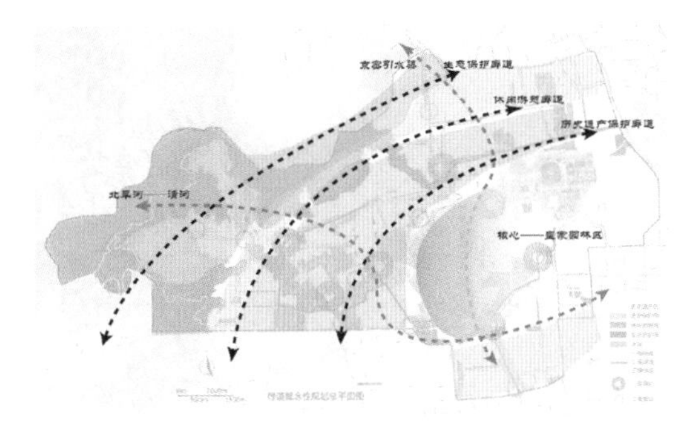

结构分析图
Structure of Green Ways

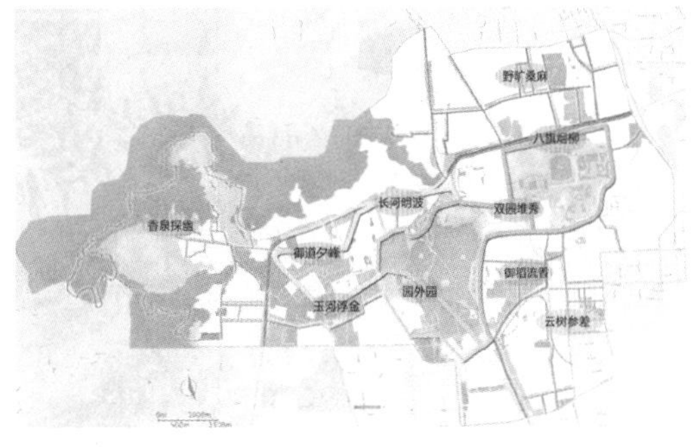

景点分布图
Distributions of Scenic Spots

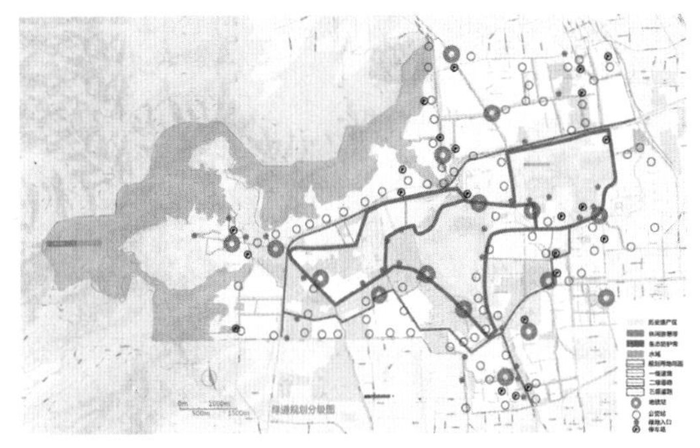

道路交通规划图
Road Traffic Plan

慢行系统规划图
Slow Traffic Plan

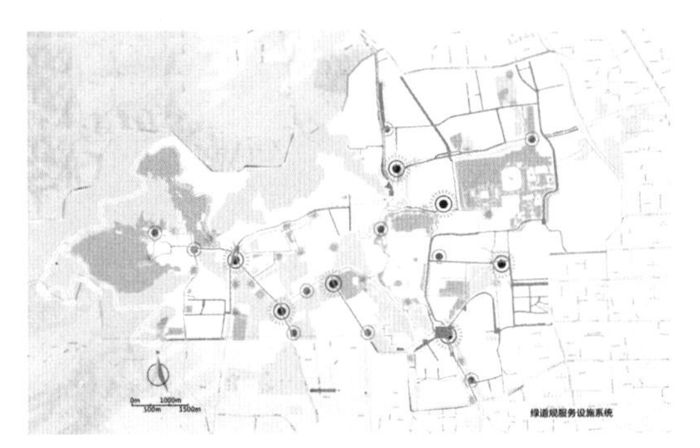

服务设施规划图
Service Facilities Plan

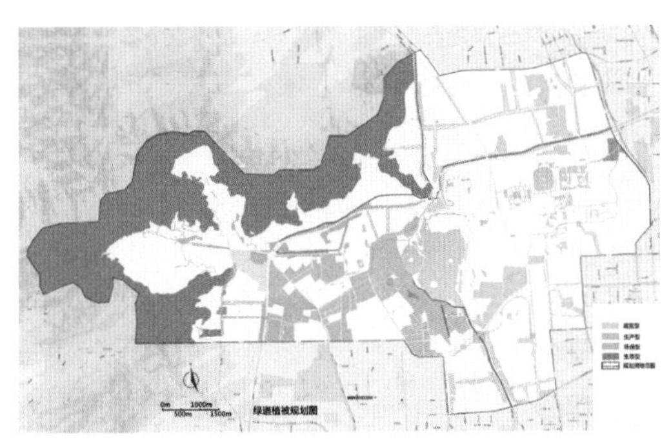

植被规划图
Planting Plan

水系规划图
Water System Plan

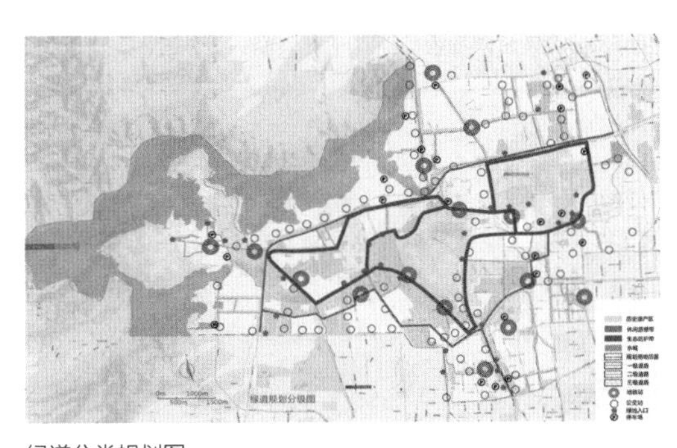

绿道分类规划图
Categories of Green Ways

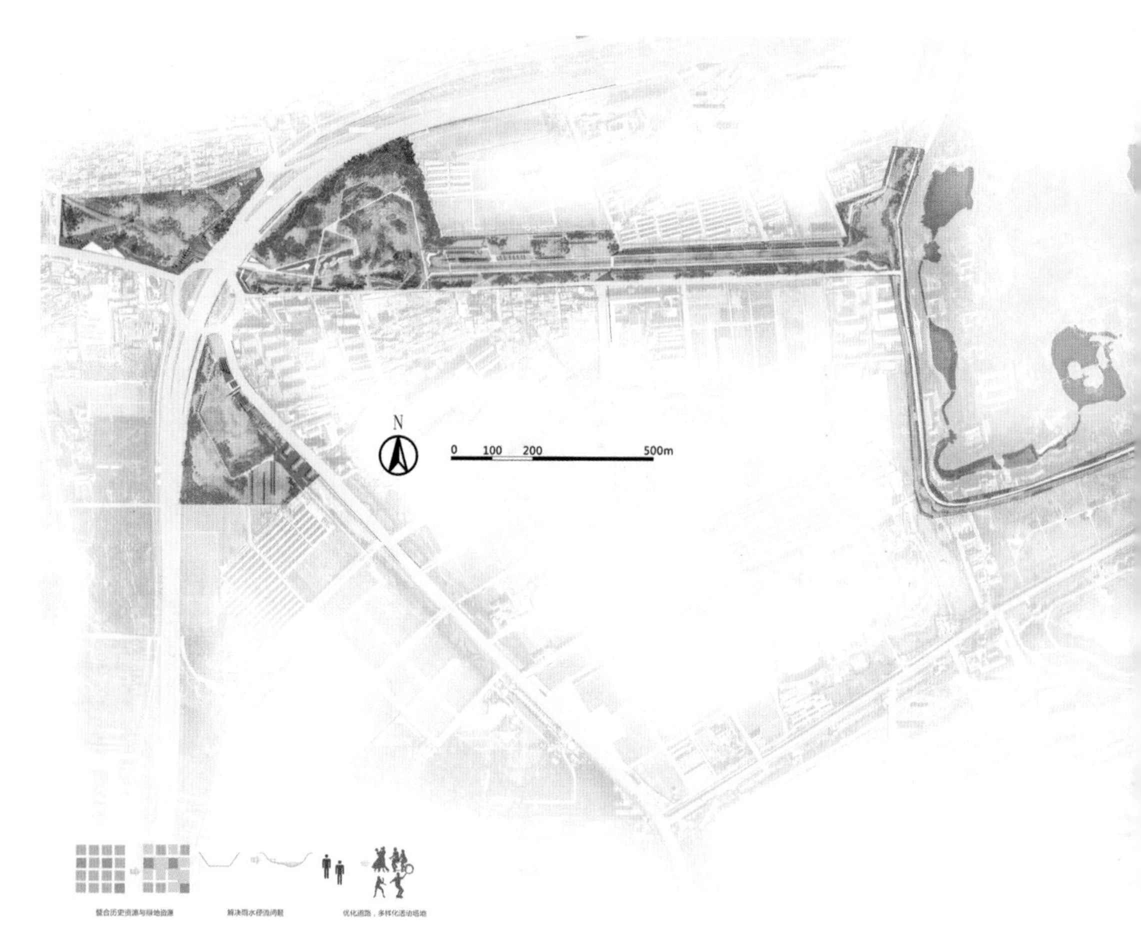

香泉环岛 – 颐和园片区设计总平面图
Site Plan

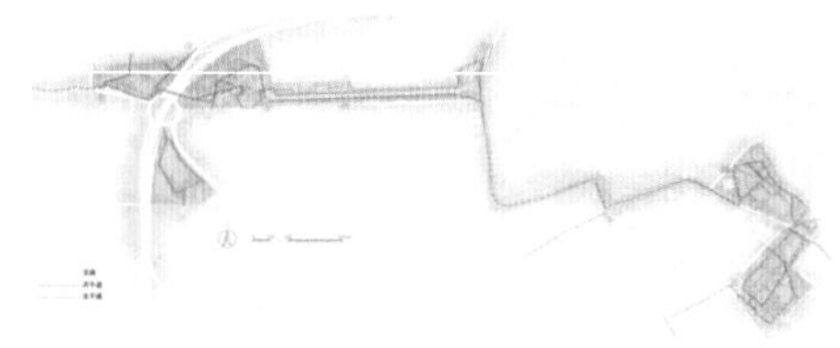

道路分析图
Road Analysis

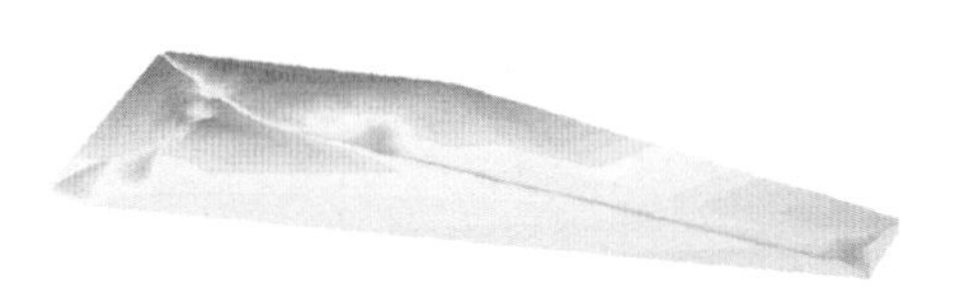

高程分析图
Elevation Analysis

功能分区图
Function Zones

视线分析图
Line of Sight Analysis

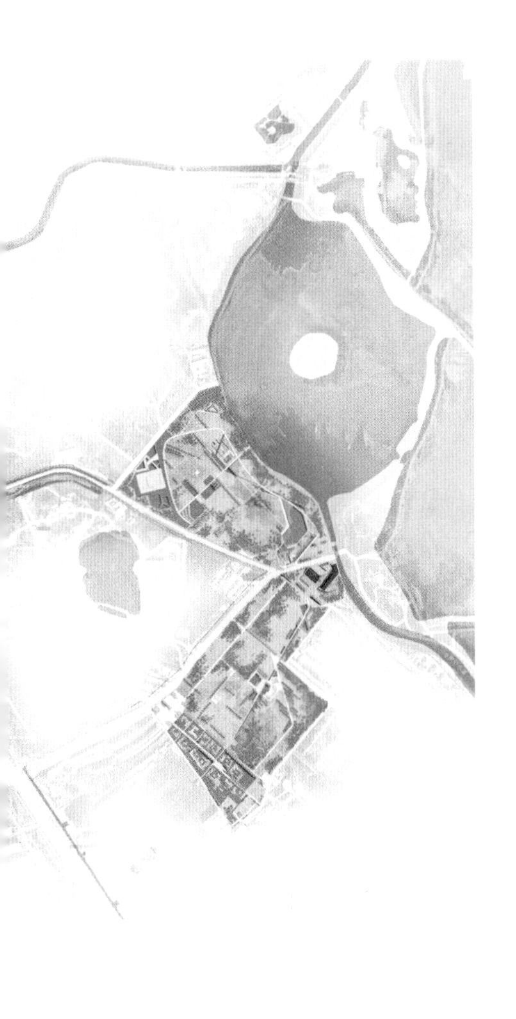

5 节点设计

廊道区设计定位：以历史文脉联系为线索，自西向东连接颐和园、玉泉山、香山等传统历史文化地段，同时沟通廊道两侧和河道等开放绿地空间。

颐和园西门区设计定位：以历史文化认知体验和休闲游憩为主的城市郊野公园，是北京西北郊郊野公园体系的一部分。设计以稻田收割等农事体验活动、园林式购物活动和砖窑制作体验为特色，吸引大量游客及居民前往，同时缓解颐和园北宫门等区域的人流压力。

场地自西向东可分为七大主要功能分区，分别为：

（1）游客集散区：靠近主干道设计大面积的硬质透水铺装，缓解人流集散压力，并引导场地与道路的水流进入河道之中；

（2）休闲游憩区：通过地形的变化与高空栈道的设计，丰富观感，同时结合丰富的游乐设施，成为游人休闲游憩的场地；

（3）生态保护区：保留原始场地植物，加以整理，形成大面积的下凹式绿地，以解决雨水压力；

（4）河道展示区：将原始的旱河加以改造，成为拥有丰富的植物造景与活动场所，并能引导雨洪的新河道；

（5）景观观赏步道：设计步道连接东西方向，成为一条以植物造景为主的景观廊道；

（6）历史体验区：设计砖窑景观游园、提供砖窑制作体验活动等方式，唤醒历史文脉，提供一个致敬历史的场所；

（7）田园观光区：曾经的京西稻田是这片土地的历史记忆，所以将这片区域打造成为宽阔的田园景观，供人游憩。

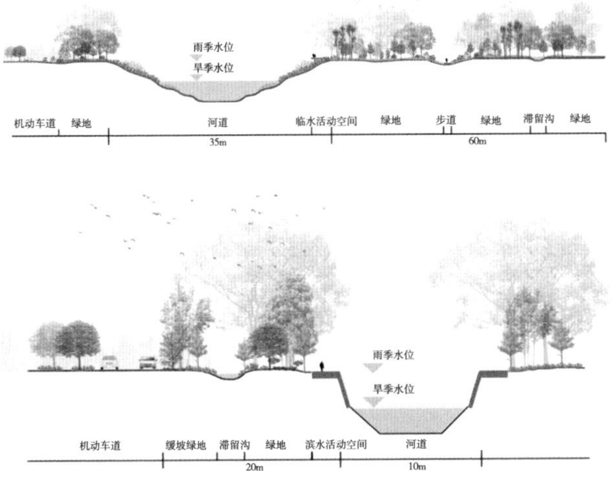

剖面图
Sections

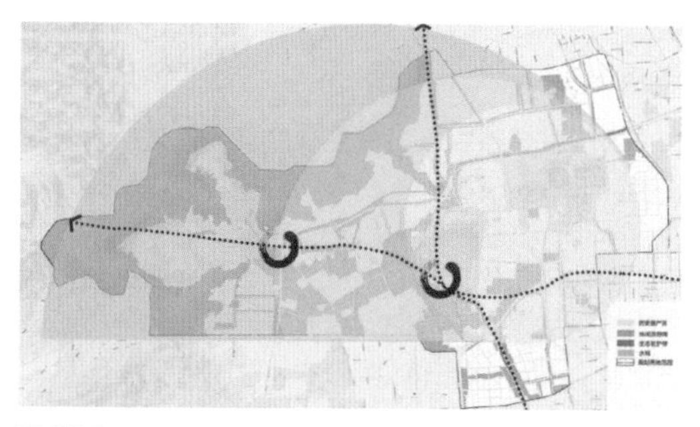

景观结构
Landscape Structure

现状道路
Existing Road System

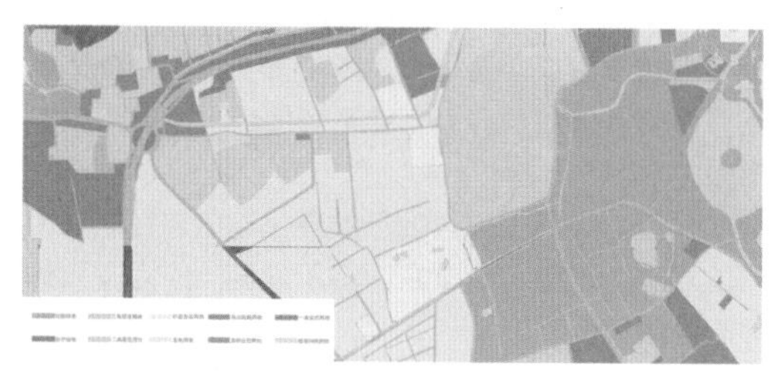

现状用地
Existing Land Use

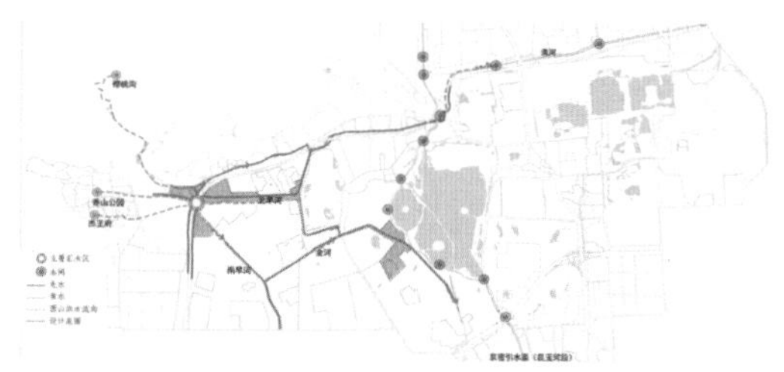

现状水系
Existing Water System

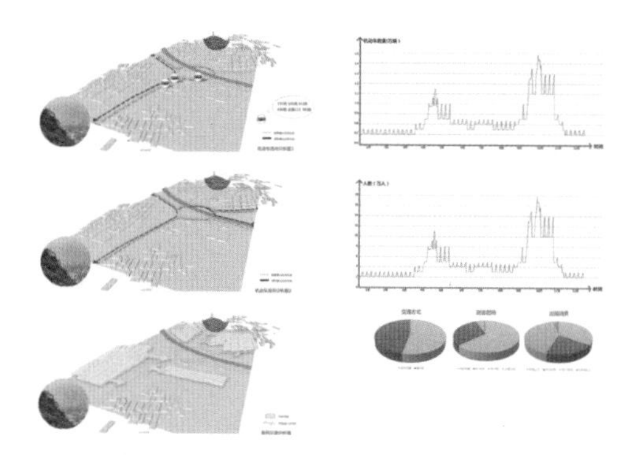

香泉环岛交通
Traffic at Xiangquan Roundabout

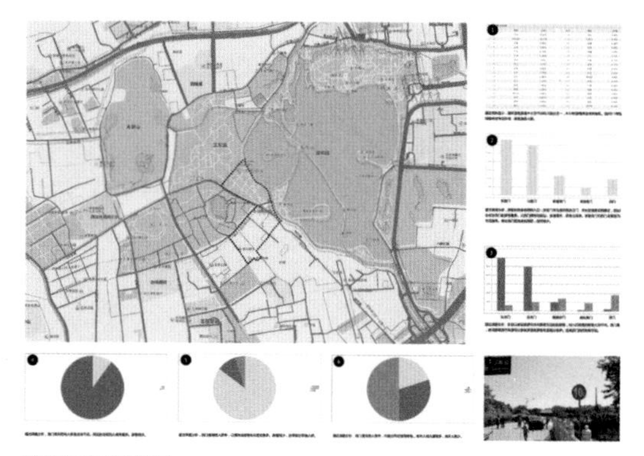

颐和园西门交通
Traffic at West Gate of Summer Palace

基于对现状整体调研做出对设计场地的分析：

现状自然资源条件：可以颐和园与圆明园等历史遗产为中心，分别划为历史遗产区、休闲游憩带、生态防护带与水域。这决定了我们的景观结构网络。

现状用地性质：可将设计区域分为公园绿地、三类居住绿地、二类居住用地、一类居住用地、行政办公用地、商业设施用地、农林用地、防护用地、文物古迹用地与教育科研用地，根据不同的用地性质，将相应进行不同的后期设计。

设计区域内分别有快速路、主干路、次干路、支路、西山隧道，不同的道路会对设计产生不同的影响。

从现状来看，香泉环岛周边居民区众多，是潜在游人数量最多的区域，其次是颐和园西门区域，临近景区，也会有一定量的游人，所以这两片区域是重点设计的地块。

用地内河流众多，为雨水设计带来了困难也提供了潜在价值。有些河道是有水的，我们可整合周边景观，将之改造为景观河道；另外一些为旱河的，我们可利用其成为新的景观水渠，以缓解周边水文压力。

香泉环岛地块交通情况复杂，机动车、非机动车流量均较大，周边地块功能复杂。设计中对该地块的行车方向与公交车站点进行了梳理。

颐和园西门周边游客量逐年增多，目前缺少一块满足游人消费需求的场地，同时颐和园皇家园林的历史记忆也需要传承。

鸟瞰图
Bird Eye's View

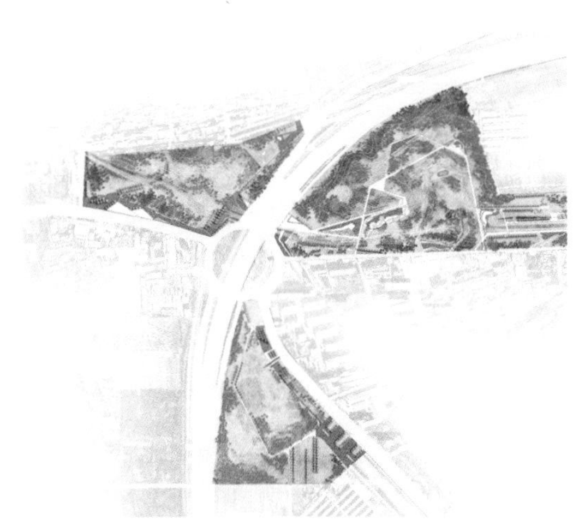

香泉环岛设计平面图
Site Plan of Xiangquan Roundabout

颐和园西门设计平面图
Site Plan of West Gate of Summer Palace

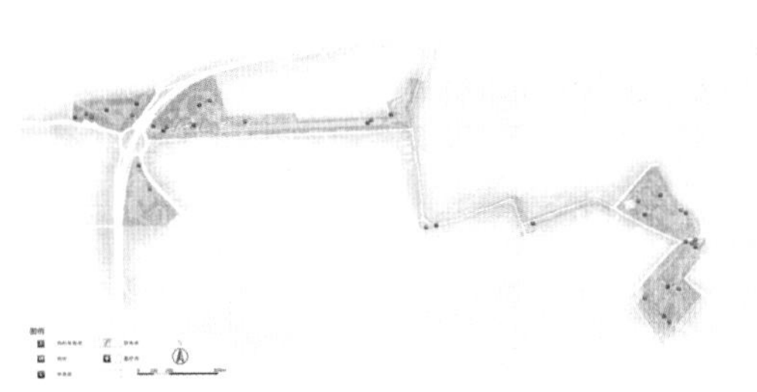

服务设施规划图
Service Facilities Plan

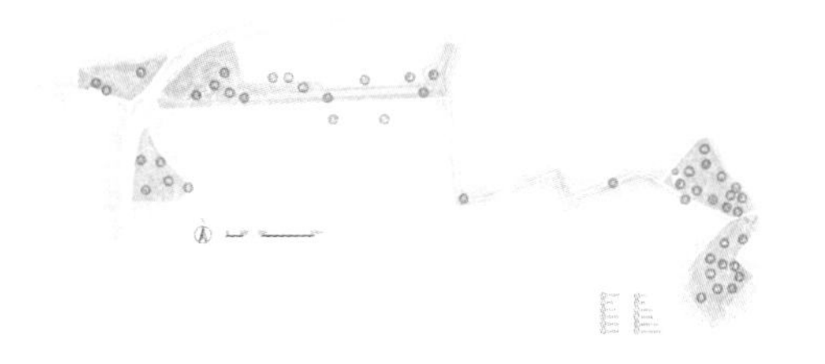

休闲活动规划图
Leisure Activities Plan

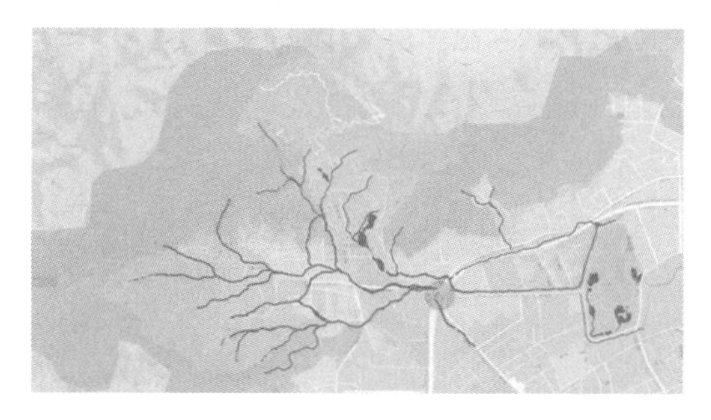

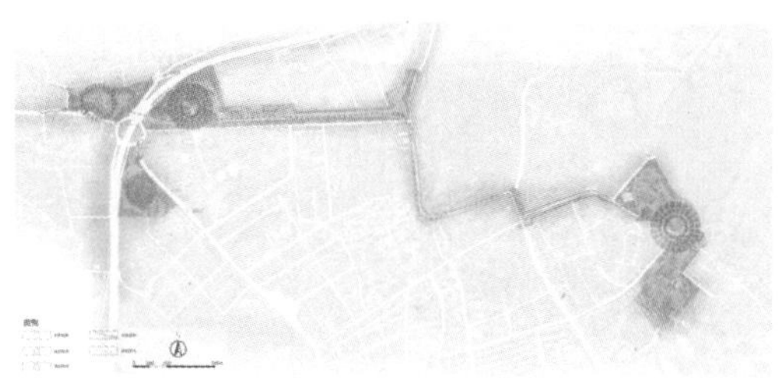

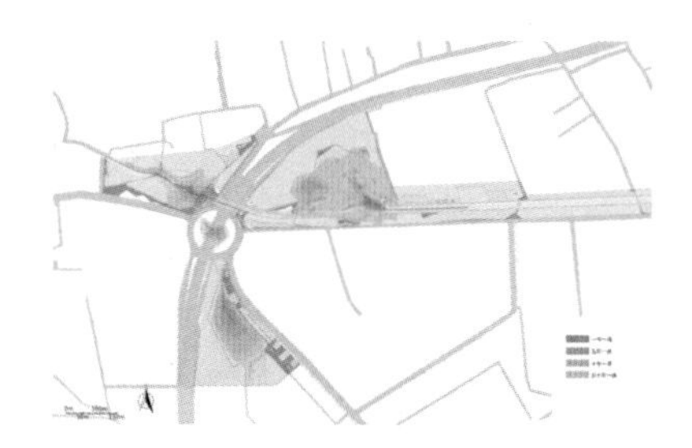

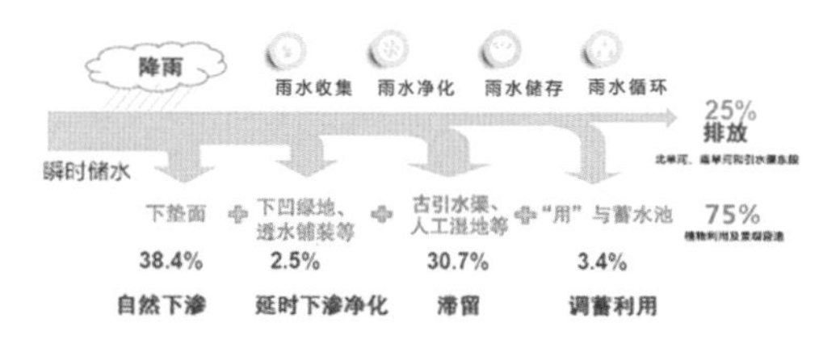

雨洪管理策略与措施
Stormwater Management Strategies

重现期（年）	1		5		10		50		100	
峰值流量（m^3/s）	3.74		17.48		23.51		37.76		43.82	
洪量（万 m^3）	4.85		25.35		38.31		71.10		86.24	
积水半径（m）	12.66		66.17		100.00		185.59		255.11	
平均水深（m）	0.10		0.30		0.36		1.48		1.80	
最大水深（m）	0.11		0.52		0.80		2.62		3.88	
积水量（Wm^3）	0.30		1.79		2.50		4.64		5.63	
保险面积（hm^2）	0.38		1.98		3.00		5.57		6.75	
建议面积（hm^2）	3.50									
场地 1	0.3hm^2	1.0m	0.5hm^2	1.5m	0.6hm^2	2.5m	0.8hm^2	3.0m	1.0hm^2	4.0m
场地 2	0.5hm^2	0.5m	0.7hm^2	1m	0.9hm^2	1.5m	1.2hm^2	2.0m	1.4hm^2	3.0m
场地 3	0.6hm^2	0.7m	0.9hm^2	1.5m	1.0hm^2	2.0m	1.5hm^2	2.5m	1.7hm^2	4.0m

由此，绿道选线与设计中以香山、万寿山、玉泉山这三山为主要的借景点 ，充分丰富景观视线，引导游览方向，打造空间结构多样性，达到步移景异的良好效果。

与其他地段所不同的是，香泉环岛地段具有复杂多样的特殊性。首先其作为重要交通枢纽，分布着众多公交车站、停车场等，担负着重要的交通组织功能，然而目前的交通现状比较混乱，人行系统被打断。其次，该地段地势较低，暴雨时极易积水，造成严重的交通压力，甚至导致交通瘫痪。通过设计，在解决人行交通畅通的基础上，整合区域内的水系统，将道路与场地的积水汇入原有的水道，以求一定程度上缓解暴雨所带来的灾害影响，同时收集雨水汇入下游成为景观水景。

颐和园作为皇家园林具有极大的旅游吸引力，周边游客众多，消费需求量大，故在场地内设计一商业片区以供需求。历史上此处为京西稻田的所在地，设计布置了大片的疏林草地与小范围稻田景观，以强化历史记忆。

服务设施：场地中分别设置了自行车租赁点、厕所、休息处、医疗所、饮水点等，以满足游客的不同使用需求。在场地的不同区域分别设计了野餐、草地排球、自行车、拓展训练、观鸟、日间露营、远眺观景、垂钓、农事体验、果树采摘、砖窑制作体验、园林式购物体验等丰富有趣的活动，给不同需求的游客提供了多样性的选择。

加强城市绿地雨洪管理，提高抗洪能力。在降雨时，通过雨水收集、净化、储存和循环等手段，将暴雨通过下垫面自然下渗，下凹绿地等进行延时下渗和净化，场地内的河流和人工湿地雨水花园等进行滞留和生态营造，并且将所存留的雨水进行利用，还对地下水和附近河流进行雨水补充，从而达到一种生态循环的过程。

整个场地中香泉环岛主要处理雨洪问题，通过旱河（引水渠）进行串联，在颐和园西门地区进行景观营造。在香泉环岛的雨水处理先通过地表径流的引导和设置暗渠在场地内进行雨水收集，通过植物和砂石进行净化，在下凹绿地和雨水花园中进行雨水储存，在整个过程中，下渗、灌溉、蒸发利用都进行了自然生态循环。

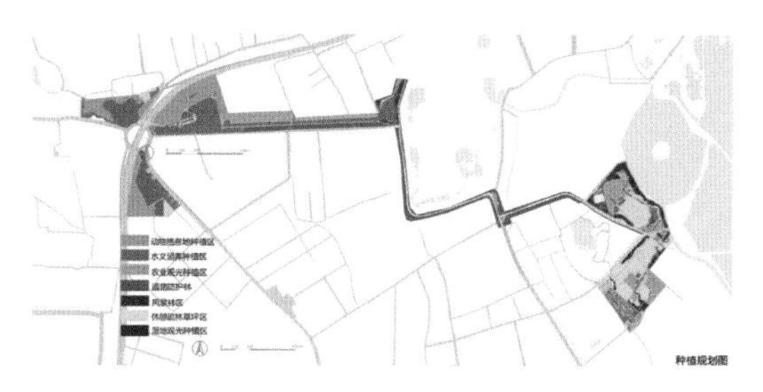

植被规划总平面图
Plan of Vegetation Planning

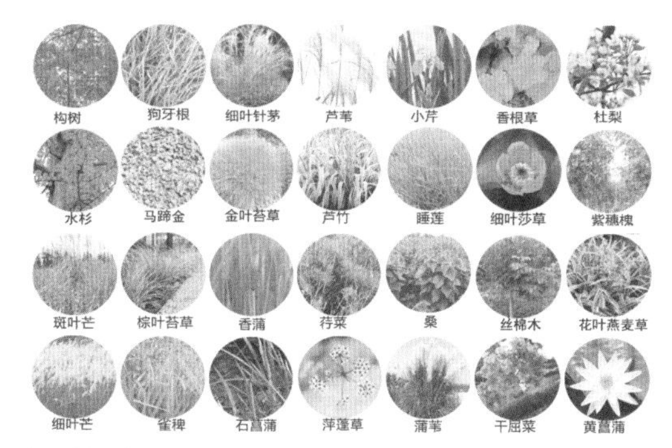

主要植物种类示意图
Schematic Diagram of the Main Plant Species

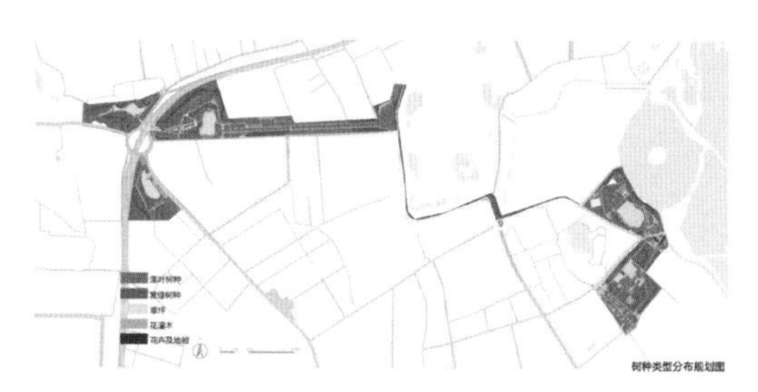

植被类型分布图
Plan of Vegetation Type Distribution

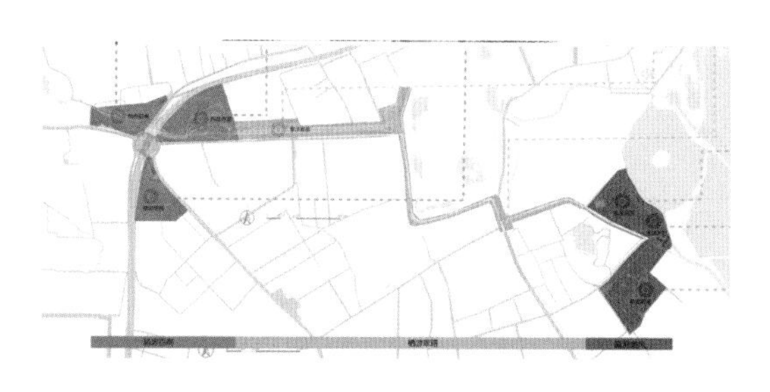

植被景观分区规划图
Plan of Vegetation Landscape Zoning

效果图 1
Rendering 1

效果图 2
Rendering 2

效果图 3
Rendering 3

6 植物分区规划

动物栖息地种植区：复层混交植被，满足动物生长环境需求。

水文涵养种植区：耐水湿植物，防汛抗涝。

农业观光种植区：狗尾草、荷、经济树种为主，营造农田体验。

道路防护林：廊道侧植常绿树，起防护作用。

风景林区：以观花、观果、观干等观赏植物为主，营造优美景观。

休憩疏林草地区：以开敞草坪结合疏林，供游人休憩。

湿地观光种植区：以荷塘湿地为主的湿地景观，再现历史荷塘风貌。

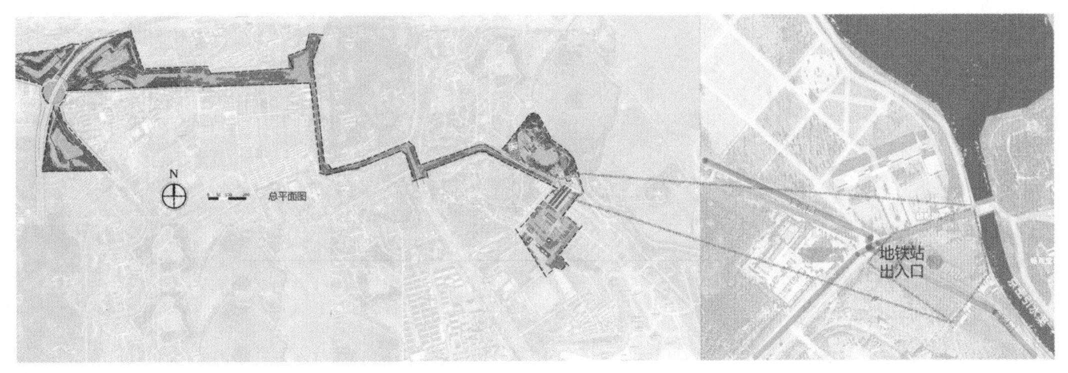

建筑区位图
Location Map

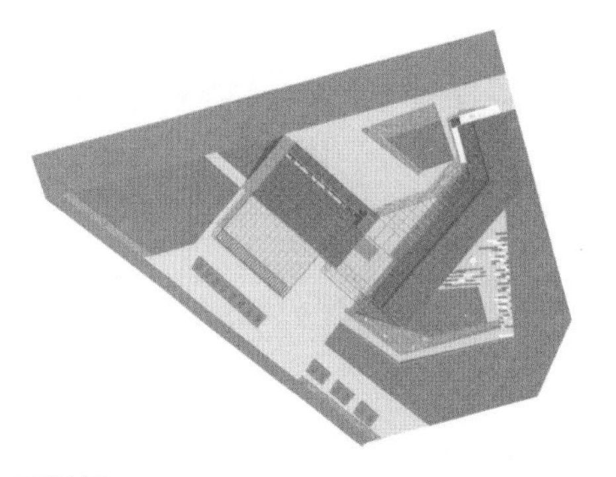

平面布局
General Layout

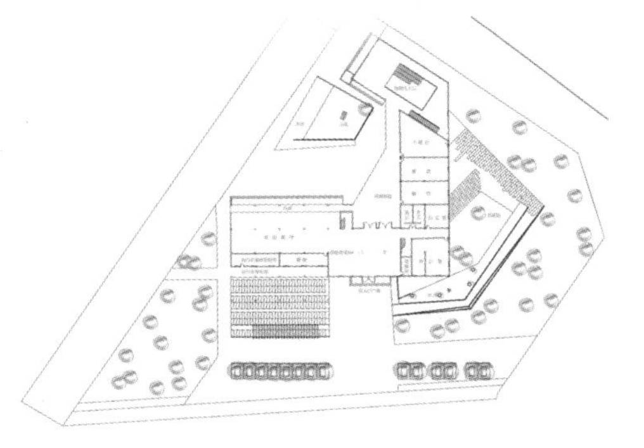

首层平面图
First Floor Plan

整体轴侧图
Overall Isometric View

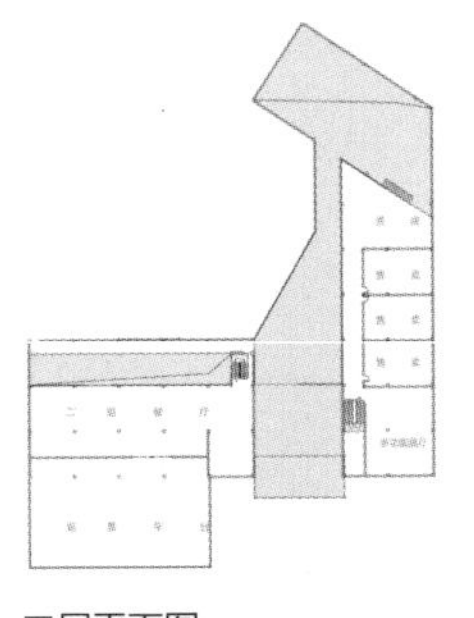

二层平面图
Second Floor Plan

建筑外立面图
Elevation

效果图 4
Rendering 4

效果图 5
Rendering 5

效果图 6
Rendering 6

建筑剖透视图
Architectural Sectional Perspective

建筑剖透视图
Architectural Sectional Perspective

7 游客中心建筑设计

游客中心建筑选址紧邻颐和园西门，绿道从场地北侧穿行而过，地铁香山线将在此设置出入口，未来将有大量人流由此进出，同时也会成为重要的绿道枢纽。

根据南北两个公园和颐和园西门的位置确定了建筑的轴线和朝向，并结合设计面积确定建筑体量。参照北京传统民居平面布局，入口—影壁（转向）—垂花门—院落—连廊—正房，布置了从颐和园西门或地铁出入口到建筑主入口的流线：出口—小品—草坪—透明廊道—主入口。

建筑外立面的色彩与元素借鉴了北京传统民居和周围环境的融合。根据北京传统民居建筑的立面材质，选取了灰瓦、青砖和红木等作为建筑外立面的主要材料。

建筑外环境的设计中提取传统园林水、树、墙、踏步等元素，结合现代手法，营造传统园林结合现代景观的空间体验。

规划方案六 及 北宫门－安河桥北片区概念设计

Plan 6 & The Conceptual Design of the Summer Palace North Gate-Anhe Bridge Northern Section

佟思明、崔滋辰、李璇、王晞月、李媛、徐慧、莫日根吉、Mosita
Tong Siming/Cui Zichen/Li Xuan/Wang Xiyue/Li Yuan/Xu Hui/Morigenji/Mosita

本规划方案中，绿道的设置除充分考虑旅游需求外，还细致地考虑了本地居民的日常生活需求，使绿道成为城市居民出行的重要选择，此外对绿道与道路、绿道与公交线路及站点以及绿道与绿道之间的衔接节点进行了分类设计，以确保整个体系的完整与贯通。

北宫门－安河桥北片区环境要素复杂且各地块受到快速交通的割裂严重。设计中，着重打造了跨越多条机动交通干道的步行景观环，使山林、滨水绿色游憩空间与城市开放空间紧密衔接，形成内容极为丰富的景观核心。

In this plan, we not only responded to the demand for tourism in the greenway design, but also carefully considered the daily life needs of the local residents, making greenway an important choice for urban residents to travel. In addition, we accomplished classification design of the connection points between greenways and roads, also greenways and public transport hubs, so as to ensure the integrity and coherence of the whole system.

The environmental elements of this area are complex and each block is severely fragmented by rapid traffic. Accordingly, we focused on building a pedestrian landscape ring spanning multiple motorized traffic arteries, so that the green recreational space of mountains and waterfront can be closely linked to the urban open space, and together they will form a landscape core with rich contents.

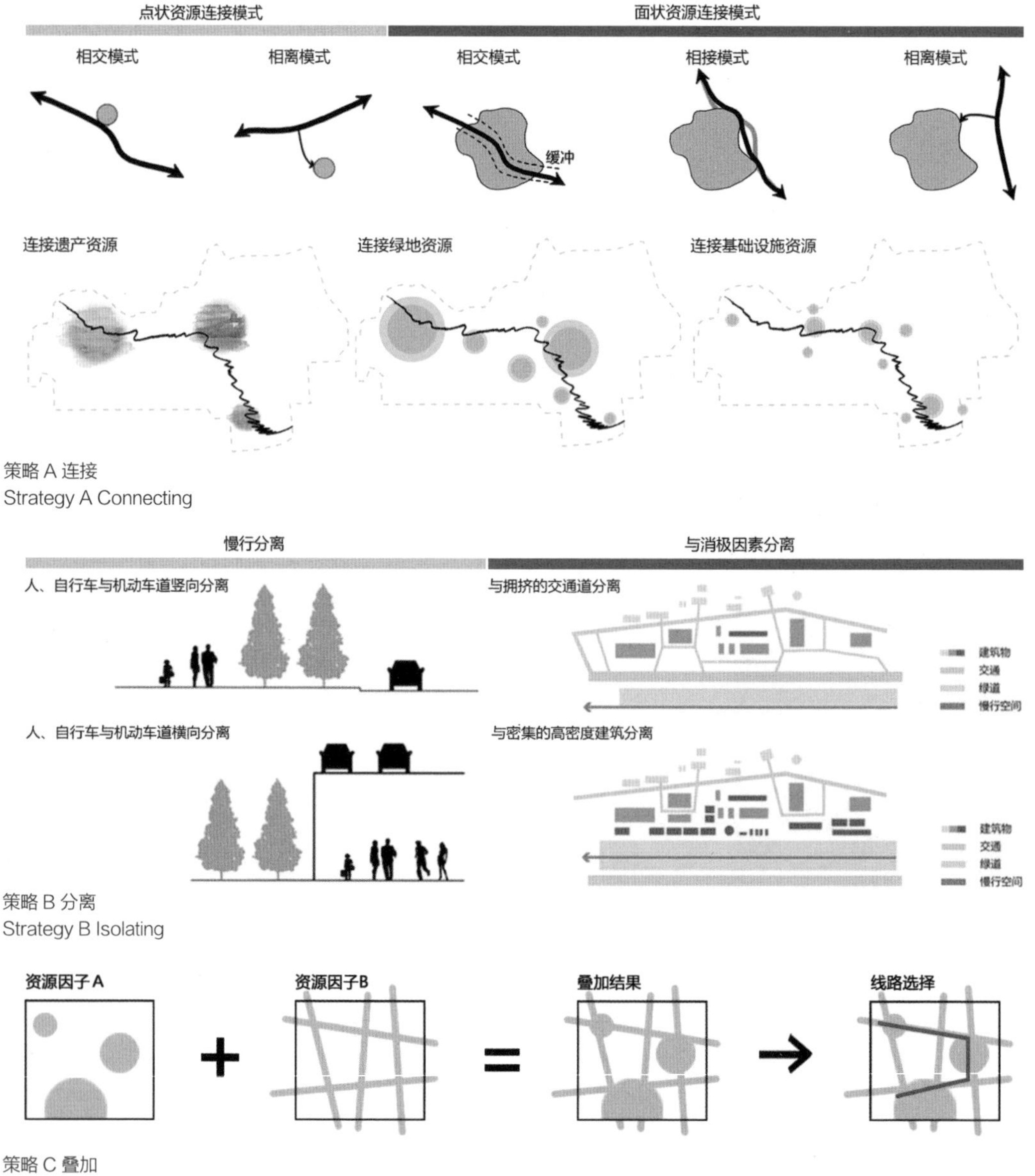

策略 A 连接
Strategy A Connecting

策略 B 分离
Strategy B Isolating

策略 C 叠加
Strategy C Overlaying

三山五园地区位于北京市西北郊，现有绿地面积 23.07km^2。三山五园区域是海淀区的重要部分，包括其功能区规划中的休闲旅游功能区和科教功能区。海淀区东西向的皇家园林景观游览轴穿过该区域。同时，三山五园地区位于北京市的生态楔形绿地之内，具有重要的生态价值。

地块内现有林地果园、现代城市公园、皇家园林、社区公园和河流等多样化的自然资源。城市用地以居住用地、商业用地、科教用地和绿地为主。

三山五园在北京城市发展中有着重要历史和公共教育意义，但在城市化进程中正逐渐被肢解成为点状景点，从而减弱发挥片区的历史文化保护和生态服务的功能价值。

本规划尝试以绿道作为载体，从历史演变和现状条件研究入手，解决这一地段发展中存在的问题。三山五园绿道除了具有普遍意义上绿道的生态保护、生活休闲功能外，同时具有历史文化遗产保护与传承、辅助旅游业发展功能。通过现场调研，对场地可利用及可恢复的资源进行整理，并通过筛选、叠加等手段，形成以文化遗产型绿道为主，兼有健康

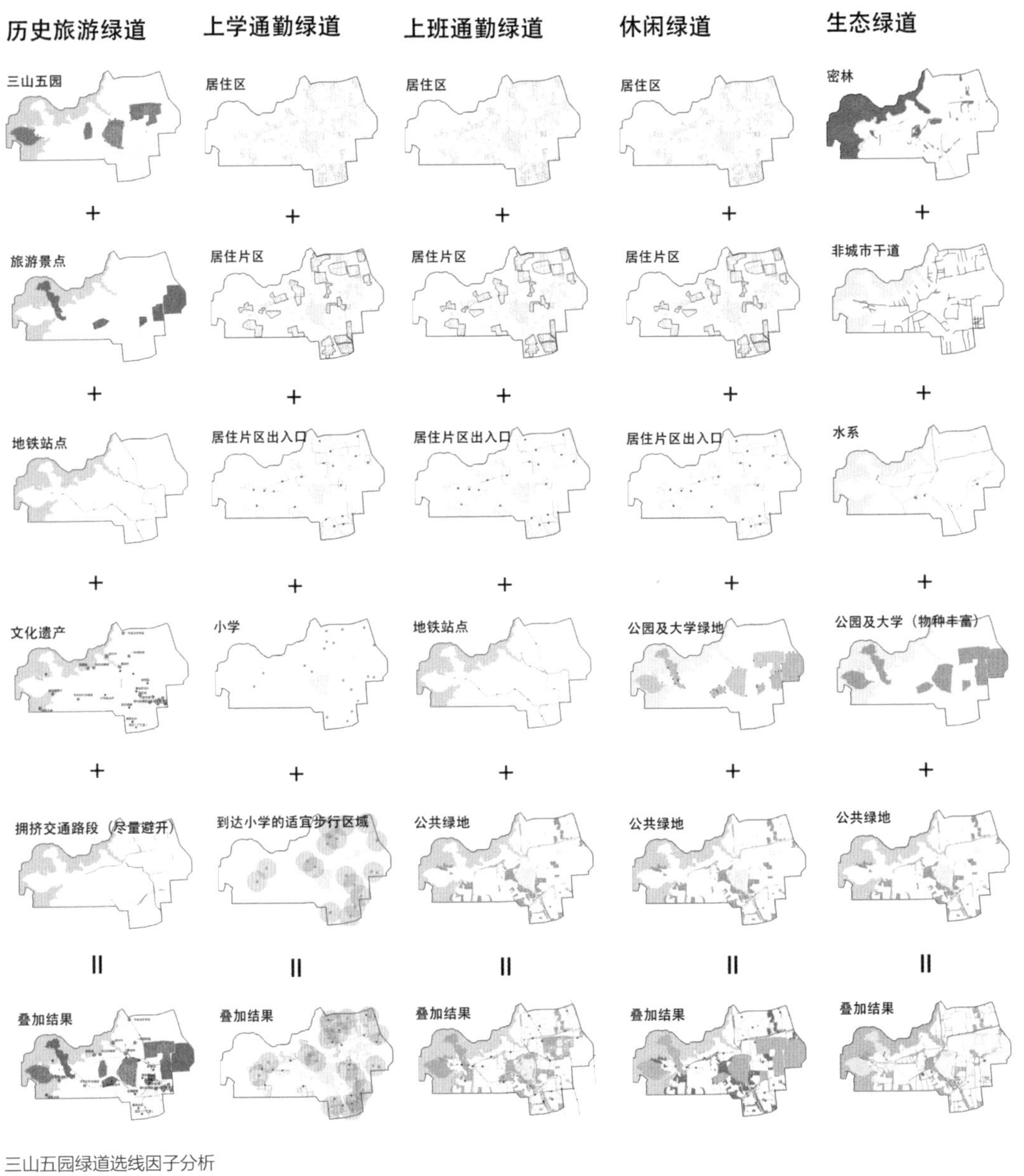

三山五园绿道选线因子分析
Factors Analysis

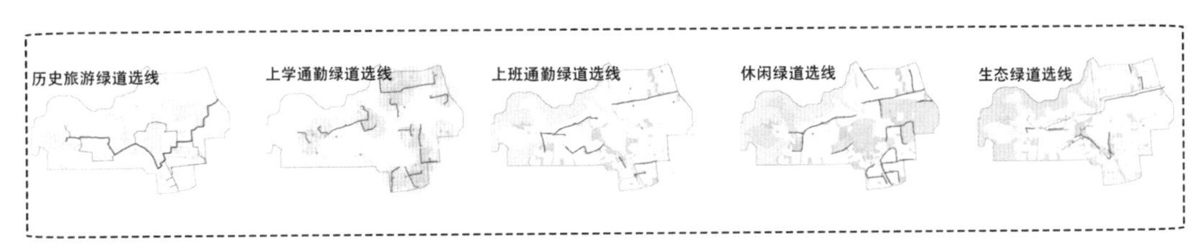

三山五园绿道选线因子叠加
Superposition of the Factors

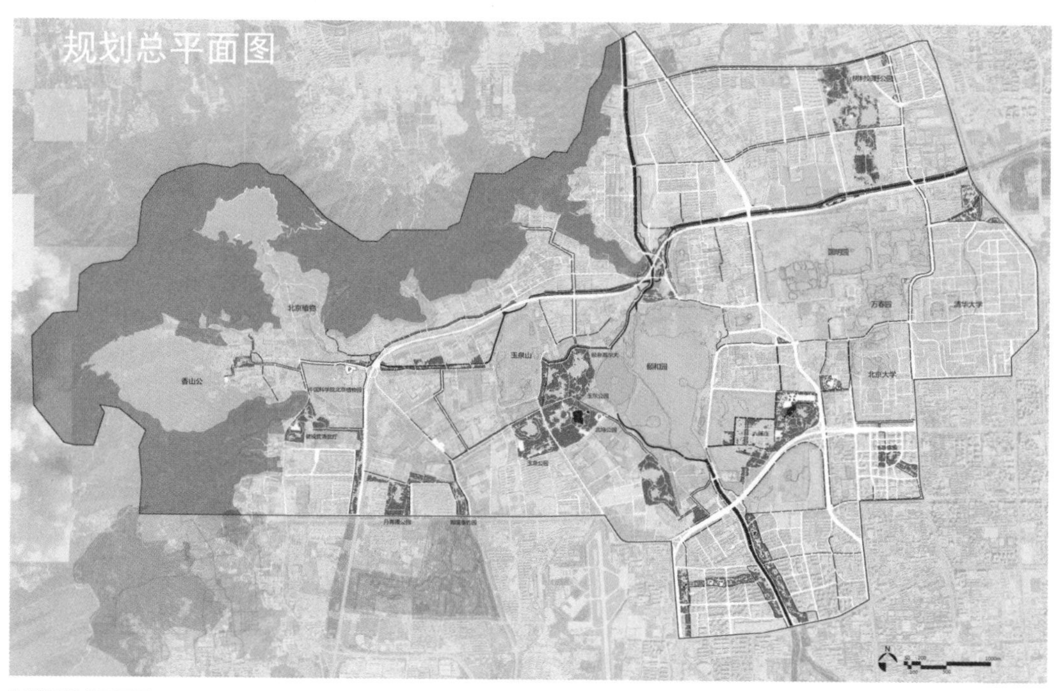

绿道规划总平面图
Master Plan

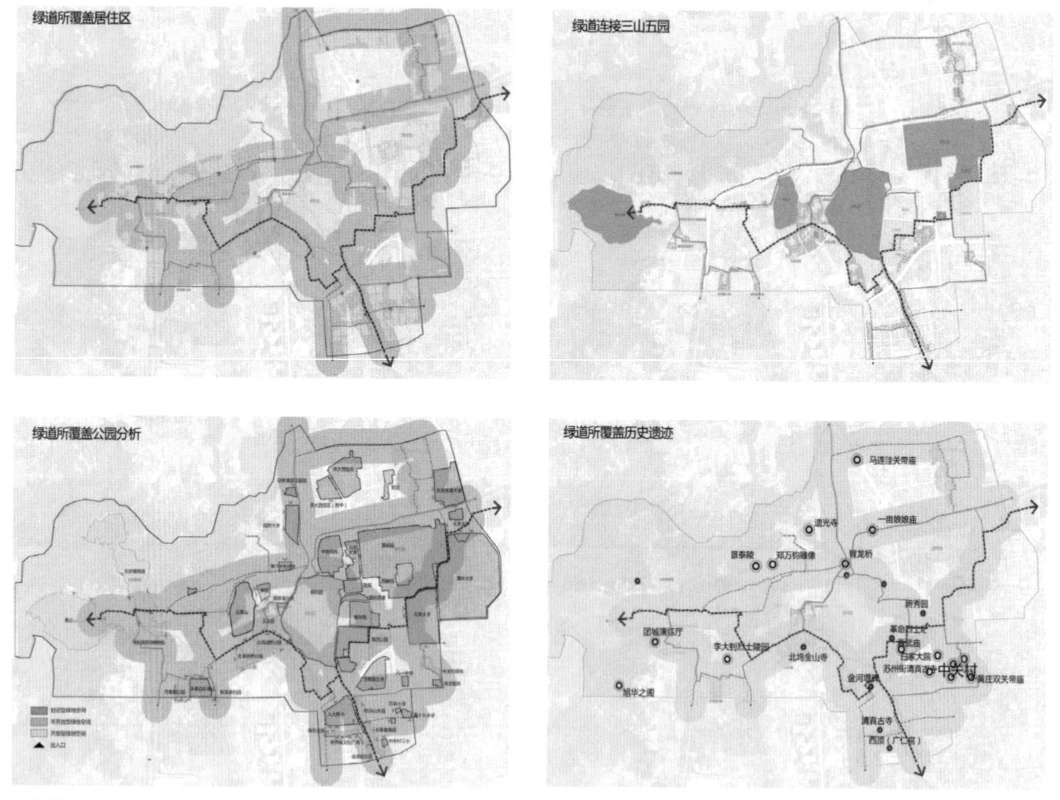

绿道价值评估
Evaluation of the Green Way System

绿道分类图
Categories of Greenways

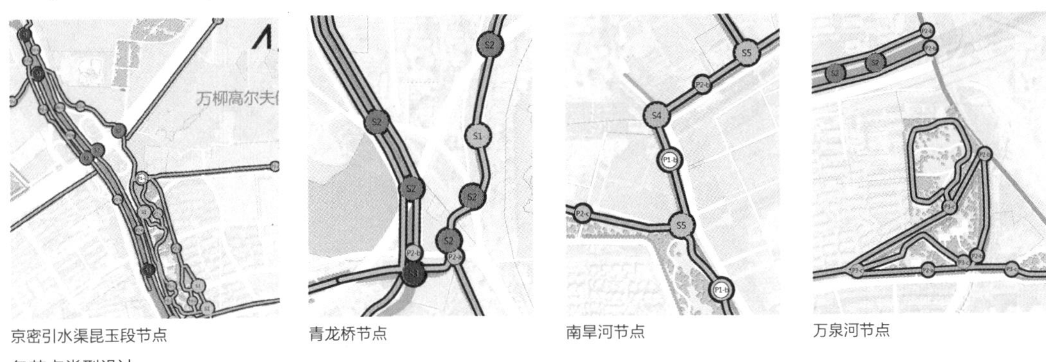

京密引水渠昆玉段节点　青龙桥节点　南旱河节点　万泉河节点

各节点类型设计
Design of Nodes

休闲、生态保护功能的城市绿道，以达到保护环境、传承历史、提升景观风貌、完善城市功能、促进经济发展及改善民生的作用。

三山五园绿道规划主要依托以下策略：

连接策略：绿道在城市尺度上是线性的开放空间，三山五园通过连接文化遗产、生态要素、公园绿地、基础设施等场地现有资源，形成绿道网络的骨架。设计通过分析形成绿道连接点状资源和面状资源的模式类型：点状资源可通过相离和相交的方式与绿道连接，绿道可以通过相交、相接、相离三种方式与面状资源连接。

叠加策略：针对绿道具有的不同功能，在街区尺度上，将绿道选线需要考虑的资源因子，如地铁站，公园出入口、居住区出入口、河道、绿地等进行叠加，从而得到最优的线路选择结果。

分离策略：包括慢行分离策略，即为了营造舒适的慢行环境，在街道尺度上，绿道规划尽可能在垂直空间或水平空间上将慢行道与车行道分离，同时采用与消极因素分离的策略，使绿道尽量远离拥挤的交通地带和高密度、高人流量的建筑群，提升绿道慢行与活动体验，同时保护绿道作为生态廊道的功能。

通过现场调研分析，将场地现状的可利用资源基底进行分类分析，主要归纳为：历史文化资源、

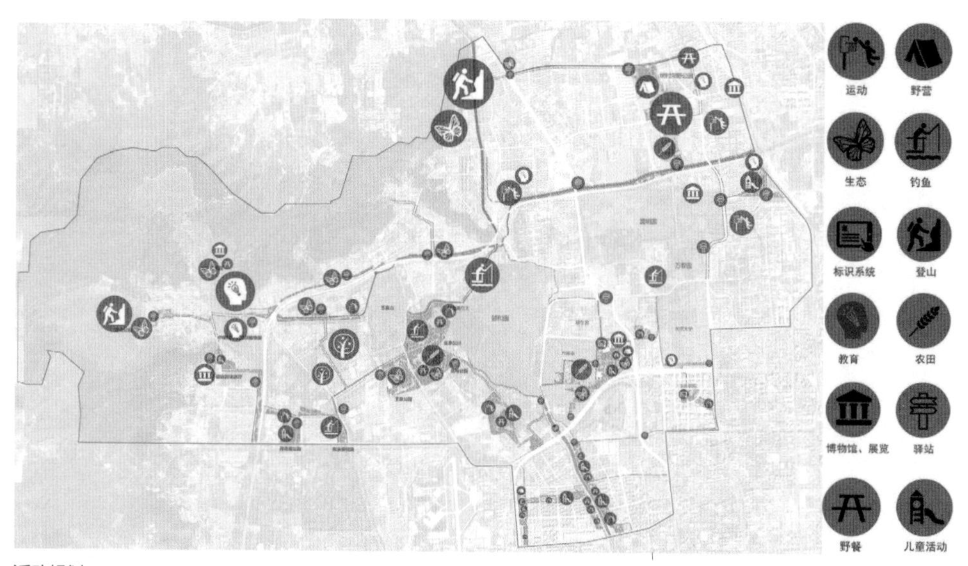

活动规划
Activities Plan

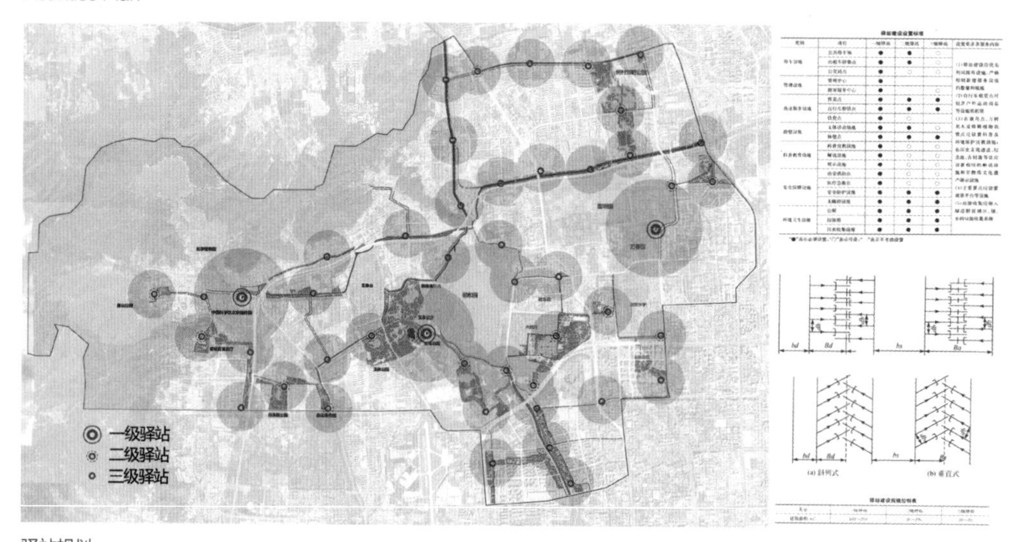

驿站规划
Service Point Plan

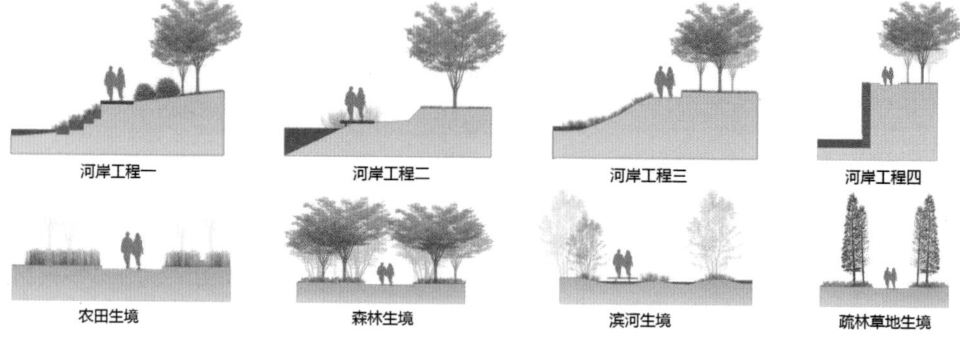

生态措施
Ecological Strategy

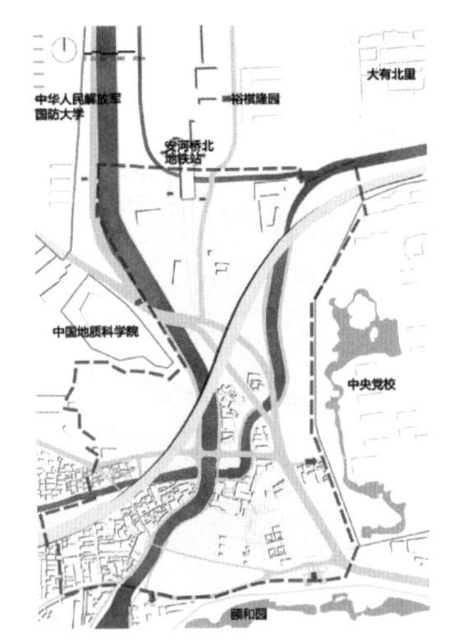

单位出入口分析
Analysis of User's Demand

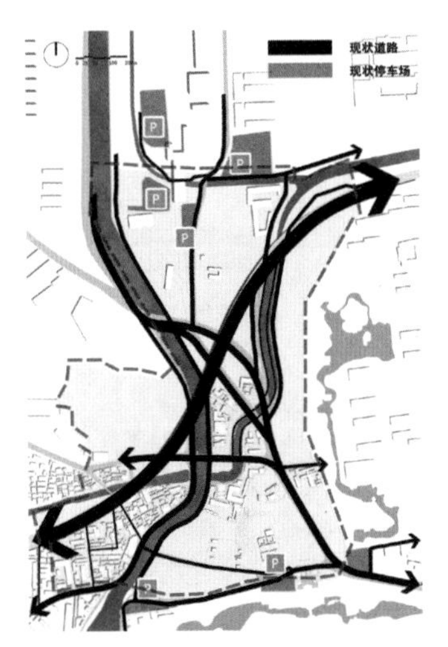

停车场分析
Analysis of Parking Lot

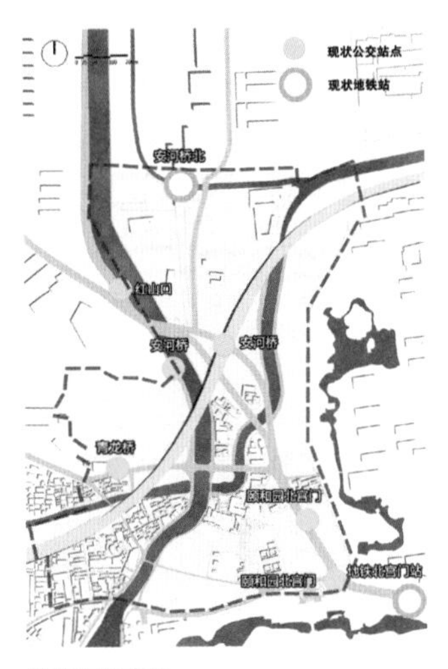

公共交通分析
Analysis of Public Transportation

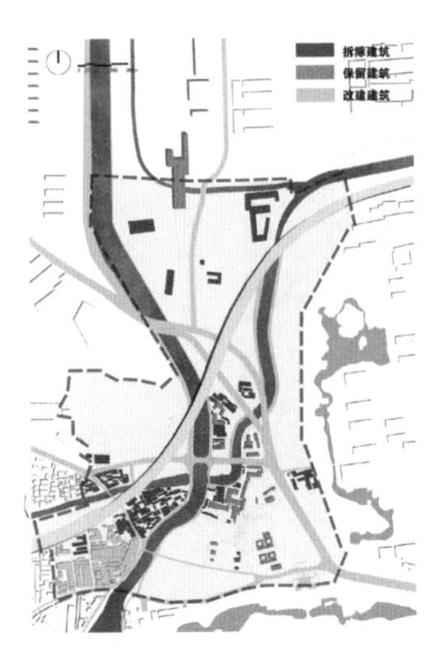

现状建筑分析
Analysis of Building

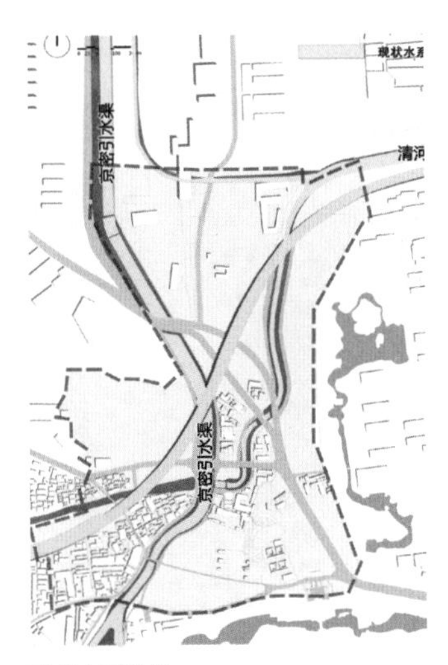

现状水系分析
Analysis of Water

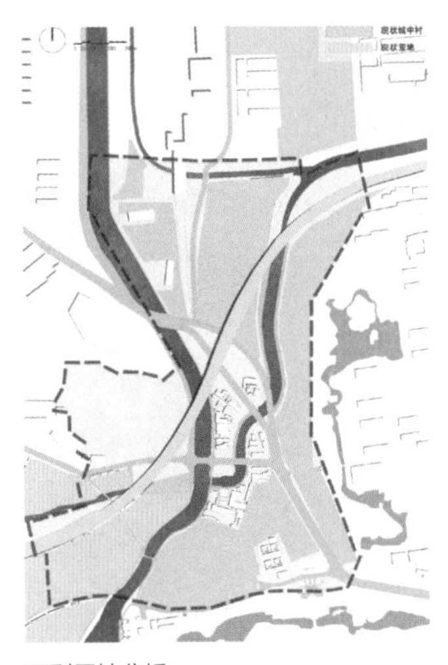

可利用地分析
Analysis of Available Land

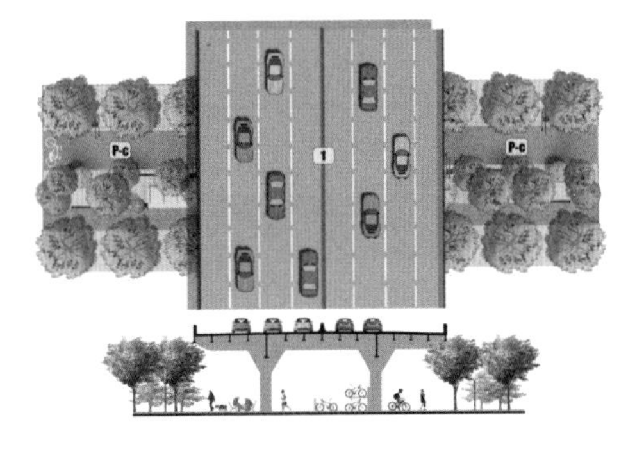

交叉口组织
Design of Road Intersections

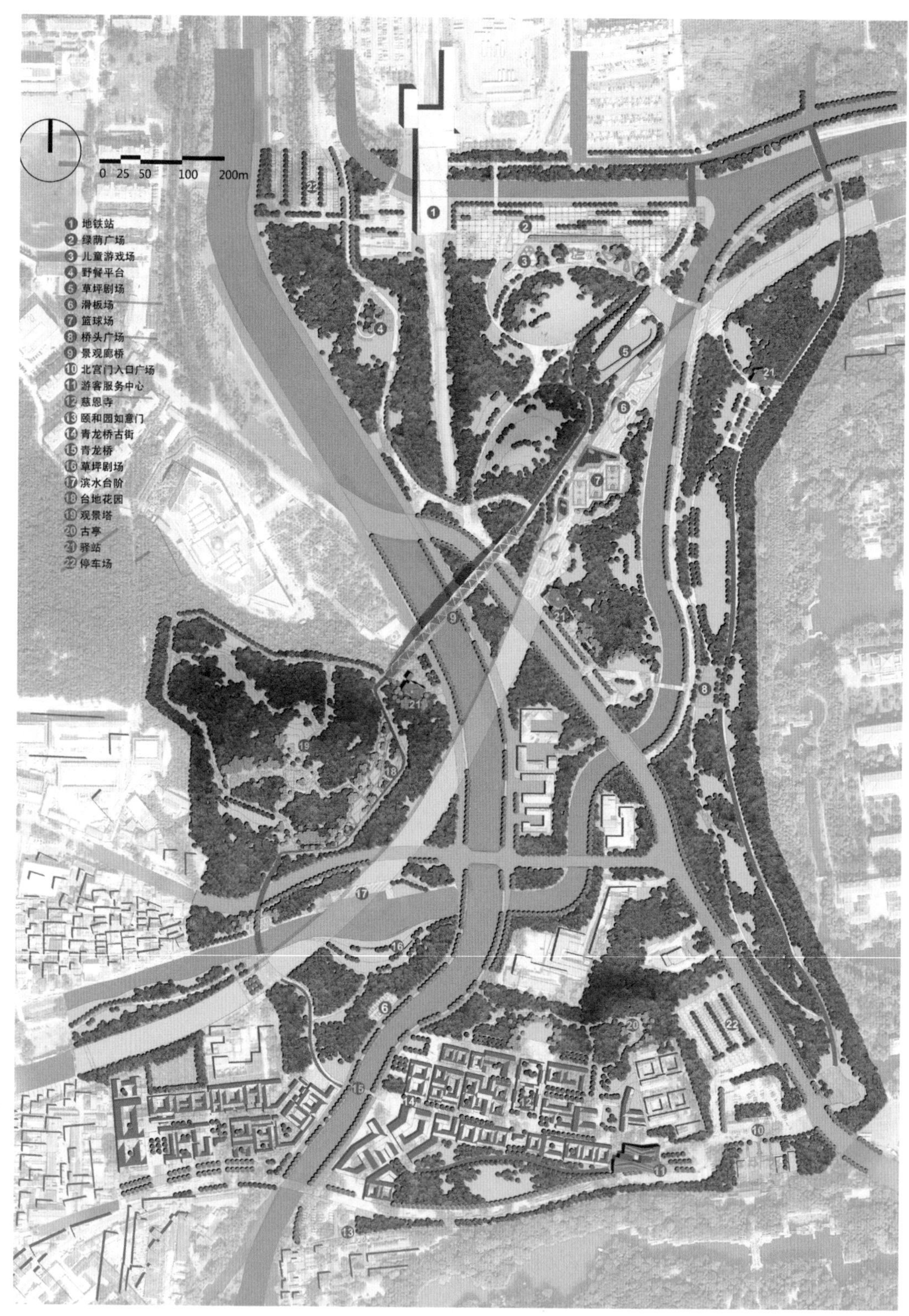

北宫门－安河桥北片区设计总平面图
Site Plan

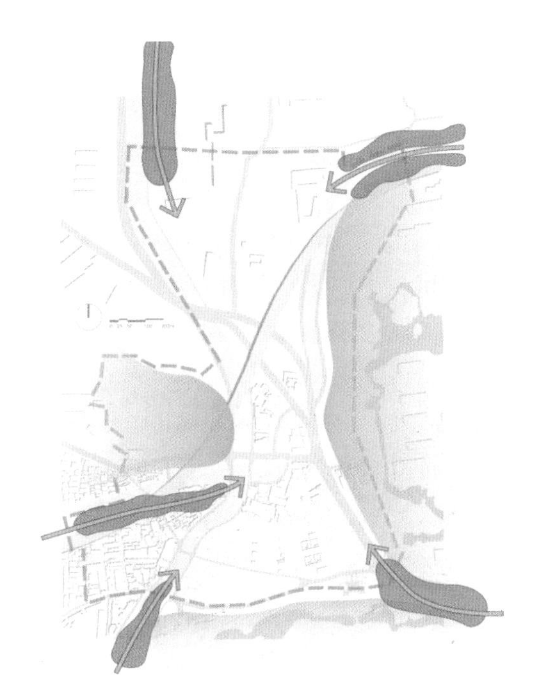

外部绿道分析
Greenway Analysis

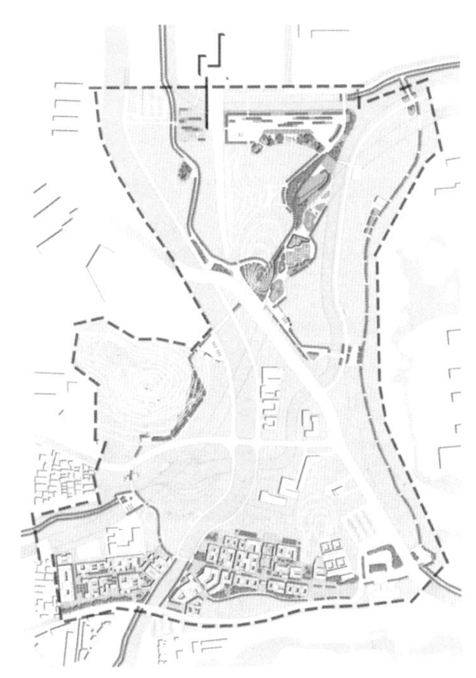

环绕绿道结构
Circle Greenway

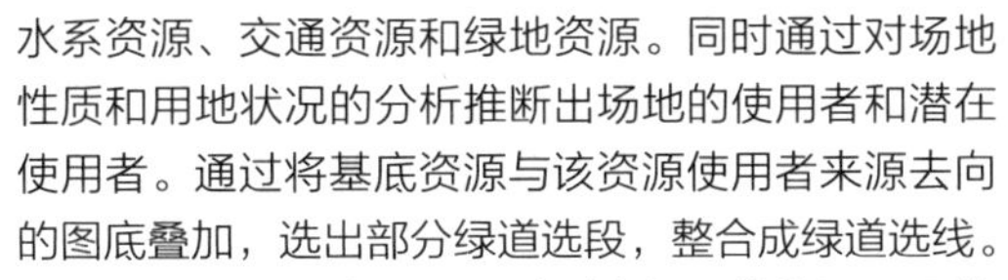

水系资源、交通资源和绿地资源。同时通过对场地性质和用地状况的分析推断出场地的使用者和潜在使用者。通过将基底资源与该资源使用者来源去向的图底叠加，选出部分绿道选段，整合成绿道选线。

绿道路径确定后，通过对路径沿线资源及现状条件的分析将绿道进行分类分级，包括沿道路绿道路径、沿水系绿道路径和沿绿地绿道路径。同时，根据绿道可利用宽度将每类绿道路径分为两级，形成绿道路径规划图，并对各级各类的绿道设计制定导则，以指导各区段详细设计。

规划完成的三山五园绿道全长 78.8km，总面积 557hm^2。三山五园绿道贯穿现有城市生态廊道，囊括现有生态资源，成为城市的蓝绿色基础设施。同时，三山五园绿道形成完善而舒适的慢行系统，为周边居民提供日常的休闲活动和通勤的绿色空间。在古迹保护传承方面，三山五园绿道贯穿北京西郊的旅游资源，构建了旅游资源框架，完善了历史文化资源的建设，并为未来发展奠定基础。

绿道选线完成后，分析了绿道与居住区、皇家园林、历史遗迹、城市公园的关系。沿绿道选线以 500m 为半径划出绿道的服务范围。经评估，设计绿道服务范围囊括了此区域 90% 的居住区出入口，连贯所有皇家园林，覆盖了 84% 的历史遗迹点和 100% 的城市公园出入口。

重要地段设计场位于颐和园北宫门至安河桥北片区，总面积 48.6hm^2。周边有玉泉山路、马连洼西路等城市主干道。地块在颐和园以北、中央党校以西。场地交通复杂，人行舒适度低，交通功能与绿地功能矛盾冲突。

以三山五园绿道规划为基础，选择具有代表性、矛盾突出的地块作为详细设计的场地。所选场地基础条件复杂，地块割裂、交通混杂，五条绿道在场地中汇聚。针对现状条件，创新性地采用外围环绕绿道作为解决策略，将周边绿道环通，同时最大程度地利用零散地块。

设计区域内部有多条快速路、城市主干道及高架穿越，使用者对交通需求量大。场地北边临近安河桥北地铁口，西边临近中国地质科学院出入口，南边临近颐和园北宫门。场地北部安河桥北地铁口附近有多处停车场，南边靠近颐和园，有专用停车场。场地内部有多处公交站点，北部临近安河桥北地铁站，地铁北宫门站位于场地东南向。场地中部有多处桥梁，如安河桥、青龙桥。京密引水渠贯穿场地南北，清河呈东北 - 西南走向。

总体上，场地现状具有以下特点：

场地优势：①可利用空间多；②临近居住区、院校、历史遗迹，周围文化气息和生活氛围浓厚。

场地劣势：①现状交通复杂，多条城市道路聚

青龙古街鸟瞰图
Bird Eye's View of Qinglong Traditional Street

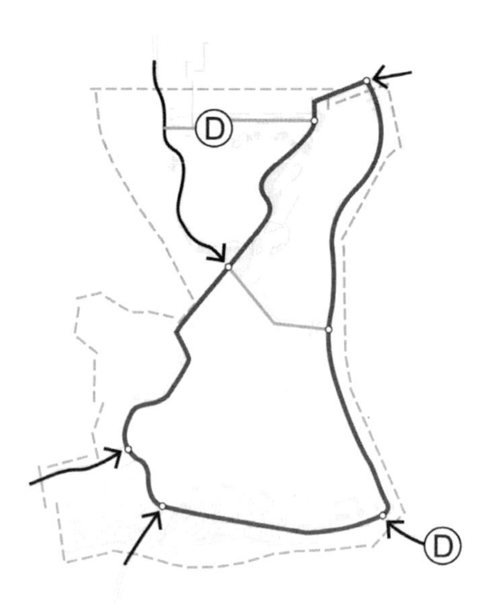

一条慢行环线
A Circular Slow Road

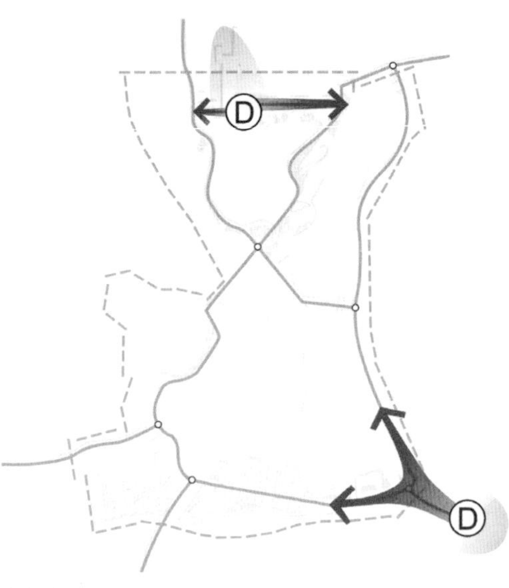

两处门户
Two Gateways

四片功能区
Four Func

集在场地中部；②地块被道路分割，连接性差，利用程度低；

机遇：整合可利用空间，传承历史文化，丰富生活体验。

挑战：面临机动车交通对活动空间的干扰。

在总体规划中，有五条绿道深入场地内部，如何结合场地的复杂交通，将五条绿道联系起来，是设计要解决的主要问题。充分利用场地四周的空地，并设计丰富的活动场地。

原场地地块破碎，周围游赏空间被机动车道阻隔，人行步道接洽处的可利用空间小。设计策略为通过局部调整机动车道，扩展外围可利用空间面积。同时，设环形慢行绿道串联四周可利用空间，减少机动车对慢行道的干扰，营造舒适的步行环境。

慢行环线设计：通过营造这条环形宜人舒适的慢行道，增强与区域间的联系；通过塑造环线上的活动节点，提升居民和游客的生活、游览品质；同时，环线与外围五条绿道联通，将场地外围的文化与活力潜入场地内，以发挥这个地块作为城市连接点的作用，改善其难以利用的交通点现状。

两处门户：包括位于场地内北侧的安河桥北地铁站及场地外东南侧的北宫门地铁站，是绿环上的两大门户，这里人流量大，机动车交通复杂，是进入环线的主要入口。

四片功能区：场地北侧为居民区、东侧为学校（中央党校）、南侧为文化遗产（颐和园）、西侧为山体，结合周围环境，绿环北、东、南、西四面功能定位为：

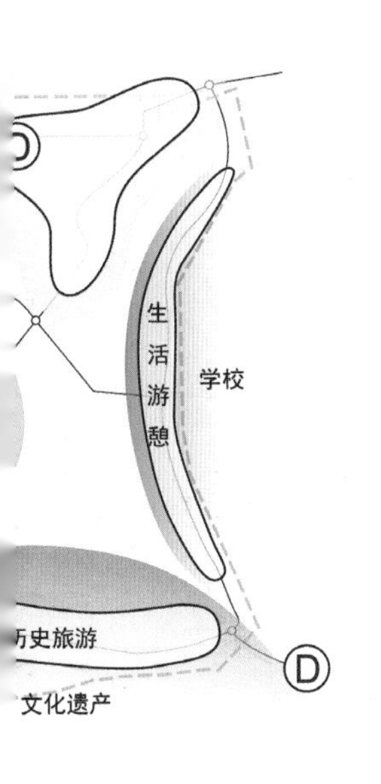

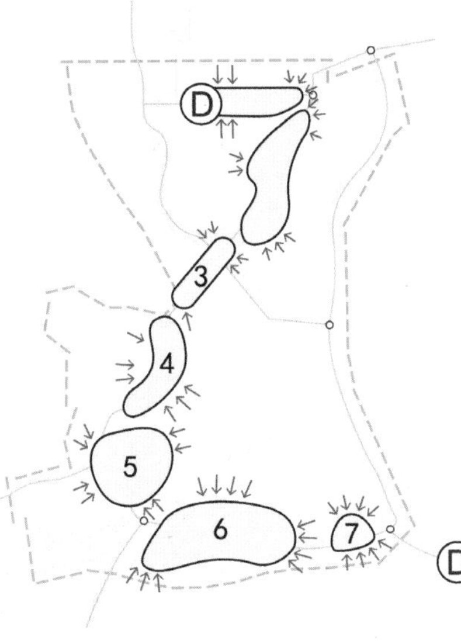

七个活力节点
Seven Energetic Nodes

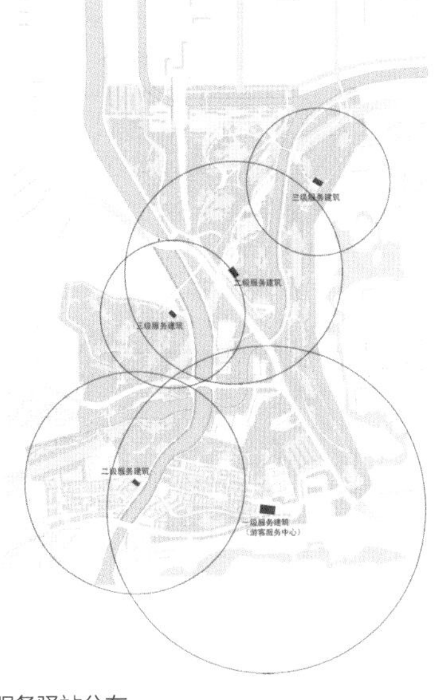

服务驿站分布
Service Points

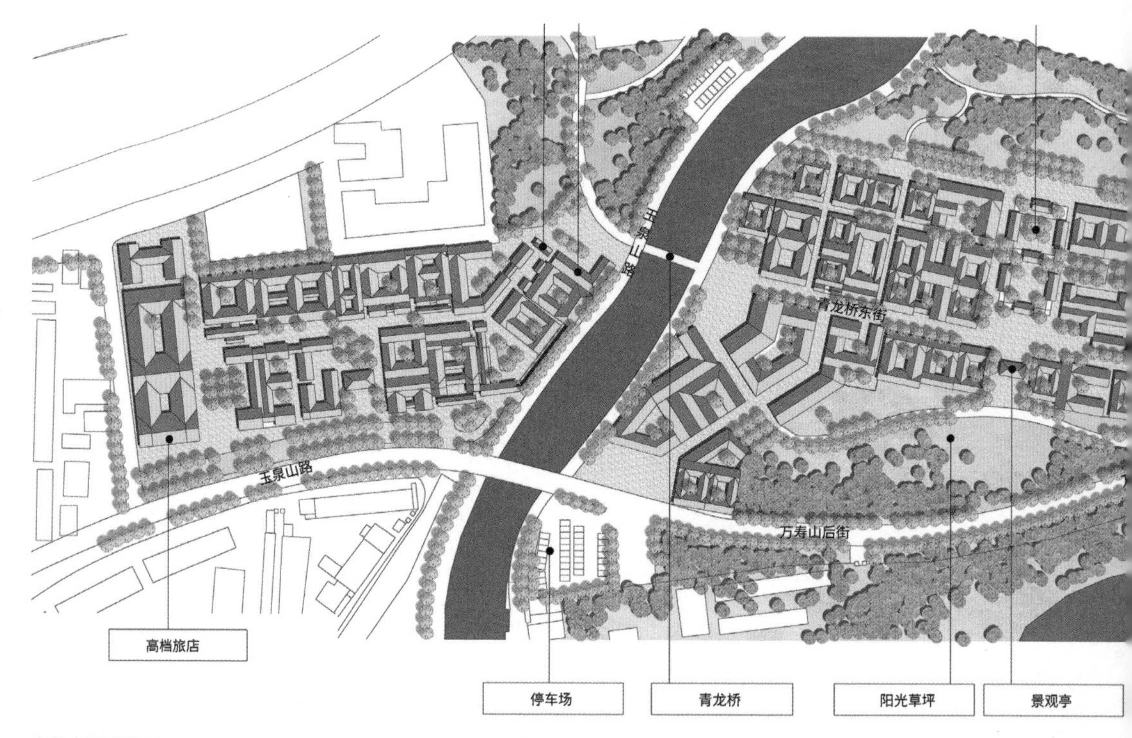

青龙古街平面图
Qing Long Historical Area Site Plan

青龙古街鸟瞰图
Aerial View

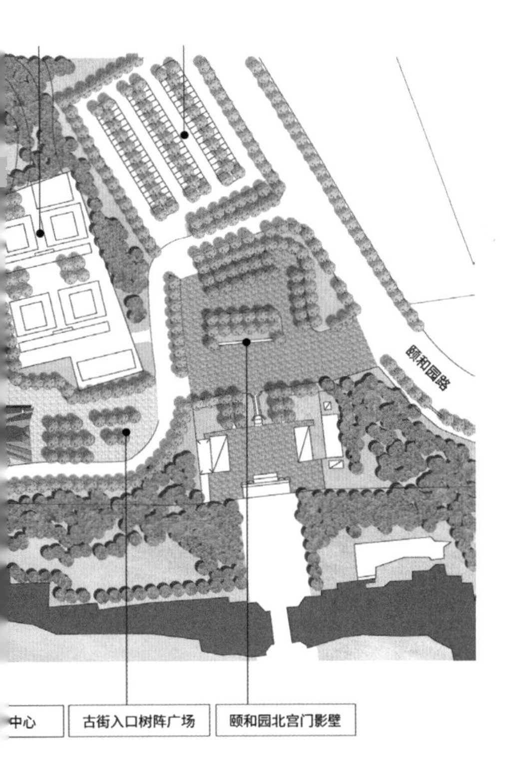

青龙古街区位
Location

青龙古街效果图
Qing Long Historical Area Rendering

生活运动、生活游憩、历史旅游、生态休闲。

七个活力节点：对应四个功能区，场地从北到南依次为安河桥北地铁站区、活力运动区、连桥、生态山林区、滨河休闲区、古街风情区、北宫门前广场。

服务驿站：场地设置一级驿站 1 处，二级驿站 2 处，三级驿站 2 处。服务范围覆盖整个地段。

节点一：青龙古街

位于地块南部，南临颐和园北宫门。场地以历史上的青龙古道和古桥为主结构，结合颐和园北宫门的交通和功能需求，设计为可供休闲游览的古街

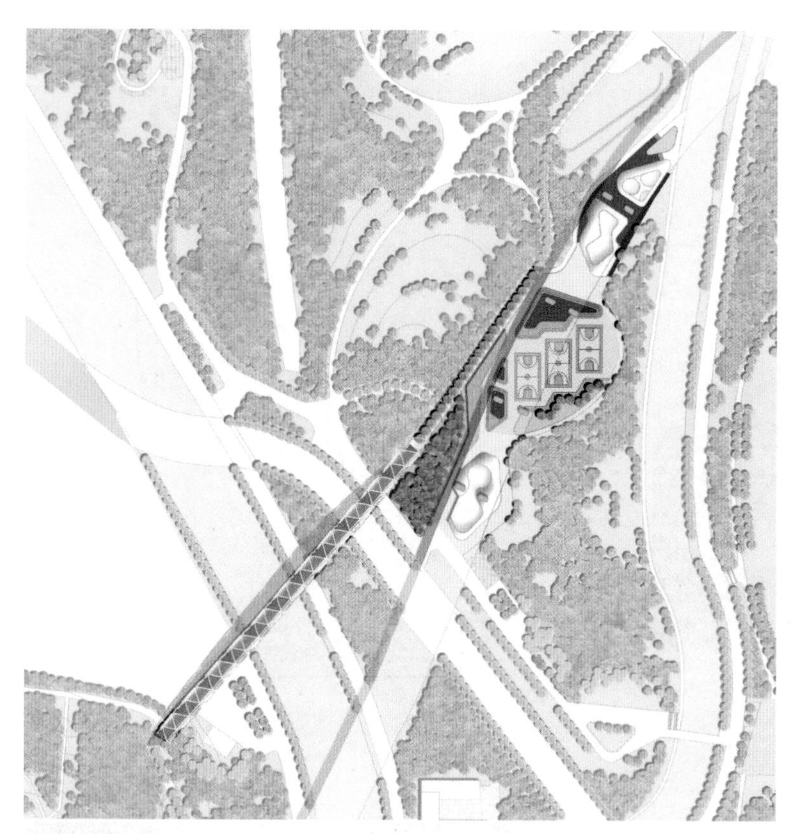

观光连廊和活力运动场
Corridor and Underside Courts

效果图
Impression drawing

巷，街巷内包含商铺旅店、文物古迹、公共空间等。设计后的颐和园北宫门外区域车行、人行分流，停车空间充裕、活动内容丰富多样，再现历史风貌。

现状交通混杂，停车混乱，商贩杂乱，人行舒适度极低。然而，这里历史底蕴浓厚，有青龙古街、青龙桥、慈恩寺等历史文化遗产，由于位于颐和园北宫门外，还具有人流集散、游客接待、文化展示、旅游业拓展等功能需求。

设计将该地段改造为可供休闲游览的古街巷，街巷内包含商铺旅店、文物古迹、公共空间等。街巷内机动车限时通行，营造舒适的步行环境，街巷外规划有车行道。设计后的场地人车分行，停车空间充裕、活动内容丰富多样，再现历史风貌。

以历史遗迹“慈恩寺”为轴，青龙桥西街为主

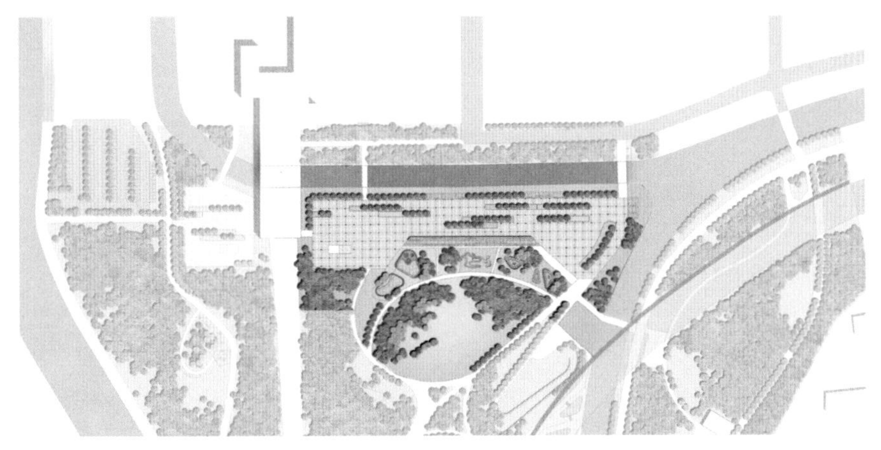

城市公园
City Garden

效果图
Impression Drawing

街，后设有小街，并通过青龙桥与河西侧街道相连，古街轴线上设有多处小空间，形成开合的空间节奏，丰富街巷体验。功能上河西古街以旅店、茶室、餐饮功能为主，建筑体量较大，河东侧古街以零售、小卖为主。

颐和园北宫门两侧不设停车场，而将停车场整合于其北侧空地，扩大颐和园前广场范围，成为可供游客和旅游团使用的集散广场。在原青龙桥南侧设一新桥，联系两侧车行道，保证青龙桥的人行需求。

节点二：观光连廊和活力运动场

环形绿道行至这里，由于京密引水渠及其两侧的车行道被阻隔，无法在地平层面衔接。因此，我们采用高架的景观桥进行连接，跨越不易穿越的京密引水渠并与机动车道连接，形成空中步道，同时步道也起到鸟瞰全区的功能。

高架的五环路下空间不宜种植植物，考虑周边居民区较多，缺少运动场地，这里设计桥下青年运动场，场地内设有篮球场和滑板场，并结合地形设计座椅和攀援场地，为市民提供多样的活动内容。

节点三：城市公园

场地位于安河桥北地铁站外，设计改造了现有机动车道，使场地不被机动车穿行和分割。调整后的完整地块内设计有广场、儿童活动场地、休憩设施等，同时与清河的优美景观结合，营造供市民游客停留等待的舒适空间。

规划方案七 及
颐和园南片区概念设计

Plan 7 & The Conceptual Design of the Summer Palace Southern Section

于静，洪卫静，乔丽霞，乔菁菁，伊琳娜，朱青，丁宁
Yu Jing/Hong Weijing/Qiao Lixia /Qiao Jingjing/Yi Linna /Zhu Qing /Ding Ning

本规划方案中，着重强调整个绿道体系中的活力注入，通过绿道沿线历史遗迹的整理和功能空间的完善，实现了整个绿道体系景观对三山五园恢宏历史进行抽象化地再现和重塑。

颐和园南片区是颐和园水系的出口，南水北调水利工程的终点。设计中将地铁西郊线（地面段）、公共展示场馆及体育活动场地均布置在场地外围，内部打造为自然郊野氛围的绿色空间，成为整个三山五园绿道体系中的一个生态调节枢纽。

In this plan, we emphasized the vitality injection in the whole greenway system, and achieved the reconstruction of the history in the 'Three Hills and Five Gardens' area in an abstractive way, by reorganizing the historic sites along the greenway and improving the functional space.

The southern section of the Summer Palace is the export of the Summer Palace water system, also the end of the 'South Water to North' water conservancy project. In the design, we arranged the western suburban line of subway (the ground segment), the public exhibition hall and the sports field on the outside of the site, and created a green space with natural countryside atmosphere, so as to make it an ecological adjustment hub in the whole greening system of the 'Three Hills and Five Gardens' area.

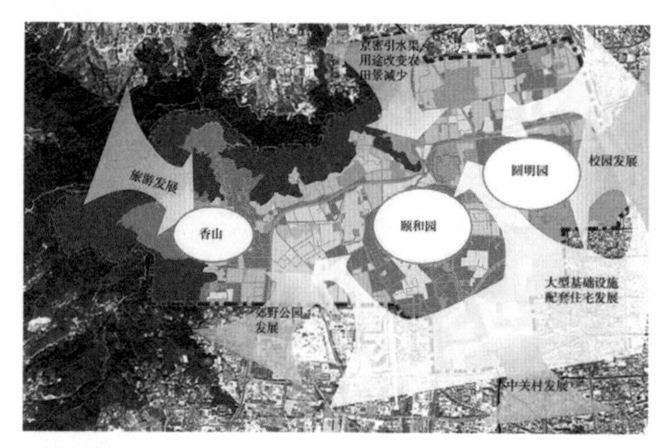

现状问题
Existing Problems

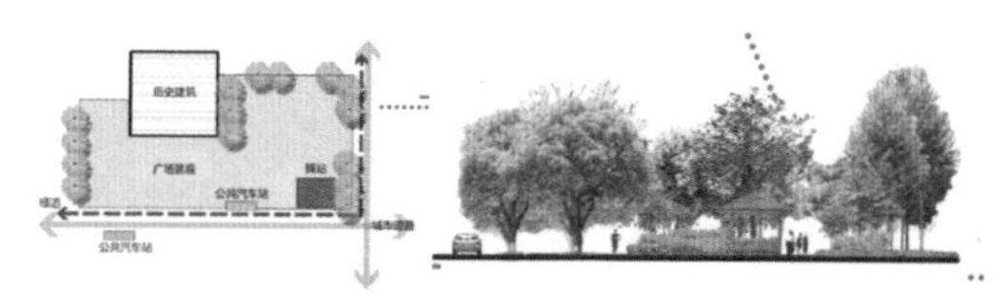

节点型游览空间广场景观示意图
Square Landscape Sketch Map

节点型游览空间与绿道衔接示意图
Convergence of Green Road

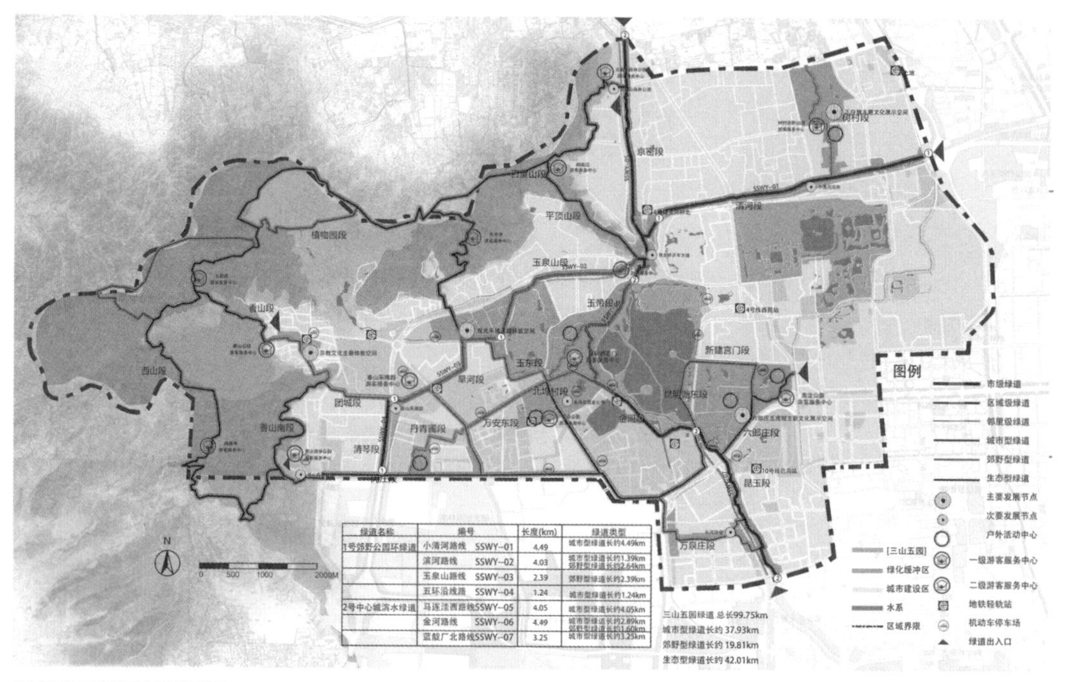

绿道名称	编号	长度(km)	绿道类型
1号郊野公园环绿道	小清河路线 SSWY--01	4.49	城市型绿道长约4.49km
	滨河路线 SSWY--02	4.03	城市型绿道长约1.39km 郊野型绿道长约2.64km
	玉泉山路线 SSWY--03	2.39	郊野型绿道长约2.39km
	五环沿线路 SSWY--04	1.24	城市型绿道长约1.24km
2号中心城滨水绿道	马连洼西路线SSWY--05	4.05	城市型绿道长约4.05km
	金河路线 SSWY--06	4.49	城市型绿道长约2.89km 郊野型绿道长约1.60km
	蓝靛厂北路线SSWY--07	3.25	城市型绿道长约3.25km

三山五园绿道规划总平面图
Master Plan

本绿道规划设计名为“唤活之道”，希望以注入活力、创造生境、唤醒历史的方式，依托绿道的建设将整个三山五园地区在历史上的恢宏以一种耳目一新的方式再现和重塑。

在城市快速发展的进程中，三山五园地区的生态环境不可避免地被城市化所影响。曾经的皇家园林或复建或重修或为遗址或不存或被占用，不能表现统一完整的景观风貌。同时三山五园现仅存三园，其全盛时期的轴线结构系统遭到极大地破坏。而且外围城市建设与园林环境极不协调，核心皇家园林面临巨大的环境压力。

在规划过程中，我们将三山五园绿道定位为保护都市文化遗产为主的保护型绿道，旨在突出三山五园皇家园林历史文化。在整个历史文化主题下，根据不同地段的属性，对绿道的建设采取分级分类的建设方式，依托于现有的破碎化绿地，提高绿地的连接性。

节点型游览空间是指沿着绿道分布的历史文化景观。类型有历史建筑、历史博物馆、城市广场或公园等多种不同的形式，具有景观形态清晰、功能明确和停留性强等特点，是绿道网络中的重要开敞空间。

线性游览空间是指游览绿道的游览空间。绿道本身就是一种线性空间，在构建绿道与历史文化景

规划年限示意图
Construction Phases of Greenways

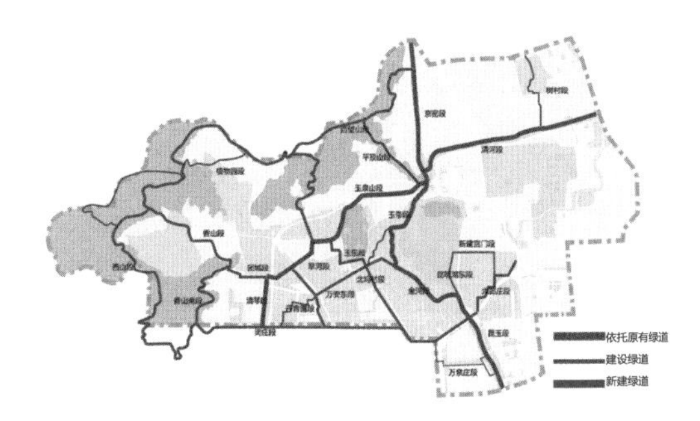

建设方式示意图
Construction Types of Greenways

用地变动示意图
Land Changes

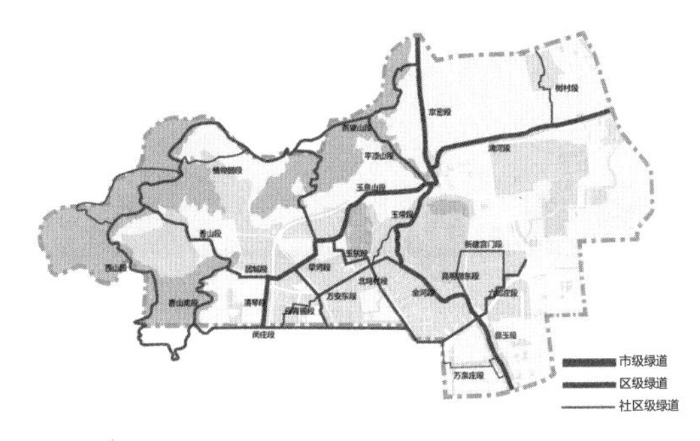

分级示意图
Levels of Greenways

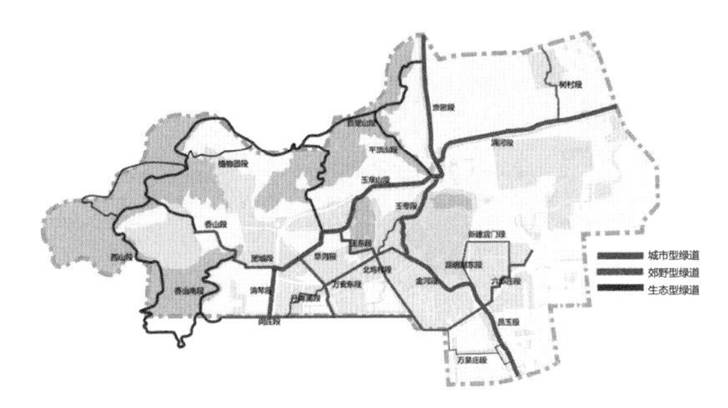

慢行系统
Non-motorized Traffic System

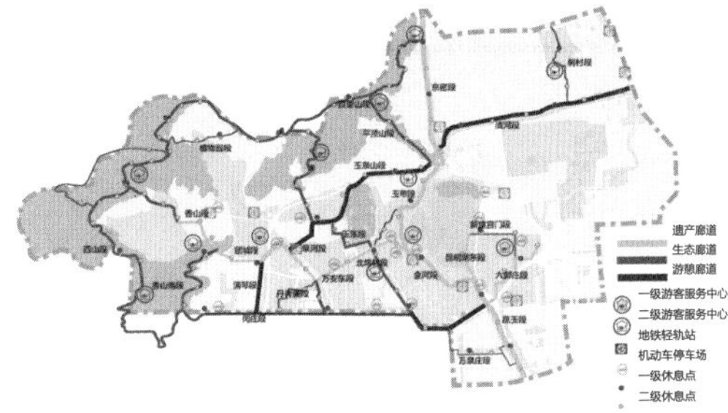

服务设施系统
Service Facilities System

观关系的时候，线性空间是最基本的形态，除了作为联系各个历史文化景观单元的纽带，本身也能成为历史文化景观形态的一种。

在规划设计中应遵循以下五个原则：

（1）顺应自然肌理，畅通生态廊道。首先尊重西山阡陌纵横、山水绵延的自然本底，充分利用地形、植被、水系等自然资源，结合市域生态廊道、生态隔离绿地、环城绿带和农田林网等构建城市绿道，使分散的生态斑块得以有机连接，从而构建和维护完整、安全的区域生态格局。

（2）串联发展节点，体现特色底蕴。充分发挥城市绿道对各类发展节点的组织串联作用，以三山五园风景名胜区、郊野公园、郊野公园、采摘园以及人文遗迹、非物质文化遗产、历史村落、传统街区等自然、人文节点为依托，尽可能多地发掘并展示本地具有代表性的特色资源，实现“在发展中保护，在保护中发展”。

（3）契合城乡布局，引导空间发展。一方面，城市绿道应契合城市的空间结构与功能拓展方向，有效发挥三山五园绿道在城乡之间、城镇之间以及城市不同功能组团之间的生态隔离功能，引导城乡形成合理的空间发展形态；另一方面，城市绿道应连通玉东公园、树村郊野公园、中关村广场、柳浪游泳场馆、中关村商业街、滨水休闲带等公共空间，

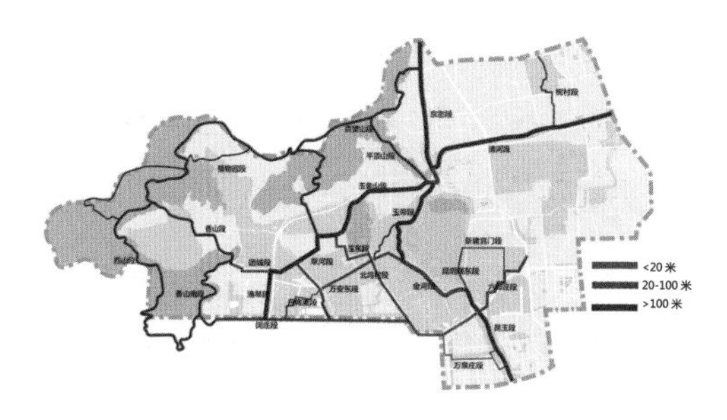

绿道绿廊系统
Greenway System

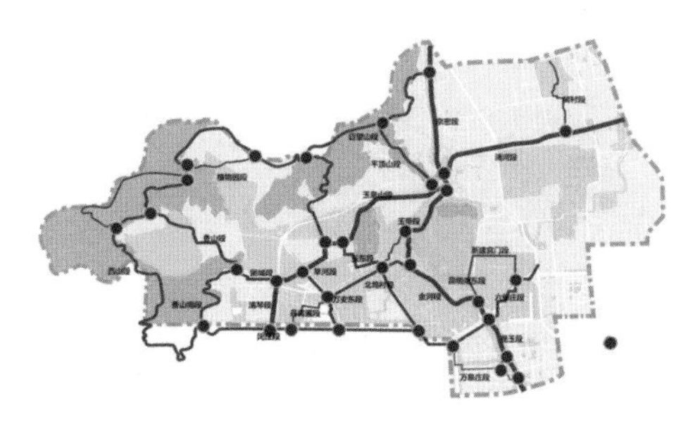

交通衔接系统
Traffic Connection System

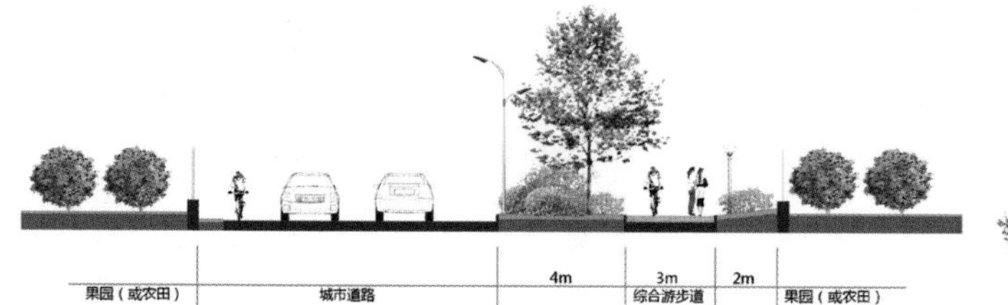

郊野型慢行系统断面示意图

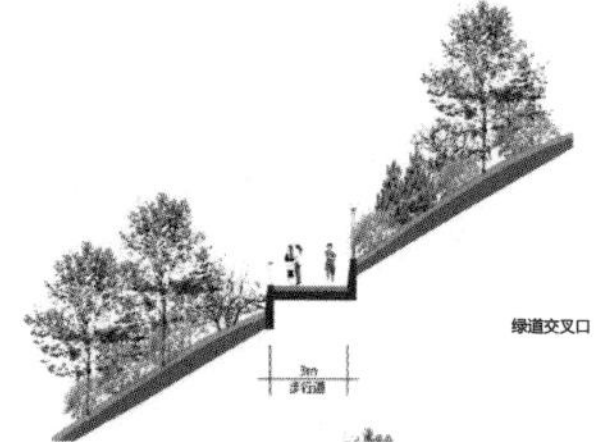

生态型慢行系统断面示意图

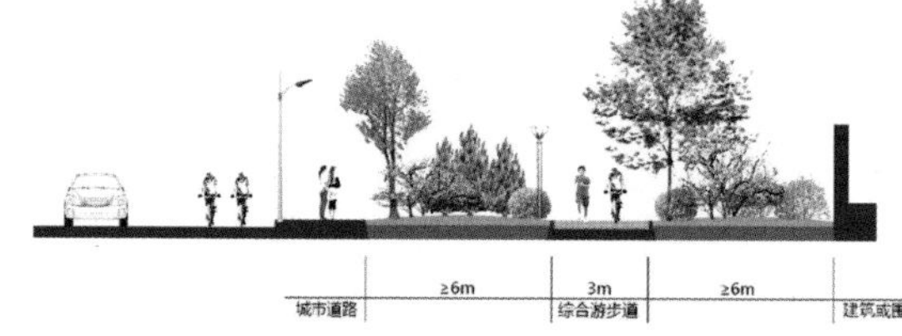

城市绿地型慢行系统断面示意图

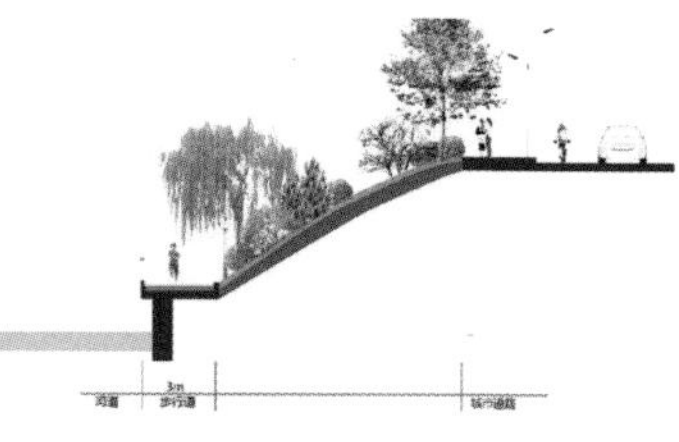

城市滨水型慢行系统断面示意图

慢行系统断面示意图
Cross Sections

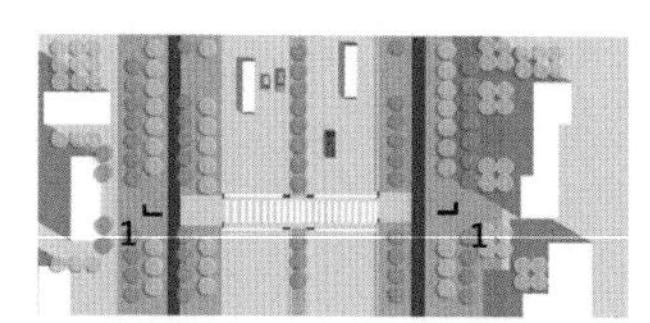

共面相交型

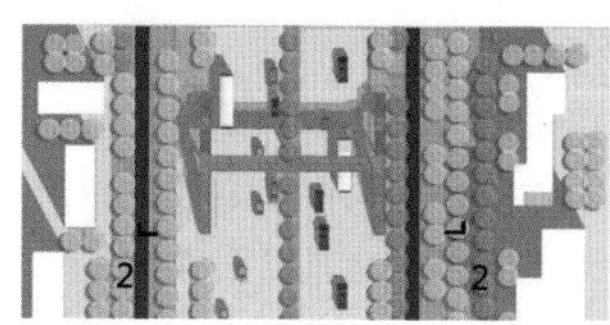

天桥型

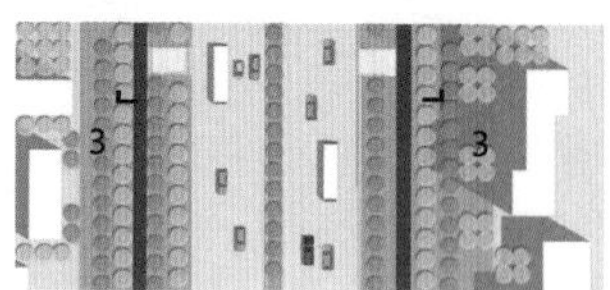

隧道型

交叉口示意图
Design of Intersections

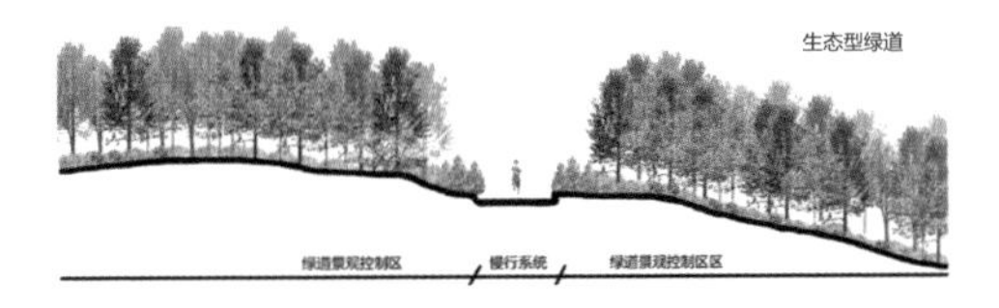

绿廊系统断面示意图
Cross Section

城市型廊道（六郎庄段—北坞村段—闵庄段）（京密段—金河段— 昆玉段）（清河段）
郊野型廊道（玉泉山段—玉带段 + 团城段—香山段）
生态型廊道（西山段—香山段—植物园段）(百望山段）

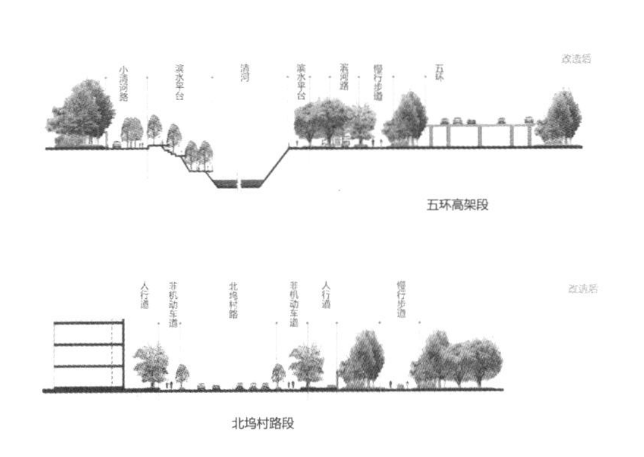

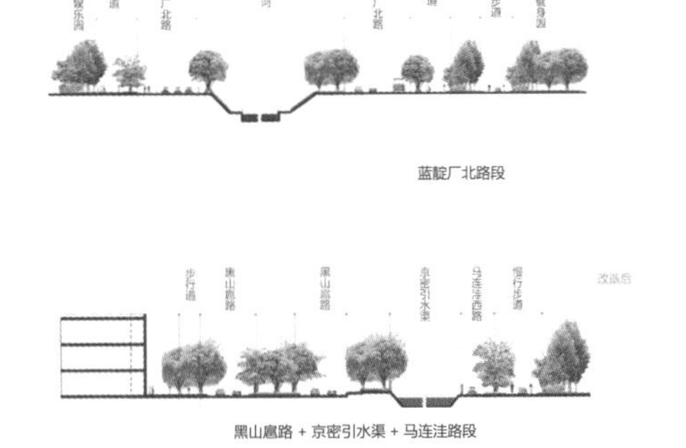

绿廊系统断面示意图
Cross Section

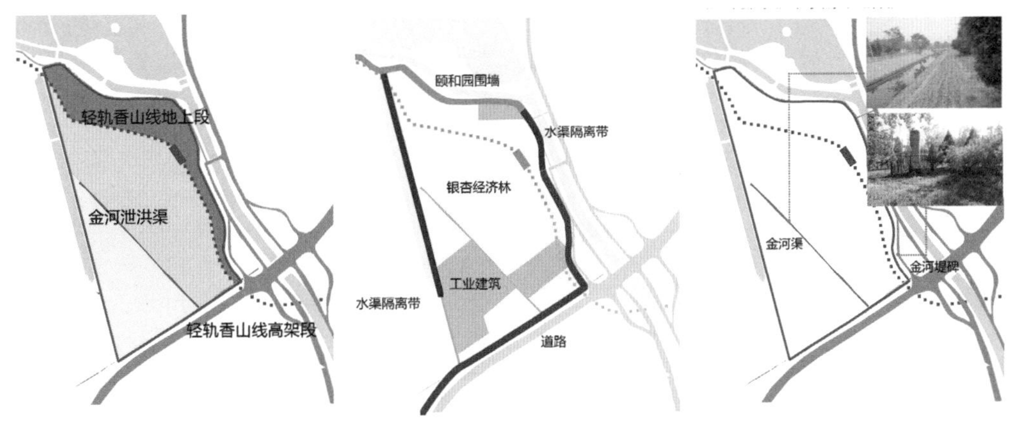

场地现状分析图
Design Analysis

成为公共空间的联系纽带，孕育城乡居民多样的公共生活空间，促进和谐社会建设。

（4）利用交通廊道，集约利用土地。城市绿道布局要尽量避免开挖、拆迁、征地，应充分利用现有的废弃铁路、村道、田间道路、山林游道等路径，在保障绿道使用者安全的前提下，集约利用土地，降低建设成本。

（5）衔接市级绿道与慢行系统，倡导绿色生活。发挥承上启下的作用，与相邻城市绿道同步对接，加大绿道网密度，并重点向中心商业区、居住社区、

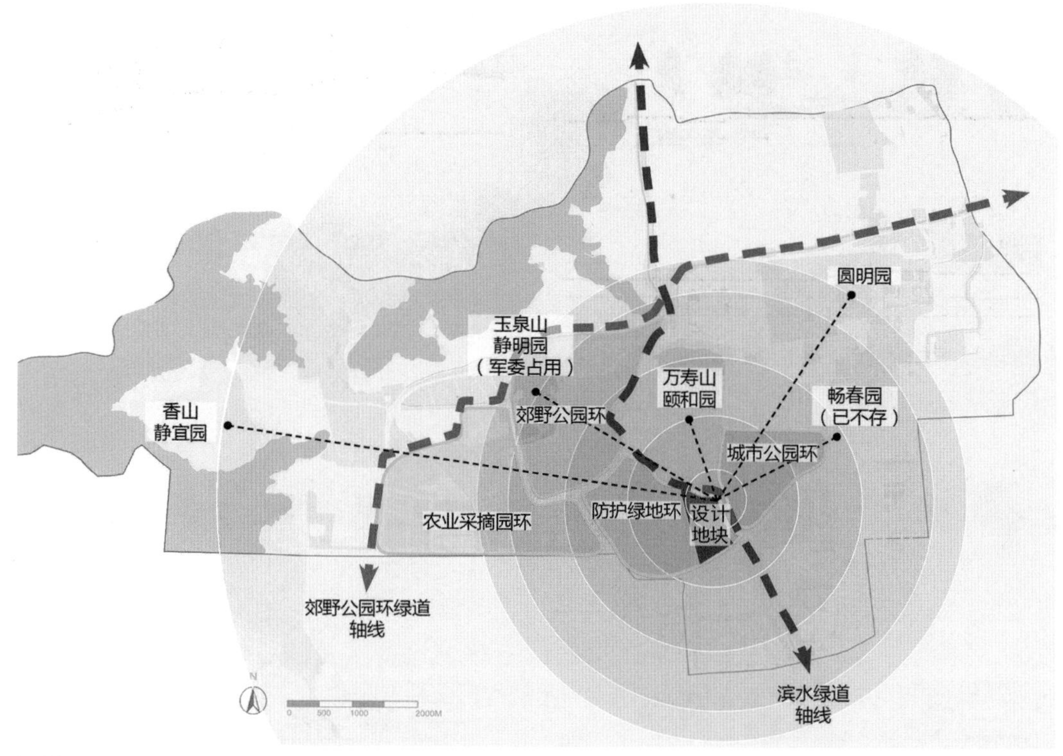

场地与绿道规划的关系
Site location

公共交通枢纽以及大型文娱体育区等人流密集地区延伸，与城市慢行系统共同构成连续、完整的绿道生活网络，丰富市民出行方式，引领“公交优先、方便慢行”的绿色出行模式。

由于建设条件限制，绿道建设的宽度与实际的建设条件相结合，尽可能地满足绿道最小建设要求9m。借助现有绿地系统规划的绿地进行建设，对于城市建成度较高的区域而言，建设绿道的难度加大，局部为保障绿道的连接性，会出现一些连接性的道路，但总体不能超过绿道长度的10%。

在绿道建设过程中为保证绿道的连接度，必要时需要与上位规划相结合新建部分绿色开放空间，能最大限度地将破碎化的绿地纳入绿道系统中，既保证绿道与市政交通换乘的连通性，也要保证绿道与城市建设有所隔离，不被较多的干扰，使得三山五园绿道在尊重和利用现状条件下形成整体性的绿道。

本次的重点设计地块位于北京市绿化隔离地区和重要的水源保护区，用地范围45.54hm^2，属于南水北调工程北京段终端地区建成景观区，在三山五园历史风貌区和世界文化遗产颐和园外围文物缓冲区范围内。本设计延续“三山五园”的历史文脉，建设符合这一地区皇家园林风貌的水源保护地。设计地块紧邻滨水绿地轴线，该部分绿道属于市级遗产廊道，且紧邻颐和园和昆玉河，对于遗产保护意义重大。并且该地块属于规划四环的中心——防护绿地环，生态防护职能突显。

该场地内历史文化资源缺乏保护，金河泄洪渠和轻轨割裂了场地，造成皇家风貌的断裂。同时该场地未能合理建设利用，缺乏活力，水源保护绿地的职能未能得到充分体现。基于这些问题，我们得出了“延续皇家园林风貌、传承历史文化资源、建设水源防护绿地、联系用地周边区域、营造社区活力空间”的项目定位，并以此指导我们的设计策略，使该场地成为我们“唤活之道”绿地规划设计的点睛之笔。

设计地块用地西界是南水北调明渠，北界是颐和园，东界是长河。南临四环，东临昆明湖东路和蓝靛厂北路，东南角为火器营桥。场地内部有金河路贯穿，东北侧有在建的地铁香山线贯穿，此外附近还有地铁十号线。

现有金河泄洪渠和在建轻轨香山线将用地分为3个部分。用地东北角将建设轻轨香山线和颐和园

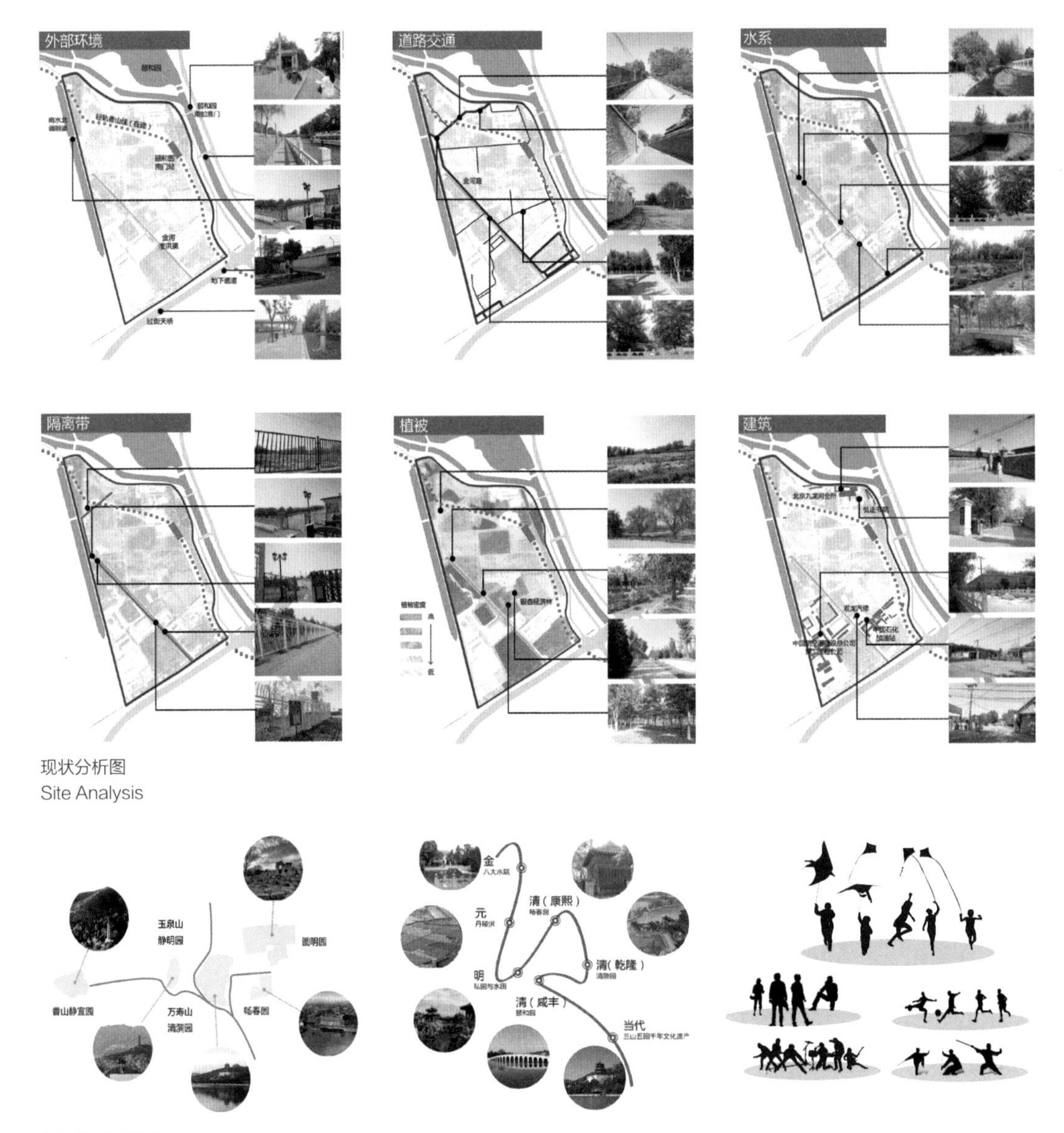

现状分析图
Site Analysis

设计构想分析图
Design Ideas

南门站。同时地块紧邻滨水绿地轴线，该部分绿道属于市级遗产廊道，且紧邻颐和园和昆玉河，对于遗产保护意义重大。

北部和东西两侧以绿地为主，包括南水北调水源保护地、颐和园和封闭的高尔夫球场。设计地块南部和绿地外围以居住用地为主，并有少量商业和文教用地。设计地块是颐和园的延伸、市民的活动空间和高尔夫球场与水源地的生态过渡。

场地现状问题总结为金河泄洪渠和轻轨对场地的割裂，用地被工业建筑和林地占用，京西稻田风貌不复存在，金河水渠污染严重，护坡粗糙生硬，水利遗迹缺乏识别度。设计用地紧邻颐和园南墙，位于颐和园风貌控制的缓冲区。该区域的空间序列自北向南依次是颐和园佛香阁、昆明湖、银杏经济林、工业建筑、四环高架。该序列中设计用地的位置存在空间的断档、皇家园林风貌的断裂。

设计用地内部充斥着银杏经济林和工业建筑，外围被颐和园围墙、水渠和道路包围，场地少有人工痕迹，未能合理建设利用，整体呈现较为荒凉的面貌。水源保护绿地的职能未能得到充分体现。

颐和园南片区设计总平面图
Site plan

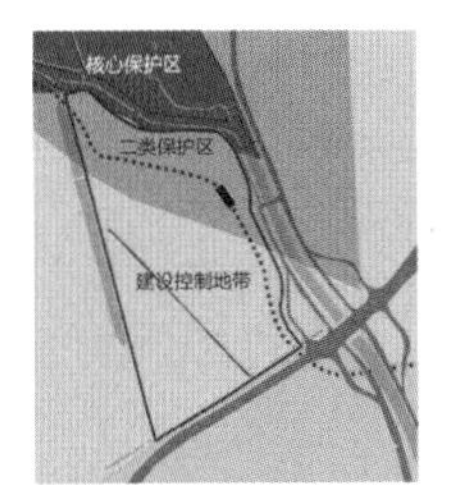

风貌控制
Style control

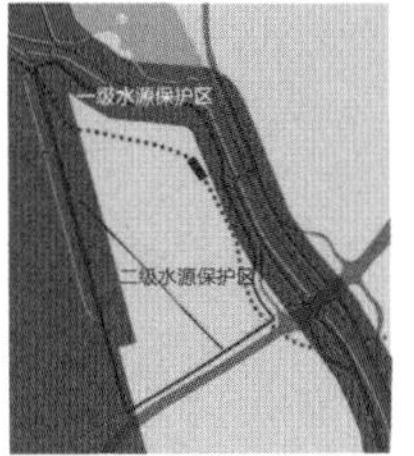

水源保护
Water protection

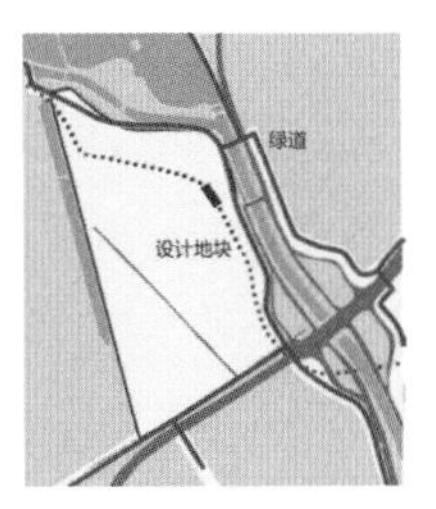

绿道节点
Greenway joint

竖向规划图
Vertical plan

植被规划图
Vegetation plan

植物种植分析图
Planting Plan

场地现状分析图分别是场地外部环境现状、场地内部交通现状、场地内部水系现状，场地东侧有颐和园南门和过河人行桥，南侧有过街地下通道和人行天桥，北侧和西侧是封闭的颐和园围墙和南水北调保护区域。场地内部有一些内部小路和金河路、金河渠，金河路和金河渠两侧有栏杆隔离。现有植被以荒地和苗圃纯林为主。场地内部建筑主要有南部计划拆迁的厂房和西北角计划保留的会所和服务性建筑。

针对不同的现状问题，设计过程中总结出相应的策略以完善整体设计。打破金河泄洪渠并架设桥梁，使得道路从轻轨使两侧在交通上取得联系。借助场地文脉对京西稻田风貌还原、西郊千年历史变迁的展示以及非物质文化遗产活力再现。通过使用代表皇家园林寓意的植物种类来营造场地的绿色空间。在园林构筑物和雕塑小品中加入皇家园林常用的体量（规模宏大、体量敦厚、场景深远）、色彩（色彩富丽，如红色宫墙、黄色瓦顶、汉白玉栏杆）和形式（轴线和制高点的控制、传统中式纹样装饰），同时在场地内塑造地形建设高地并保留透景线，在视线上使场地与颐和园取得联系。通过建设绿道联系三山五园，在用地南侧保留开敞空间，吸纳周边的居民活动。在靠近南水北调明渠的位置设置绿化隔离带，使用生态水过滤系统处理金河水污染。以上策略相互结合，将场地打造成为延续皇家园林风貌、传承历史文化资源、建设水源防护绿地、联系用地周边区域、营造社区活力空间的景观。

颐和园于 1998 年被列为世界文化遗产，设计地块位于其南侧，用地东北部分为遗产地二类保护区，西南部分为建设控制地带。该用地的建设要彰显北京西郊皇家园林风貌。

地块西侧是南水北调工程（北京段）明渠，东北侧为京密引水渠，近水渠两侧 100m 为一级水源保护区，用地其他部分为二级水源保护区。本绿地也是水源保护绿地。

设计地块目前主要被银杏纯林和厂房所占据，场地与周边联系较少，缺乏活力。该用地的建设强调加强西郊皇家园林与居民之间的联系，提升文化保护水平，吸纳周围居民的游憩活动。

竖向设计中，将设计地块北部地形提升 4m，结合场地使游人可以透过围墙眺望颐和园景色。南部扩大金河渠，形成水面与建筑相结合并软化金河渠驳岸。山水之间以缓坡活动草坪和稻田作为过渡。场地西南角结合微地形形成富有活力的活动场地。

种植设计中，按照地形和空间使用功能，规划不同的乔、灌、草类型，部分改造和丰富现有植被。因场地特殊性质划分有几处特色植被分区，包括以银杏、油松、玉兰等中国古典园林常用植物为主的

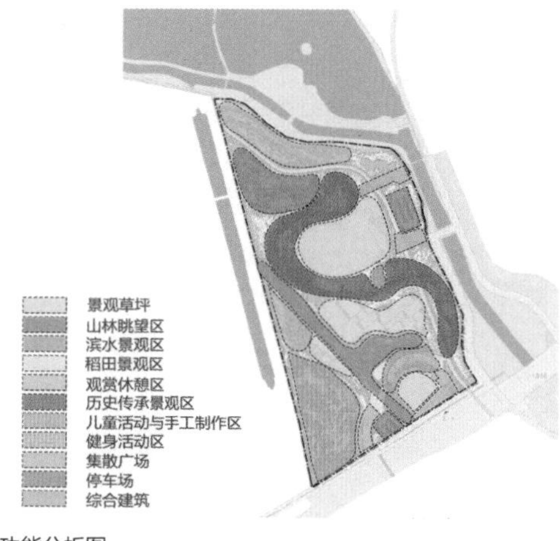

功能分析图
Functional Zones

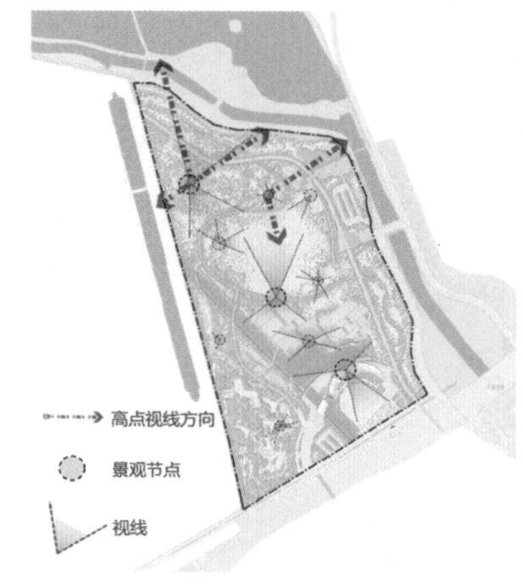

视线分析图
Sight Lines

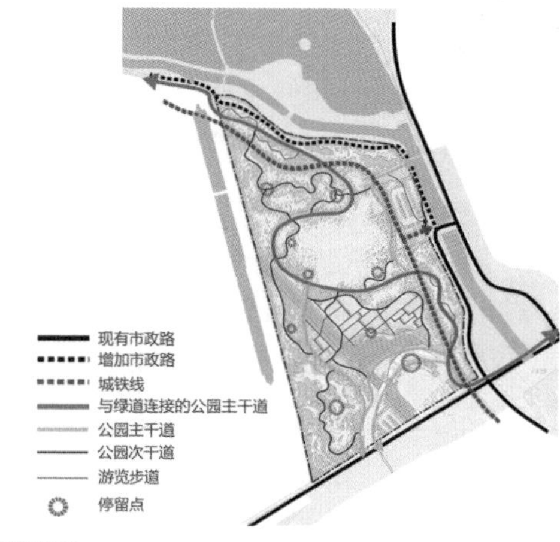

交通分析图
Traffic System

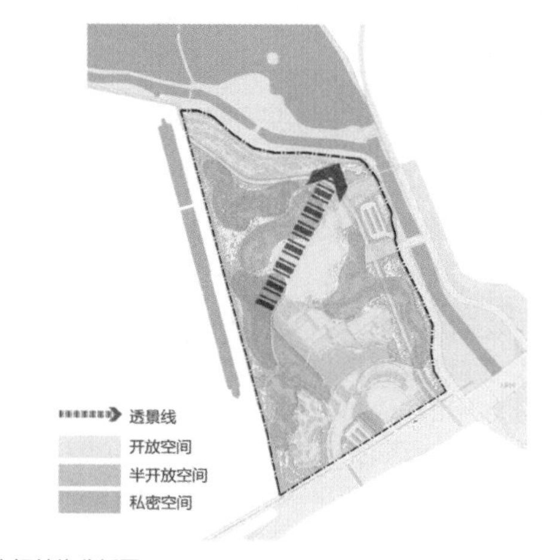

空间结构分析图
Spatial Structure

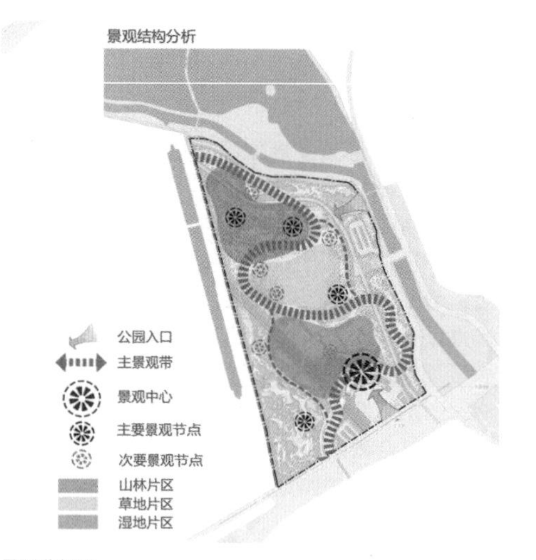

景观分析图
Scenic System

历史文化区；以刺槐、紫穗槐、苜蓿等具有水源涵养、水土保持功能的植物为主，以复层混交的形式防止游人穿行的水源防护林；以千头椿、火炬树、紫丁香等抗性强、茂密的树种隔离烟尘和噪声，兼顾观赏的道路轻轨隔离带；以京西稻配以少量农业防护林常用乔木，营造乡村气息的稻田区；以芦苇、花叶水葱、狼尾草、玉带草等水生、湿生观赏草为主，与稻田区相映成趣的湿地草甸区等。

同时在靠近水源保护地的区域使用水源绿地典型的植物配置群落和林分结构。选用具有水源涵养、水土保持功能的树种，以生态价值为主兼顾观赏，以复层混交的形式防止游人穿行。选用抗性强、茂密的树种，隔离烟尘和噪声，兼顾观赏的植物。

场地功能分区包括景观草坪、山林眺望区、滨水景观区、稻田景观区、观赏休闲区、历史传承景

金水河改造前水质
Analysis of Water Quality Before Reconstruction

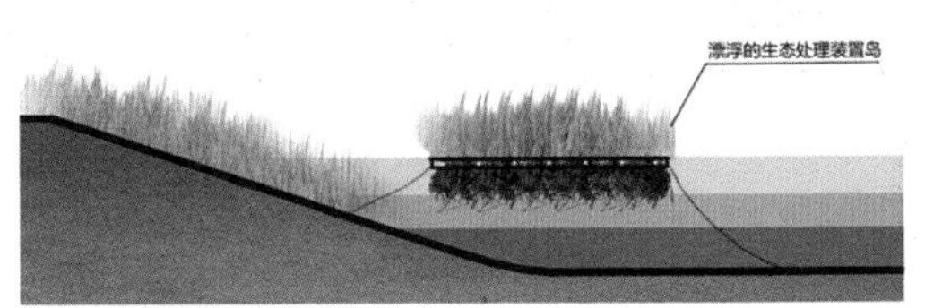

金水河改造后水质
Analysis of Water Quality Before Reconstruction

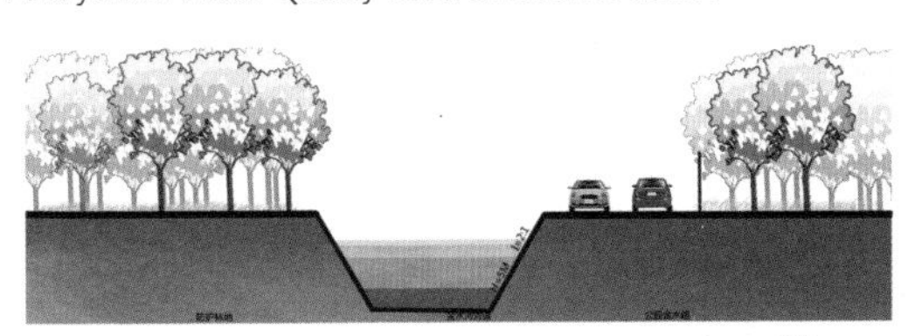

金水河现状断面示意图
Current Section of Jinshui River

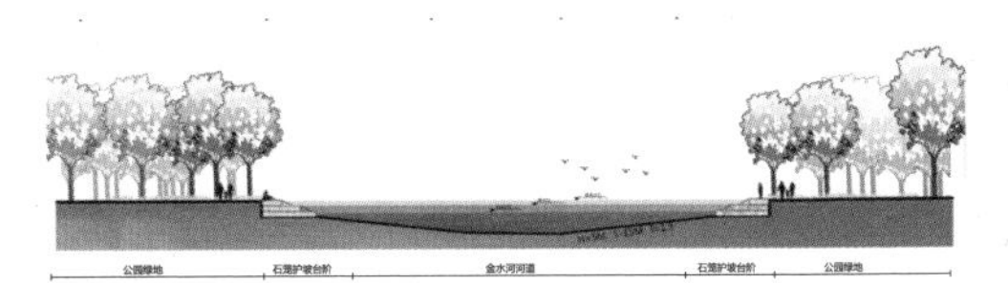

金水河改造后断面示意图
Post Reconstruction Section of Jinshui River

观景廊架视线分析图
Viewing Gallery View Analysis

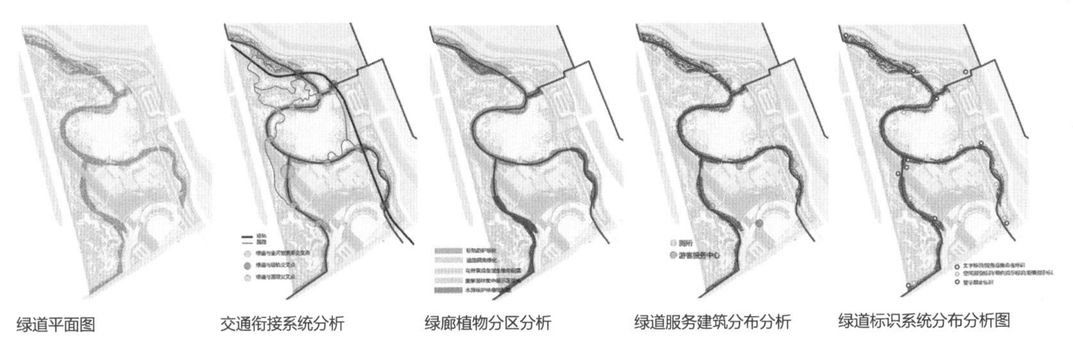

绿道平面图　交通衔接系统分析　绿廊植物分区分析　绿道服务建筑分布分析　绿道标识系统分布分析图

绿道分析图
Greenway analysis

观区、儿童活动与手工制作区、健身活动区、停车场、集散广场和建筑区。场地交通分为穿过设计地块的规划绿道、主干道、次干道和游览步道。沿主要游览路线和重要景观节点设置停留点。场地视线分为三种：从制高点向四周眺望、草坪稻田区开阔的视野和小型景观节点、休息区被阻挡的视线，营造立体的观赏体验。活动场地分为开放、半开放、私密三种类型，靠近南水北调明渠的部分较为郁闭，小型活动场地和过渡空间以半开放空间为主，大型活动场地为开放空间。并有面向颐和园的透景线。场地景观结构为“一带、一心、三面、多节点”。主景观带绕过场地内三个主要景观片区，沿线布置多种景观节点。

金水河水质在改造前较差，各种污染物质超标，在改造后，通过在水中增加漂浮的生态处理装置岛等措施，改善水质。在改造后，水质清澈，适合鱼类等生物的生存。改造后，通过减缓金水河驳岸的坡度，同时在驳岸上种植植物，软化驳岸。在驳岸两侧增加石笼台阶，给人们亲水创造机会。通过在场地西北侧增设竖向地形设计、在地势相对较高的位置设置眺望台、在植物种植上留有一定的透景线，加强场地内部和颐和园的沟通。

04 研究团队

RESEARCH TEAM

核心研究团队的组成充分体现了北京林业大学园林学院以风景园林为特色的人居环境学科群的研究复合度，采用了以风景园林设计方向教授领衔，城市规划、建筑学、园林工程、植物景观等多方向教师共同支撑，并特邀资深一线规划设计专家和校外学者为顾问团队的形式。数十名研究生共同全程参与了持续数月的辛勤研究工作，确保了整个研究过程的深入与全面。

The composition of our core research team fully embodies the compound degree of the research community of human settlements subject with the characteristics of landscape architecture in Beijing Forestry University. Our team is guided by professors in the direction of landscape architecture design, and is supported by multiple directional teachers who expert in urban planning, architecture, landscape engineering, planting landscape, etc. We have also invited senior front-line planning and design experts and off-campus scholars as our consultants. Dozens of graduate students took full participation in our hard work which have lasted several months, ensuring the thorough and comprehensive research progress.

核心研究团队

Core Researchers

团队负责人 Principal

■ 王向荣 Wang Xiangrong

1963年生，北京林业大学园林学院院长，教授，博士生导师，研究方向：风景园林规划设计与理论。

Born in 1963, Dean of School of Landscape Architecture, BJFU. Professor and Ph.D supervisor . Research Fields: landscape architecture design and theory.

主要成员 Key Researchers

■ 刘志成 Liu Zhicheng

1964年生，北京林业大学园林学院副院长，教授，博士生导师，园林设计教研室主任，研究方向：风景园林规划设计与理论。

Born in 1964, Vice Dean of School of Landscape Architecture, BJFU.Professor and Ph.D supervisor,Director of landscape design department, Research fields: landscape architecture design and theory. architecture design and theory.

■ 林箐 Lin Qing

1971年生，北京林业大学园林学院教授，博士生导师，研究方向：园林历史，现代风景园林设计理论，区域景观等。

Born in 1971, Professor School of Landscape Architecture, BJFU. Ph.D supervisor. Research Fields: garden history, modern landscape design theory, regional landscape, etc.

主要成员 Key Researchers

钱云 Qian Yun

1979 年生，北京林业大学园林学院城市规划系副教授，研究方向：城市风景环境规划设计，住房与社区研究。

Born in 1979, Associate Professor of Department of Urban Planning, School of Landscape Architecture, BJFU. Research Fields: Urban landscape planning and design, housing and community studies.

郑小东 Zheng Xiaodong

1977 年生，北京林业大学园林学院副教授，北林风景园林规划设计研究院风景建筑研究中心主任，中国一级注册建筑师，研究方向：风景园林建筑、地域性建筑。

Born in 1977, Associate Professor of School of Landscape Architecture, BJFU. Director of Landscape Architecture Research Center of Beilin Landscape Architecture Planning and Design Institute, and class 1 Registered Architect(PRC). Research Fields: Landscape Architecture, Regional Architecture.

王沛永 Wang Peiyong

1972 年生，北京林业大学园林学院副教授，风景园林工程教研室，研究方向：风景园林工程与设计，海绵城市建设及城市绿地用水的可持续设计研究。

Born in 1972, Associate Professor of Landscape Engineering Department, School of Landscape Architecture, BJFU. Research Fields: Landscape Engineering and Design, Sponge City Construction and Sustainable Design of Urban Greenland Water.

李倞 Li Liang

1984 年生，北京林业大学园林学院副教授，研究方向：现代风景园林设计理论与实践、景观基础设施。

Born in 1984, Associate Professor of School of Landscape Architecture, BJFU.Research Fields: Modern Landscape Design Theory and Practice, Landscape Infrastructure.

尹豪 Yin Hao

1976 年生，北京林业大学园林学院副教授，研究方向：现代园林设计理论，植物景观营造、生态规划与设计。

Born in 1976, Associate Professor of School of Landscape Architecture, BJFU.Research Fields: Modern Garden Design Theory, Plant Landscape Design, Ecological Planning and Design.

张云路 Zhang Yunlu

1986 年生，北京林业大学园林学院副教授，研究方向：风景园林规划与设计。

Born in 1986, Associate Professor of School of Landscape Architecture, BJFU.Research Fields:Landscape Architecture Planning and Design.

特邀专家

Invited Experts

- 陈珊珊——北京清华同衡规划设计研究院总规中心总工
  Chen Shanshan - Chief Engineer of Research Center for Master Planning, THUPDI.

- 高巍——北京交通大学城市规划系主任
  Gao Wei - Director of Department of Urban Planning, Beijing Jiaotong University

- 郭竹梅——北京北林地园林规划设计院规划所所长
  Guo Zhumei - Director of Beijing Beilin Landscape Architecture Institute.

- 李建伟——EDSA 东方总裁兼首席设计师
  Li Jianwei - President and Chief Designer of EDSA Orienta

- 朱育帆——清华大学建筑学院教授
  Zhu Yufan - Professor of School of Architecture, Tsinghua University.

- 李战修——北京创新景观园林设计有限公司总经理兼首席设计师
  Li Zhanxiu - General Manager and Chief Designer if Beijing Topsense Landscape Design Ltd.

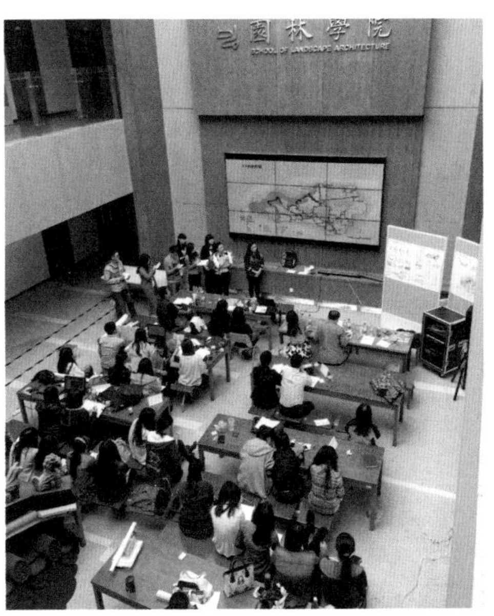
園林學院
SCHOOL OF LANDSCAPE ARCHITECTURE

研究生团队

Postgraduates

孟宇飞 Meng Yufei	张　俞 Zhang Yu	朱友强 Zhu Youqiang	何二洁 He Erjie	罗亚文 Luo Yawen
施　莹 Shi Ying	甘其芳 Gan Qifang	施　晨 Shi Chen	王训迪 Wang Xundi	孙　津 Sun Jin
尚尔基 Shang Erji	李娜亭 Li Nating	吴晓彤 Wu Xiaotong	周珏琳 Zhou Juelin	高　琪 Gao Qi
莫斯塔 Mosita	李婉仪 Li Wanyi	葛韵宇 Ge Yunyu	张　芬 Zhang Fen	胡盛劼 Hu Shengjie
李雅祺 Li Yaqi	赵　欢 Zhao Huan	邓力文 Deng Liwen	刘加维 Liu Jiawei	吴明豪 Wu Minghao

王 茜
Wang Xi

刘 童
Liu Tong

张琦雅
Zhang Qiya

夏 甜
Xia Tian

韩 冰
Han Bing

王博娅
Wang Boya

丁小夏
Ding Xiaoxia

郑 慧
Zheng Hui

张司晗
Zhang Sihan

郭 沁
Guo Qin

刘健鹏
Liu Jianpeng

佟思明
Tong Siming

王晞月
Wang Xiyue

李 璇
Li Xuan

李 媛
Li Yuan

崔滋辰
Cui Zichen

徐 慧
Xu Hui

莫日根吉
MorigenJi

丁 宁
Ding Ning

于 静
Yu Jing

洪卫静
Hong Weijing

伊琳娜
Yi Linna

乔菁菁
Qiao Jingjing

乔丽霞
Qiao Lixia

朱 青
Zhu Qing

本书编辑工作

排版校对　王长悦　胡　玫　贾家妹
英文翻译　彭丹麓